Megadrought in the Carolinas

Archaeology of the American South: New Directions and Perspectives

Megadrought in the Carolinas

The Archaeology of Mississippian Collapse, Abandonment, and Coalescence

JOHN S. CABLE

THE UNIVERSITY OF ALABAMA PRESS
TUSCALOOSA

The University of Alabama Press
Tuscaloosa, Alabama 35487-0380
uapress.ua.edu

A Dan Josselyn Memorial Publication

Typeface: Caslon

Cover image: Jacques Le Moyne de Morgues (1533?–1588), *Floridae Americae provinciae recens . . .* (detail); Library of Congress
Cover design: Michele Myatt Quinn

Cataloging-in-Publication data is available from the Library of Congress.
ISBN: 978-0-8173-2046-1
E-ISBN: 978-0-8173-9276-5

Contents

Illustrations

Figures

Tables

Preface

This book is the result of a personal research project that I undertook with the goal of understanding what happened to cause the collapse of numerous agriculturally dependent, complex societies in the southern latitudes of the continental United States during a period of megadrought and other extreme climate events in the first half of the last millennium. The germ of the project was planted in my mind when I was 11 years old, soon after my family moved from Kansas to Scottsdale, Arizona, in 1960. One Sunday afternoon my dad loaded us up in our black Rambler sedan and drove out to visit Pueblo Grande Museum, where the sole above-ground monument of Hohokam civilization in the Phoenix metropolitan area—an adobe-walled platform mound—had been preserved and protected from urban development. At the time, the museum stood as a dusty and dilapidated remnant of the public works programs of the 1930s and 1940s. This was of little concern to me, however, as I was fascinated by both the mound, the top of which could be accessed by deck stairs, and the exposed ball court next to it that was encased in a crumbling veneer of cement. My dad told me that such ruins were at one time spread all across the valley and that the Hohokam had once thrived in the desert by virtue of networks of canals that watered vast agricultural tracts. Naturally, I asked whatever happened to these people, to which he replied that no one really knew, that they had just disappeared. Unsatisfied with his reply, I tucked this mystery away in my memory, vowing one day to solve it.

Shifting forward in time, I became an archaeologist, but my childhood enterprise remained dormant because opportunity dictates its own goals and subject matter, especially in cultural resource management. I conducted excavations at a stratified Archaic site on the Haw River in North Carolina and then returned to Phoenix to investigate and collaborate on a range of Hohokam sites, not least of which Pueblo Grande itself. It was not until I returned to the Southeast and became engaged in long-term regional research in the Francis Marion National Forest on the South Carolina coast that the mystery sprang forth again in an entirely new light. During my years of surveying the swamplands and coastal fringes of the forest, I became gradually aware that the prehistoric inhabitants of this region had also experienced societal collapse and regional abandonment. Moreover, this happened at virtually the same time as the Hohokam event, in the early to mid-fifteenth century. David Anderson had already demonstrated a contemporaneous abandonment scenario for the nearby middle and lower Savannah River valley, which led me to believe that more was afoot here than simple coincidence. I then set out to see just how big this zone of abandonment on the Southeast Coast might have been by reviewing the occupation sequences of adjacent regions. This was a difficult task, but after a long period of collections reanalysis and inter-regional correlations I was able to ascertain that a contiguous zone of roughly 34.5 million acres along the coast from the Cape Fear River in North Carolina to the Ogeechee River in Georgia had been abandoned or depopulated during this same time frame.

I reasoned that kindred events of such grand scale and synchronicity must surely have had a common root cause, and it was my hunch that it was linked to the global patterns of climate change that were manifested during what paleoclimatologists identify as the Medieval Climatic Anomaly, or MCA (AD 1000 to 1500). This was a period of warmer temperatures and extended droughts that would have exerted extreme pressures on the productivity of prehistoric farming systems and other natural resources and could well have led to societal collapse, regional abandonments, and migrations. The idea was not new, as Anderson had suggested previously that droughts were a principal factor in the collapse of the Savannah River chiefdoms; other scholars had already proposed that the prehistoric metropolis of Cahokia at St. Louis may have been the victim of extended drought. However, Cahokia had collapsed in the thirteenth century, much earlier than the events I was specifically interested in and interjecting noise into an otherwise neatly ordered synchronicity. A similar contradiction to my thesis was presented by the large-scale abandonments and settlement reorganizations that occurred on the Colorado Plateau and Mimbres regions of the southwestern United States at about the same time as Cahokia was collapsing. Undeterred, I consulted the paleoclimatic literature to see whether there might be geographic and chronological patterns of extreme drought that could explain these differences. Sure enough, there were indeed different geographic patterns of megadrought on a continental scale, but it was Greg McCabe's study of twentieth-century drought distributions that brought the whole abandonment scenario into fine focus. From his maps, I could see that there are two geographic patterns of drought associated with global warming, of which the first is caused by hot sea surface temperatures in the Atlantic Ocean and cool sea surface temperatures in the Pacific. This pattern is most intense/frequent over Cahokia and the Colorado Plateau. The other pattern is caused by hot temperatures in both oceans and its effects are most intensely felt along the South Atlantic coast and the Ohio River valley, another region that experienced depopulation during the fifteenth century. If one more leap of faith is to be taken, the late prehistoric collapse and abandonment scenarios, then, can be explained by chronologically offset patterns of megadrought conditions, with one scenario corresponding to the twelfth and thirteenth centuries and a second occurring in the fifteenth century.

Eureka, synchronicity is re-established and explained by contrasting patterns of global warming! I left out one small detail—the fifteenth-century climate pattern presupposes a period of wet conditions for the Hohokam. It should be appreciated that the consequences of global warming are not always drought. They are sometimes manifested by other extreme climate events, and in the Hohokam case there is strong evidence that their civilization was brought down by alternating periods of drought and megafloods in the late fourteenth and early fifteenth centuries due to the teleconnected relationships of hot Pacific sea surface temperatures and the El Niño-Southern Oscillation pattern. I spend much of this book's last chapter defending this hypothesis as originally formulated by Donald Graybill and colleagues.

It is not the charge of anthropologists and archaeologists to reconstruct climate and I hope those who know much more about the subject will not be too critical of my foray here. In our field, we are most interested in the implications of climate change patterns for explaining the context of social change and transformation. I endeavor in this book to rearrange what we know about the late prehistoric landscapes of the continent so that we can begin to formulate new questions about the demographic shifts of migration and immigration in times of great climatic upheaval. I demonstrate that such processes were in play within and between the regions examined here, but the hard work of piecing the puzzle together and understanding how the coalescence of different cultural groups acted as a catalyst for culture change remains to be done.

Acknowledgments

A project built over decades could not have been accomplished without the contributions of many people too numerous to thank for their individual efforts. Here I will acknowledge those who have had the most direct impact on the final product. First, I thank the reviewers of the various renditions of this manuscript, including Charlie Cobb, Chris Rodning, and Michael Trinkley. Their suggestions and advice were particularly instrumental in improving the presentation of the arguments and inferences made; I am especially grateful to Charlie Cobb for his important feedback throughout the process of writing and rewriting. Wendi Schnaufer of the University of Alabama Press is to be thanked for guiding me through the publication process and offering advice on how to present the subject matter. My copy editor, Irina du Quenoy, is to be thanked for expertly improving the flow of the text and ensuring a clearer presentation of the ideas and arguments offered herein. I benefitted from discussions with my longtime friend and colleague, Chester DePratter, who commented on certain aspects of the book and delivered his thoughts to me in the form of one of his signature three-ring binder notebooks, a treasure I will curate and always appreciate in my personal library. I would like to thank Steve Davis for steadfastly and earnestly responding to my many queries about late prehistory in North Carolina and granting me early access to the digital archive housed at the Research Laboratories of Archaeology. Others who made important contributions include, in no particular order, Joe Herbert, Chuck Cantley, Brett Riggs, Mark Williams, Gary Huckleberry, Gail Wagner, Chris Judge, Greg McCabe, Jeffery Clark, and Brett Hill. I would also like to acknowledge the support of the many individuals who sweated out the fieldwork with me in the swamps and remote areas of Francis Marion National Forest. In particular, the efforts of George Price, Ray Talley, Ken Pinson, Jason Smith, and Sean Taylor will always be appreciated and fondly remembered. Thanks are extended to Chris Judge and Gail Wagner as well for providing me the opportunity to work on three wondrous and magnificent sites in the Wateree Valley. A special note of gratitude is extended to my friend Bob Morgan of the United States Forest Service for his sponsorship and encouragement of my effort to understand regional abandonments in South Carolina. I am grateful to my son Patrick for contributing most of his free summer months while in college to help manage my collections and databases. Finally, I would like to thank my wife, Nita, whose loving support ensured that my project came to fruition.

Abbreviations

ADT	Agricultural Demographic Transition
AMO	Atlantic Multidecadal Oscillation
AMS	Accelerator Mass Spectrometry
BR	Brainerd-Robinson
ENSO	El Niño–Southern Oscillation
GDO	Greater Desert of Ocute
INSF	Inferred Nuclear Sherd Family
MCA	Medieval Climatic Anomaly
PDO	Pacific Decadal Oscillation
PDSI	Palmer Drought Severity Index
RDR	Ring of Drought Resilience
SAM	South Appalachian Mississippian
SSA	Seriated Subassembleges
SST	Sea Surface Temperature
TU	test units

Megadrought in the Carolinas

Introduction

We learn by rearranging what we know.
—attributed to Ludwig Wittgenstein

This book represents the culmination of a personal research quest that began nearly three decades ago with a series of cultural resource investigations into the archaeological record of the Francis Marion National Forest located just north of Charleston, South Carolina. Although the forest has never been the site of large-scale excavations, survey coverage of systematic shovel testing has been extensively and consistently implemented, yielding an unusually clear picture of prehistoric settlement and occupation in a sizeable geographic area scarcely duplicated anywhere in the United States. Throughout the years, as I became more familiar with this record, I began to suspect that the late prehistoric Mississippian occupation was abruptly terminated sometime prior to European contact. I knew that several neighboring coastal regions were also suspected to have been abandoned during this same time frame, most notably the middle and lower Savannah River valley so admirably described and documented by David Anderson (1990a, 1994, 1996). He attributed the Savannah River abandonment to a combination of the natural instability of Mississippian chiefdoms (i.e., chiefly cycling) and mid-fifteenth-century drought as reconstructed from tree-ring sequences. Adding the Mississippian occupation of the Francis Marion National Forest (henceforward to be referred to as the Central Coast region) to the mix, however, suggested to me that we were looking at an event that was macro-regional in scale and whose primary cause was most likely an external factor such as prolonged drought and not the normal instability of chiefly rivalry and factionalism. Indeed, such organizational oscillations had occurred frequently over several centuries prior without leading to regional evacuations. Moreover, there was no good evidence that the Central Coast Mississippian population had ever achieved chiefdom-level organization, meaning that it would probably not have been subject to the destabilizing effects of chiefly cycling, and yet, it was included in this macroregional abandonment event.

In order to prove my thesis, however, I needed first to demonstrate rigorously that the Central Coast had indeed been abandoned at the same time as the middle and lower Savannah River valley. Ceramics provided the vehicle for this endeavor because there is not a more sensitive analytical method for assigning relative ages to occupations. When this information is combined with associated absolute dates such as radiocarbon ages and further integrated into a system of correlated ceramic sequences from adjacent regions, it is transformed into an even more powerful instrument for precisely dating archaeological deposits. Of course, this simple solution was not that easy to execute, because it required satisfying several levels of closure. First, I needed to understand which attributes of the ceramic assemblage would be most conducive to chronological analysis. This was accomplished by consulting the models of decorative change developed by Chester DePratter and Chris Judge (1986) and Tony Boudreaux (2005) for the closely related cultures of the middle Wateree and upper Pee Dee valleys. Next, I needed to ensure that the relevant attributes were consistently recorded in the survey collections that I had earlier analyzed as well as in those reported by other researchers who were conducting investigations in the Francis Marion National Forest. Finally, I needed to obtain radiocarbon dates from selected contexts that would provide a basis for assigning absolute date ranges to the ceramic phases I had constructed. Once these matters were ironed out over a period of six or seven years, I was able to demonstrate rigorously that the Central Coast was abandoned, or at least severely depopulated, between AD 1425 and 1450, coinciding with the Savannah River abandonments.

The first three chapters of this book focus on the structure of the Central Coast Mississippian occupation. Chapter 1 describes the subsistence base and settlement pattern of the Central Coast Mississippian system, which is representative of a marine adaptation dependent on estuarine resources as opposed to the intensive corn agriculture economies of the classic Mississippian systems of the interior Southeast and Midcontinent. Chapter 2 presents the steps I took to construct a cultural phase chronology for the Central Coast and to develop an interregionally correlated model of ceramic decorative change capable of undergirding a reconstruction of the occupation history of the Central Coast. Chapter 3 demonstrates statistically that the Mississippian system that had flourished in the region for some two centuries was abruptly terminated in the early fifteenth century and that its sizeable population either perished or out-migrated. Since it is my thesis that this event was forced by prolonged drought, I consider how dry climatic conditions might have adversely affected the productivity of the estuaries on which the Central Coast Mississippian populations depended.

Reasoning that similar abandonment events might have occurred at this time across other coastal regions of the South Atlantic Slope, I set out to examine the archaeological record of these adjacent areas to see what they might be able to tell us about the geographic scale of the hypothesized disaster. This endeavor was supported by the model of ceramic decorative change developed earlier and presented in chapter 2. Although the evidence is spotty and not conducive to statistical demonstration, this review of the archaeological record led me to determine that there is a high likelihood that fifteenth-century abandonment events occurred over a vast contiguous area of the South Atlantic coastline, from the Cape Fear River in North Carolina to the Ogeechee Valley in Georgia, covering roughly 140,000 km^2 or 34.5 million acres. The Soto entrada encountered the western boundary of this unoccupied and tangled zone as the Spaniards pushed forward between the native chiefdoms of Ocute and Cofitachequi in 1540; the entrada's chroniclers were so impressed by its deserted condition that they called it the "Desert of Ocute" in reference to its deserted condition. Building on this convention, I named the larger zone the Greater Desert of Ocute (GDO).

To more fully understand how megadrought might have caused the abandonment of the GDO, I consulted available studies of paleoclimatic reconstructions for North America. The timing of the prehistoric abandonment events of the South Atlantic Slope coincides with the Medieval Climatic Anomaly (MCA), which is best described as a period of recurring multidecadal droughts (megadroughts) associated with increased global temperatures occurring between AD 1000 and 1500. Within this period, the continental distributions of megadrought shifted in accordance with the teleconnected patterns of sea surface temperatures in the Atlantic and Pacific oceans (B. Cook et al. 2014). Although these paleoclimatic distributions have not been precisely mapped, related modern studies have identified two distinct drought concentrations in the Southeast and Southwest regions of the United States (McCabe et al. 2004, 2008) that correlate with major prehistoric abandonment events. The abandonments of Chaco Canyon in New Mexico, the Four Corners region of the Colorado Plateau, Cahokia, and the middle Mississippi valley during the eleventh through thirteenth centuries correspond to a pattern of concentrated drought frequency associated with high Atlantic and low Pacific sea surface temperatures. For their part, the abandonments of the lower Tennessee and Ohio River valleys and the Greater Desert of Ocute in the fifteenth century conform to similar patterns brought about by high sea surface temperatures in both oceans. Clearly, there is more to these correlations than mere coincidence and the fact that local tree-ring sequences identify the fifteenth century on the Atlantic Slope as a period of increased drought intensity strongly supports the inference that megadrought caused the depopulation of the GDO.

Chapter 4 presents a paleoclimatic overview of MCA megadrought across the continental United States and discusses the geographic correspondence of this information with the major prehistoric abandonment events examined in this book. Chapter 5 details the state of knowledge concerning the settlement histories of the GDO. Both archaeological and ethnohistorical evidence is reviewed, including the sixteenth-century landings of the Ayllón voyages along the Southeast Coast, in which the colonists' concerted search for human occupation resulted in only sparse evidence of the latter.

Another objective of this project was to strengthen the overall argument by examining the archaeological record of the regions immediately surrounding the GDO, a composite area that I call the Ring of Drought Resilience (RDR) because it was less affected by drought due to its fortunate geographic position and more favorable geophysical structure relative to evapotranspiration rates. I reasoned that if populations were out-migrating from the GDO, then they were moving to zones already occupied, for which the RDR is a prime candidate because we know that this area teemed with native populations at the time of the Soto entrada. I further proposed that the arrival of refugees from the GDO should be evidenced by crowding, conflict, and unusually high population growth rates in the host territory. I examine this proposition in chapter 6. Although I limited my search to three regions within the RDR, in each case there was ample evidence of expanding population and increased hostilities marked by the construction of palisade fortifications. Together these factors satisfy the conditions conducive to the formation of coalescent societies as described by Stephen Kowalewski: "heavy and continuing population loss, warfare, people on the move as refugees and migrants, and abandonment of large areas" (2006:118). Most important to the work I have outlined here is Kowalewski's hypothesis that the process of coalescence can result in a weakening of hierarchical authority structures in societies such as chiefdoms and a transformation of the prevailing social structure into one that is more communal or collective in nature. I return to a consideration of this hypothesis in the last chapter.

Chapter 7 explores the possible impacts of indigenous drought-related disease epidemics that might have afflicted Mississippian systems prior to and during initial European contact. I focus this discussion on the ethnohistoric and archaeological details related to the pestilence that purportedly beset the great town of Talimeco just prior to the arrival of the Soto entrada in the chiefdom of Cofitachequi on the Wateree River. The origin of the disease, if indeed the Soto accounts describe an actual event, has always been an issue of great controversy. I show that the criteria for an abrupt abandonment of a great town in the 1530s appear to be met if the Mulberry site is assumed to be the archaeological manifestation of Talimeco. This argument runs counter to the prevalent view that Mulberry in fact represents the town of Cofitachequi where Soto met the casica of the chiefdom, who in turn directed him to the abandoned town of Talimeco (cf. DePratter 1989). My reasons for contesting this interpretation are discussed at length, based on a review of both the ethnohistoric accounts and archaeology of the Mulberry site. Speculating as to what kind of disease outbreak might have caused the pestilence, I draw on the work of Acuña-Soto and colleagues (2002), who have argued that the *cocoliztli* epidemics that raged during the sixteenth century in central Mexico were caused by an indigenous form of hemorrhagic fever. Contagions of this type are transmitted to concentrated human populations by rodent reservoirs that expand during wet spells that succeed prolonged droughts. A wet spell linked to a prior prolonged drought occurred in central South Carolina in the 1520s and 1530s, a climatic episode that could well have exposed Talimeco to an outbreak fueled by a growing rodent reservoir centered in the agricultural fields and town environs. Certainly, these coincidences by no means demonstrate on their own that an indigenous epidemic caused the abandonment of Talimeco, but if this was a chronic condition of dense human population aggregations under drought cycles, then such diseases might have been endemic to Mississippian towns with previously long, continuous occupation histories like Talimeco.

The eighth and final chapter provides an opportunity to, in Glenn Davis Stone's (2001:164) turn of phrase, "square the chickens." That is, to step back, hit the pause button, and ask what parts of the narrative as presented do not fit within the nice neat box I have constructed to contain it and what additional factors might require consideration if we are to arrive at a more satisfactory explanation. As I see it, four topics emerge that need further elaboration. First, I would like to know more about why some regions in North America seem not to conform to the expectations for abandonment based on the PDO/AMO model of chrono-geographic drought intensity that I developed in chapter 4. I further consider this issue in relation to the Hohokam Core Area of southern Arizona, where a fifteenth-century collapse and abandonment scenario occured under wet conditions. Second, my inference that Talimeco was abandoned because of an indigenous disease outbreak would seem to imply that this might have been an endemic problem for Mississippian towns with highly concentrated and large populations, especially in an age of megadroughts. This led me to search for correspondences between dry-wet periods and other seemingly abrupt town abandonments. I show that the abandonment of Moundville in Alabama at the turn of the fourteenth century is a good candidate for this type of event scenario and suggest some ways we might go about confirming the occurrence of viral diseases, which rarely manifest their symptoms on human bone because of the short interval between contraction and death.

The third topic concerns Kowalewski's (2006) hypothesis, which is only tangentially addressed in earlier chapters but demands further development. He expects that coalescence should result in breakdowns in centralized authority structures as clans reassert their corporate power. In the Southeast, his model has been used primarily to explain the processes that led to the collapse and transformation of Native American cultures in the context of the depredations of European contact in the seventeenth and eighteenth centuries (Beck 2013; Ethridge 2009; Ethridge and Hudson 2002). What I am suggesting, along with some others (for example, Regnier [2006, 2009, 2014] and Kowalewski himself), however, is that disruptions of similar magnitude were already underway in the Southeast prior to European contact, in the fifteenth century, and that these changes most likely involved similar circumstances of coalescence. In this case, however, I argue that the external driver was megadrought. This interpretation extends the chronological depth of the "fracture zone" (Ethridge 2009) well into the prehistoric era and provides a more informed basis for understanding the structure and condition of Native American societies at the time of first European contact. As a starting point, I examine what evidence there might be to support the hypothesis that the hierarchical authority structure of late prehistoric and protohistoric chiefdoms was weakening.

Fourth, I consider how we might go about structuring the study of coalescent contact and assimilation on the South Atlantic Slope. I begin by examining some archaeological instances of coalescence and discussing the stages of assimilation that such culture contact situations produce. In doing so, I rely on the excellent archaeological models of migrant-host relations generated from years of investigation by other scholars in regions of the southwestern United States, along with Amanda L. Regnier's book (2014) on the remarkable coalescence of three distinctively different cultures in the Alabama River valley in the fifteenth and sixteenth centuries. Based on these findings, I provide an initial strategy for studying these processes in places where material culture differences are much more subdued, such as sites related to the Mississippian culture of the South Atlantic Slope.

Before closing this introduction, it will be beneficial to describe the larger cultural context in which this story plays out. The Mississippian settlement of the Central Coast, the other groups of the GDO, and many of the groups within the RDR are part of a culture area referred to as South Appalachian Mississippian (SAM). In his original formulation, W. H. Holmes (1903:130) defined what would later be called SAM as a pottery tradition of distinctive stamped designs found in a province covering the states of Georgia and South Carolina and contiguous portions of Alabama, Florida, North Carolina, and Tennessee. Holmes was referring more specifically to the archaeological groups that inhabited the South Appalachian Province after about AD 1000, whose members typically stamped the exteriors of many of their pots with unique geometric motifs and built temple mounds in their main towns. Historically, this province was home to ethnographic tribes belonging to three linguistic stocks: Siouan, Iroquoian, and, most prominently in Holmes's opinion, Muskogean, suggesting that the ancestral macroculture was to some degree multiethnic in composition. We know today that the geographic boundaries of the South Appalachian Province supported a long sequence of earlier pottery-producing cultures, some of which exhibited ceramic traits suggesting continuity with the culture Holmes described. To distinguish the later cultural pattern from these earlier associates and ancestors, Leland Ferguson (1971:11), following James B. Griffin (1967), promoted the concept of South Appalachian Mississippian. By this construction, he meant a broadly defined cultural system that shared the material attributes of complicated stamped pottery and temple mounds (and raised elite residences) but also incorporated a wide range of other traits reflecting the Mississippian adaptive complex evident throughout much of the eastern United States in late prehistory (cf. Muller 1997). These included corn agriculture, large villages, high population densities, palisade fortifications, and "Southern Cult" ceremonialism, or what is referred to today as the Southeastern Ceremonial Complex (see King 2007) or Mississippian Ideological Interaction Sphere (Reilly and Garber 2007:3–4).

A modified version of Griffin's (1967) map of the larger Mississippian macrocosm shows the boundaries of the various types of Mississippian culture as they stood fifty years ago (Figure I.1). Based on the current state of knowledge, the northern boundaries of the SAM culture area should be adjusted slightly northward to include the upper Catawba valley. Mississippian culture spread eastward over a relatively time span from its origins in the American Bottom on the middle Mississippi River to its maximum geographic extent on the Atlantic coast. Anderson (1999:226) mapped this spread at the turn of the millenium. Modern radiocarbon calibrations suggest that the specific times of arrival are about 100 to 120 years later than those depicted on his map, but the rate of progress remains unchallenged. This process of spread, or Mississipianization, began at about AD 1050 in and around Cahokia, reached the Black Warrior River valley (Moundville) at around AD 1120 and emerged at Etowah some 20 to 40 years later. Adam King (2003) would position the beginnings of Mississippian in the Etowah Valley as much as 70 to 150 years earlier, but I agree with Pauketat's (2007:115–116) argument that the radiocarbon dating King presents as evidence is too late to support his chronology. If the calendrical ages of the dates are accepted at face value, a beginning date for Mississippian at Etowah is more likely to fall around AD 1150, which is consistent with what is understood to be the timing of Mississippianization's spread eastward. If we take an estimated start date of AD 1150 for fully Mississippianized Etowah, then the calibrated radiocarbon dates from the Central Coast to be discussed herein suggest that SAM's

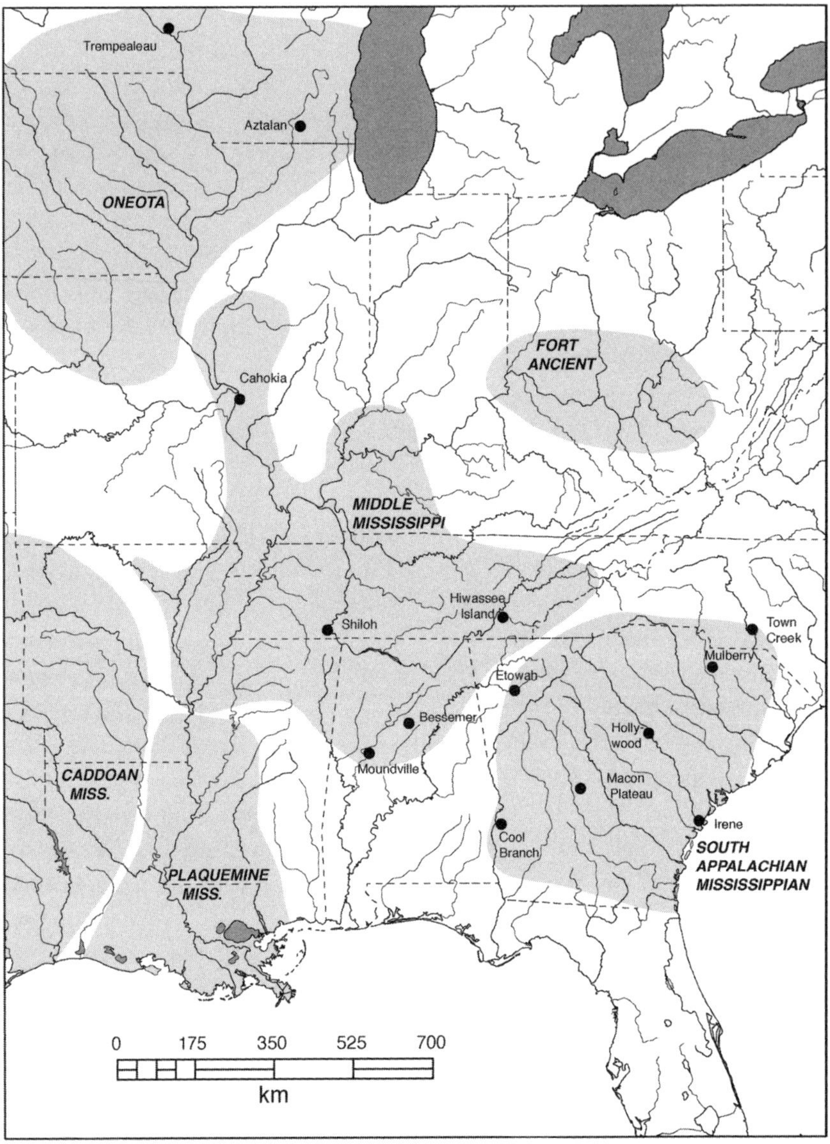

Figure I.1. Distribution of Mississippian culture areas (adapted from Griffin 1967).

march to the sea was complete by AD 1275 to 1300, some 125 to 150 years after it was initiated at Etowah.

Most SAM societies appear to have been chiefdoms of varying complexity and geographic reach (see Anderson 1990; Hally 1993; Widmer 1994). From a cultural evolutionary perspective (Sahlins 1963; Service 1962), chiefdoms are situated within the broad adaptive niche of sociopolitical evolution bridging simpler egalitarian tribes and complex archaic states such as those that emerged in Peru, Mesoamerica, Mesopotamia, Egypt, and south Asia. Chiefdoms share characteristics with both ends of the evolutionary spectrum and the various forms they take are limited only by the ability of individual chiefs to exert control over and manipulate diverse modes of economic production to aggrandize their status and increase personal access to wealth (see Earle 1991). Because of this intermediate position, much argument has ensued since Oberg's (1955) original formulation concerning the parameters and meaning of the concept (cf. Beck 2003; Kristiansen 1998; Pauketat 2007). As Timothy Earle (2011:31) states, chiefdoms "have little coherence as a type, except for the flexible power strategies of chiefs and their chieftaincies."

Even if one recognizes the utility of the concept, which I do, it is difficult to capture its essence in a short sound bite. Henri Claessen's (2014:7, 9) definition is succinct, but he is still forced to partition it into two dimensions. The first describes the political settlement hierarchy of chiefdoms: "two-tier sociopolitical organizations (structures) consisting of a number of local communities (villages, hamlets, and families) with their own leaders, which are subordinated to the leader of the political center." Although straightforward, this description does not capture the full essence of chiefdoms, because it fails to illuminate the way in which the system is internally structured and held together. This requires an understanding of the role of chiefs, whom Claessen describes as "hereditary sociopolitical leaders occupying political offices, having the alleged capacity to influence positively fertility, the possibility to amass goods and food with which to support the chiefly hierarchy and erect great works, and the power to eventually fight wars of defense or conquest." Given their central role in society, chiefs are deified figures that represent a direct connection to the pantheon of founding ancestors.

I would add to these dimensions Earle's (2011:31) concept of chieftaincy, not to be confused with Elsa Redmond's (1998) notion of the same name that refers to short-term, centralized political leadership exerted over autonomous villages by leaders of achieved rather than ascribed status. Earle's chieftaincy encompasses the ascribed chief's support network: the warriors, priests, village managers, and other persons who profess fealty and act as the chief's agents in collecting revenue and advancing strategies of control and power enhancement. According to Earle the primary objective of chiefs is to obtain and direct the movement of wealth (e.g., food surpluses, corvée labor, and exotic trade goods) to themselves and their loyal cadre of supporters, ensuring their ability to maintain and expand control of society (within the limits of societal reproduction). The chief's power is further enhanced by his ability to escalate surpluses, through either intensification of the local resource base or the geographic expansion of tributary pools by means of military conquest or colonization of new territory by close kin.

This version of what chiefdoms are is commonly characterized as the classic Polynesian chiefdom model, which has come under weltering attack from some scholars (cf. Pauketat 2007; Yoffee 1993). Pauketat has gone as far as to claim that the concept is merely a delusion and recommends its retirement from discussions of Mississippian social organization. However, anyone who reads the ethnohistory of the Natchez as collated by John Swanton (1911) from early eighteenth-century French accounts will readily recognize the classic authority structure of Polynesian-type chiefdoms, and consequently it would be difficult to make the case that this particular social formation did not guide the fundamental trajectory of cultural evolution in the Mississippian world system.

1

The Central South Carolina Coast

At the Margins of South Appalachian Mississippian Interaction

As we move in space from the center of the area—which, it may be noted, is not of necessity the geographical center—and extraneous cultural elements begin to intrude, we shall come at length to a place where the culture, for the ethnologist, is what may be called weak.
—Melville Herskovits, *A Preliminary Consideration of the Culture Areas of Africa*, 1924

Anthropologist Melville Herskovits's stand against heterogeneous and poorly integrated cultural expressions comes across as somewhat prejudiced, but I do not believe that he intended to cast geographic patterns of culture in a negative light. He was attempting to explain the divergent characteristics of cultures situated at the interfaces between "typical cultures" and those with vivid and extensive cultural expressions. Kopytoff (1987:5) recognized the same phenomenon when he described the groups situated along the frontier interstices between dominant sub-Saharan cultures in Africa as "not quite formed societies." What he meant by this description was that these interstices were populated by multicultural groups with diverse ethnic backgrounds that came together to form societies with new institutions and adaptations that differed in significant ways from their various parent communities. The Mississippian settlement of the Central Coast was in many ways just such an intermediate cultural expression. It was a new society, forming late in the process of Mississippianization and attempting to adapt the basic tenets of South Appalachian Mississippianism to an environment that was poorly suited for intensive corn agriculture, the prime mover of SAM's chiefdom tribute systems and demographic expansion. The abandonment of the region in the middle of the fifteenth century cut short the experiment. I show in the chapters to come that this event was not limited to the Central Coast but extended to a much larger area of the Georgia and South Carolina Coastal Plain, most likely as a consequence of megadrought. In this chapter, I present an overview of the Mississippian occupation of the Central Coast to set the stage for a better understanding of the broader implications of this event.

By far, the microenvironment most intensively exploited by Mississippian populations on the Central Coast was the tidal creek mouths of Bulls Bay and Cape Romain (Figure 1.1). The creek mouths have been investigated over a span of 24 km and extensive Mississippian occupations have been documented from as far north as Jeremy Creek at McClellanville, South Carolina, to the northern end of Sewee Bay at Andersonville. Southward, the creek mouths are submerged below sea level and this probably served as both an environmental and cultural boundary, effectively separating the groups around Bulls Bay from contemporaneous formations to the south along Copahee Bay, Hamlin Sound, and Mount Pleasant, South Carolina.

The Bulls Bay estuary was a major focus of Mississippian subsistence strategies. The shell middens that demarcate the settlements are not highly mounded, ranging between 25 cm and 80 cm in thickness, but they are horizontally extensive. None of the site clusters have received the kind of broadscale excavation that would expose house patterns, but when similar middens have been peeled away along tidal creek mouths in Beaufort, South Carolina (Mozingo et al. 2004), Georgia (Keene 2004; Pearson 1984), and North Carolina (Millis 2011), ample evidence of domestic structures has emerged. The contemporary Broad Reach site in Carteret County, North Carolina, although not Mississippian in affiliation, is particularly informative. Here, 9 acres of topsoil and shell midden were stripped away, exposing 22,265 postholes and over 100 house plans, numerous human and dog burials, and a wide range of pit types (Millis 2011:6-2). The late prehistoric shell middens along Bulls Bay are of comparable extent and likely contain similar configurations of residential features. Tibwin Creek supports two of the most intensively shovel-tested Bulls Bay clusters. Three major residential sectors have been defined in the North Tibwin tract (Cable 2002a), which together encompass 14.5 acres. The South Tibwin tract has been subjected to two intensive surveys (Brockington and Associates 2011; Cable et al. 1994) and limited test excavations (Cable et al. 2005). The two main

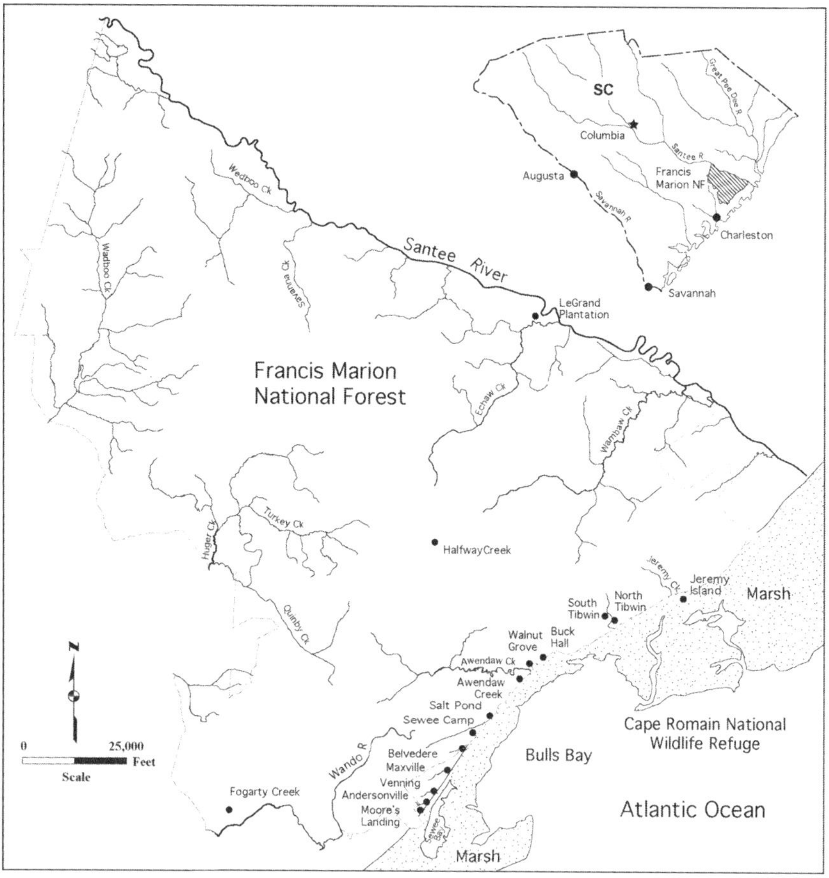

Figure 1.1. Locations of Central Coast Mississippian site clusters.

clusters there cover 11.6 acres. Smaller clusters across both tracts add an additional 2.9 acres of either residential space or special activity area. Similar aggregations occur across each of the tidal creek mouths.

Close-interval shovel testing at the relatively small shell midden site of Fogarty Creek (38BK552) in the upper reaches of the Wando River provides a view of how one residential sector within the larger coastal clusters was probably organized (Figure 1.2). The site was originally identified during a Forest Service road-grading project in 1983 that exposed a portion of a human burial and a grooved ax. Emergency salvage excavations by Dan Elliott (1984) led to the exposing of a portion of a wall-trench house as well as the recovery of artifact collections from an overlying shell midden. In addition, Elliott identified pit features scattered across the graded roadbed. Subsequent investigations of the site

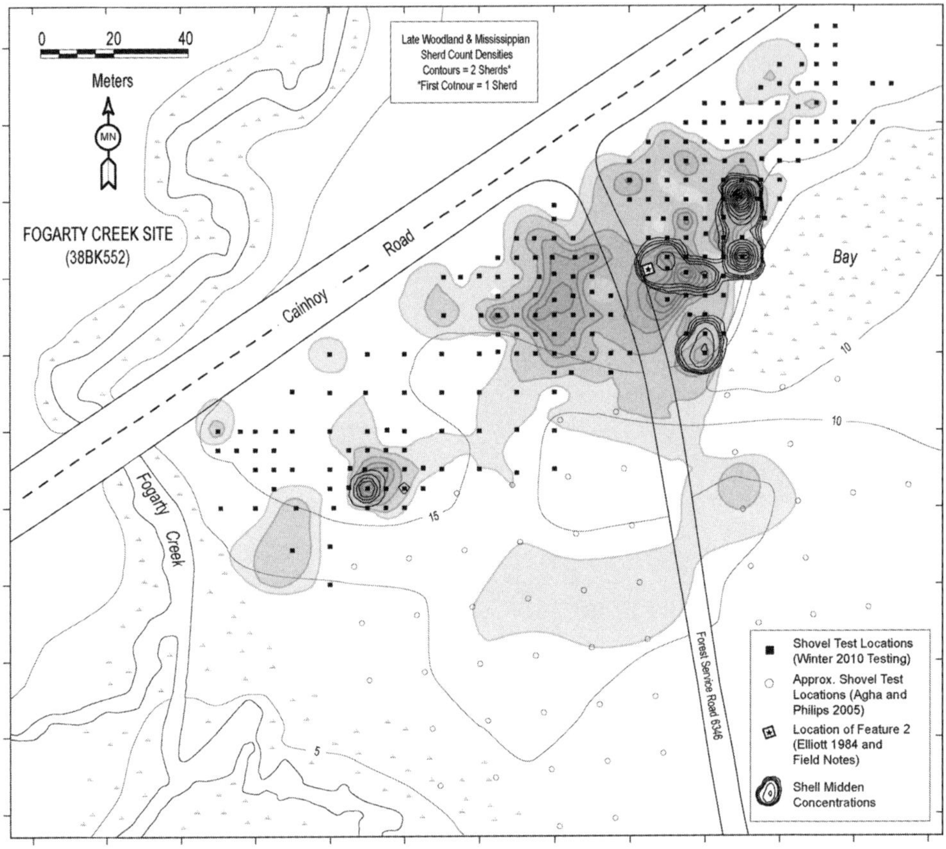

Figure 1.2. Mississippian sherd densities, Fogarty Creek site.

have focused on establishing site boundaries and further delineating artifact density distributions with 15 m and 5 m interval shovel tests (Agha and Phillips 2006; Cable 2013). Based on artifact classification and associated radiocarbon dating, a long history of late prehistoric occupation has been documented that spans the period between about AD 1200 and 1425. The site is situated near the head of Fogarty Creek, a tributary of the Wando River; the presence of dense shell midden concentrations indicates that estuarine resources were brought to the site by boat. In addition to shellfish, the midden contains a wide range of other estuarine species (hardhead catfish, gaftopsail catfish, red drum, and mullet), fresh and brackish water species (snapping turtle, mud turtle, and gar), as well as terrestrial animals (deer, opossum, rabbit, raccoon, turkey, and box turtle) (Reitz and Weinand 1993).

The main shell midden is situated on the northeastern side of the site and its rich assemblage of subsistence remains suggests strongly that it served as a community refuse dump for multiple household units. Late prehistoric ceramics are spread across a maximum site area of 3.7 acres, although the high artifact density core measures only about 1.3 acres. The core is situated directly adjacent to the main shell midden and appears to represent the site's major residential zone. Since this area was occupied over a long period, it is likely that individual household patterns, as reflected in artifact density distributions, are obscured by occupation superpositioning. The house foundation that Elliott partially exposed, for instance, was buried by later accumulations of the expanding midden. Counts of Late Woodland and Mississippian ceramics from the shovel test samples provide a rough estimate of the relative contribution of occupation components. About 13.4 percent of the ceramics represent Late Woodland types, 22.9 percent belong to the Early Mississippian occupation, and 63.7 percent are associated with components dating between about AD 1325 and 1425. Based on equivalent areas of maximum late prehistoric ceramic distributions, it can be inferred that the clusters along the tidal mouth of Tibwin Creek may have supported as many as eight multiple household aggregations like the one described at Fogarty Creek, with relatively long-lived histories bracketed by spans of 100 to 200 years.

The other site type commonly occurring in the Central Coast settlement system consists of small, dispersed nonshell midden occupations scattered throughout the interior uplands. The importance of such sites has gone unappreciated because of their low archaeological visibility, brought about by detection methods ill-suited to the effective definition of small sites with low artifact densities. To adequately begin to define such sites, a much shorter shovel test interval is required than the standard 30/15 m survey methodology routinely deployed in cultural resource management surveys in the region. In 2015 the United States Forest Service sponsored a small testing project at one of these smaller sites, the Royall site, employing short-interval shovel testing to map the spatial structure of the Mississippian occupation and plan test excavations in areas that might yield evidence of cultural features. The site is situated on a high terrace overlooking Huger Creek, a tributary of the Cooper River on the southwestern edge of the Francis Marion National Forest. Shovel testing at 5 m intervals defined a mid-fourteenth-century occupation extending over an area of 2.3 acres.

Mapping of sherd frequency distributions at Royall led to the identification of a large, high-density concentration on the eastern margin of the site and several smaller, less dense concentrations scattered across its central and western portions (Figure 1.3). This pattern closely resembles the structure of Lamar farmsteads scattered across the upland tributary creeks of the upper Oconee River valley as described by James Hatch (1995:141). Hatch had the opportunity to conduct surface collections on relatively undisturbed, but plowed, Lamar

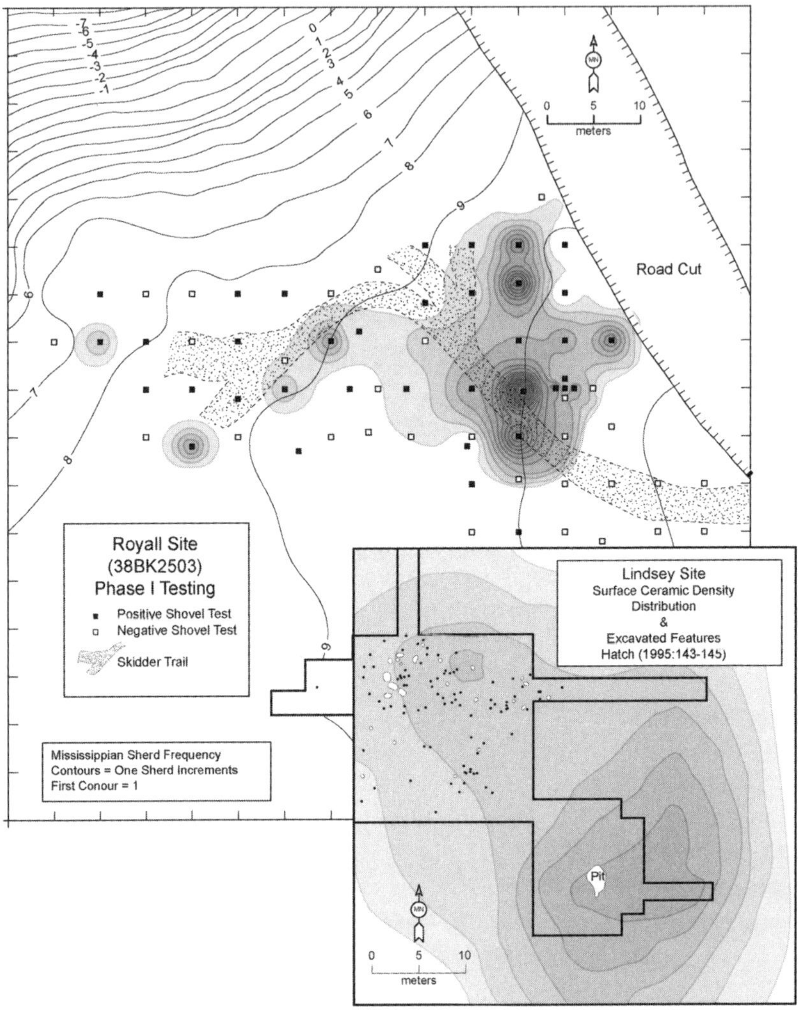

Figure 1.3. Comparison of farmstead site structure between the Royall site and the Lindsey site. AMS = accelerator mass spectrometry (adapted from Hatch 1995).

sites ranging between about 2 and 4 acres in size. The map of one of the sites Hatch investigated, the Lindsey site, may be seen in Figure 1.3. In all four of the sites he investigated, large, high-density concentrations were situated on site perimeters and generally conform to the expectations for midden refuse and debris concentrations associated with functions better controlled away from household activity space, such as pit roasting (see O'Connell 1987:83). The secondary concentrations to the side of the refuse middens contained the remains of houses and other domestic activity features. In the case of the Lamar farmsteads, architectural remains typically consist of a circular "winter house" and several rectangular "summer" houses and storage facilities that appear to have been occupied by nuclear or extended families. Although Hatch's upland Lamar sites are believed to represent permanent homesteads, it is likely that seasonally reoccupied farmsteads would reflect similar artifact patterns and yield comparable evidence of domestic habitation. In the case of the small, dispersed nonshell midden occupations found in the interior uplands, while it is theoretically possible that they represent permanent homestead occupations, the proximity of the large coastal shell midden sites suggests that they functioned as seasonal arms of the larger settlement system centered along the coastal fringe.

This possibility coincides with Larson's (1978; 1980:206–209) and Crook's (1978; 1986:17–20) models of dispersed shifting cultivation, which they argue characterized the late prehistoric and protohistoric groups occupying the Southeast Atlantic coast. These models are largely derived from the AD 1570 accounts of the Jesuit fathers Rogel and Sedeno concerning the geographic movements of the Guale surrounding the mission on St. Catherines Island. Crook's comprehensive model of the seasonal settlement cycle, which he named the Guale Annual Model, stipulates the fissioning of family units away from permanent coastal villages and into dispersed agricultural farmsteads in the interior between the spring planting and mid-summer's harvest in June or July. Numerous objections to this facet of the model have been raised, generally concerning aspects of the context and meaning of the Jesuit testimony. G. Jones (1978:191) has argued that dispersal was not related to agricultural production strategies at all but was linked to a desire on the part of the Guale to avoid the Spanish mission system. From another perspective, because the ethnohistoric accounts coincided with a period of severe drought between AD 1562 and 1571, some doubt has been cast on whether dispersal was a common feature of the Guale settlement system or merely a situational response to environmental stress (see Saunders 2002:42).

Importantly, the Larson and Crook models are not based solely on ethnographic analogy. A principal element of their argument is grounded in the agronomic character of the Coastal Plain, which is suboptimal for corn agriculture. The optimal development of corn requires maximum root penetration to depths of at least two to four feet in soils that are well drained but are also capable of delivering uninterrupted and abundant moisture uptake (Feldman 1994). A single corn plant can absorb 30 to 50 gallons of water during its growth cycle (Benson 2011; Bundy 1998). Deep loams, silt loams, and some silty clay loams are ideal for such requirements, especially along stream terraces where crucial nutrients are continually replenished by overbank flooding. The deep sandy soils of the Coastal Plain suffer from nutrient deficiency, are poorly drained in creek bottoms, and are subject to desiccation in the uplands due to excessive drainage (Helms 2000; Spivey et al. 1986), conditions that are clearly suboptimal for traditional swidden agriculture techniques (Hudson 1976:18; Larson 1980:221; see also Larson in Thomas 1993:48–49).

Production limitations on the Coastal Plain would have been felt most severely during drought periods, when pot watering would have been required to sustain proper soil moisture levels. The brackish water of the tidal creek mouths would have been ill-suited for this purpose, and for this reason it is likely that agricultural fields were commonly located on loamy soils adjacent to freshwater creeks and bay ponds in the interior. Worth (1999) presents a convincing argument based on ethnohistoric accounts that the protohistoric groups of the South Atlantic coast could grow ample amounts of corn under favorable climatic conditions and that corn agriculture was a well-established adaptation extending back several centuries. However, it is also true that corn agriculture is practiced more effectively over the long run on loamy stream terraces and floodplains along the courses of major interior river channels.

These contrasting environmental features surely presented coastal populations with a different set of strategic challenges than those faced by Mississippian groups in the interior Southeast, where wide expanses of fertile and regenerating loamy alluvial bottoms abounded. Corn played a major role in the agricultural economies of many Mississippian polities and occurs ubiquitously in the archaeobotanical records of groups that are inferred to have depended on it as a primary staple. For instance, Welch and Scarry (1995:406) found that both corn cupules and kernels were present in 72 to 100 percent of all flotation samples they analyzed from Moundville and other associated sites in the Black Warrior River valley. Moreover, the samples (n = 60) yielded an average of 136 corn elements each. By contrast, corn has only been sparingly identified in late prehistoric flotation samples on the South Atlantic coast (Thomas 1993:51–52; Wagner 1997). No confirmed incidences of corn have actually been identified in flotation samples from the Central Coast Mississippian deposits. Larger sample sizes and evaluations of differing conditions of preservation are needed to fully understand potential contrasts between the agricultural productivity of coastal and interior Mississippian systems (cf. Saunders 2002), but existing evidence suggests that corn productivity in the Coastal Plain was environmentally limited.

The archaeobotanical record from the Broad Reach site in Carteret County, North Carolina (Millis 2011), is an exception to the poorly documented records from other coastal locations and appears to provide a better measure of the importance of corn in the diets of coastal groups. Although the occupation there is represented by White Oak series ceramics and is not affiliated with the SAM ceramic tradition, it is contemporaneous with the Bulls Bay populations and likely deployed a similar intensity of agricultural production. Corn ubiquity was relatively low in the 124 samples analyzed, amounting to just over 16 percent, with an average of only 5 elements per sample. However, 3 smudge pits contained 89 percent of the corn elements and if these are eliminated, an adjusted average of only 0.6 elements per sample remains. Even when adjustments for poorer preservation are allowed, comparatively low ubiquity and density values for corn in the Broad Reach samples suggest that this coastal population was much less dependent on corn agriculture than those occupying the interior riverine systems.

Interestingly, the relative paucity of corn and other cultigens in archaeobotanical samples from concentrated residential sites may supply additional support for the existence of dispersed, shifting agricultural production in the Mississippian economies of the South Atlantic coast. It stands to reason that corn and other crops, if harvested from remote farmsteads, would be processed prior to transport back to residential bases. Such a dynamic would reduce the presence of cultigens in village deposits, whereas much higher ubiquities and densities should characterize the associated farmsteads. Scarry and Steponaitis (1997:115–117) have demonstrated this very relationship in their comparisons of corn cupule ubiquity and density in flotation samples from farmsteads and villages in the Moundville and West Jefferson systems along the Black Warrior River valley in Alabama. If we are going to understand the role that corn and other domesticates played in coastal economies, it will be necessary to investigate the small, late prehistoric sites scattered across the adjacent interior landscapes where cultivated fields and seasonal farmsteads are most likely to have been concentrated.

Available archaeobotanical and faunal data (Poplin et al. 1993; Reitz and Weinand 1993; Runquist 1980; Wagner 1997) indicate that the economy of the Bulls Bay Mississippian population corresponded to a version of what Bruce Smith (2001) refers to as "Low-Level Food Production." The model was developed to describe the economies of the earliest farming communities in the American Bottom, but he notes that its fundamental structure continued to guide the prehistoric economies of the Eastern Woodlands throughout time and space. He further (B. Smith 2009) contends that prehistoric groups in the Mississippi River valley corridor constructed resilient long-term resource niches through manipulation and modification of the landscape (e.g., agricultural land clearing, controlled burning, and agroforestry). In addition, biodiversity was enhanced through the application of land management techniques to create and maintain plant community patches at earlier stages of succession. Basic elements of the niche included (1) the exploitation of animal species with high reproductive rates that were resistant to overharvesting (e.g., white-tailed deer, migratory waterfowl, and fish); (2) the relocation and enhancement of the density of fruit-bearing trees and ground berries to the edges of agricultural plots and old fields;

(3) the broadcast sowing of starchy grains such as chenopod and goosefoot in floodplains; (4) in-place expansion and encouragement of nut and mast-bearing trees; and (5) the multicropping of domesticated plants.

Along the South Atlantic coast, the exploitation of estuarine resources—including shellfish (oysters and hard clams), fish (e.g., hardhead and gafftopsail catfish, red drum, mullet, and gar), and terrapin—can be added to this list. Given the probability that agriculture was nonintensive, dependence on estuarine productivity may have been the factor permitting sedentary settlement of the coastal regions during the late prehistoric period (see Reitz 1982; Reitz et al. 2012:77–81; Saunders 2002:43; Waselkov 2012). Reitz and colleagues (2012) found that prehistoric fishing in the Georgia Bight was selectively focused on the efficient harvesting and mass capture of a primary group of species (e.g., seatrouts, mullets, gars, Atlantic croakers, hardhead catfishes, gafftopsail catfishes, flounders, red drums, and star drums) that school or occur in large aggregations. Most of these same species are represented in the late prehistoric faunal samples from Bulls Bay and it is likely that very similar communal extractive techniques were employed.

Under the best of conditions, the Central Coast was suboptimal for the long-term survival of a Mississippian adaptation based on intensive corn agriculture. No platform mounds have been identified within the boundaries of the Central Coast Mississippian settlement area and although several occur across the surrounding South Atlantic Coastal Plain, they are not numerous and do not occur as multiple mound complexes as is commonly characteristic of town centers in the interior. Most of the Mississippian mound constructions along the coast in fact are rounded burial mounds (see Pluckhahn and McKivergan 2002) and not the rectangular substructure mounds that are typically understood to be the locations of elite and chiefly residences and temples. For instance, the Irene site, which is one of the most substantial coastal town centers, had a single, modest platform mound that was covered over after AD 1300 by a much larger, rounded burial mound (Caldwell and McCann 1941). A smaller burial mound was also present at the site, the initial construction of which was probably contemporaneous with the earlier platform mound, although its use extended into the fourteenth century. Present also was a large rotunda or council house associated with the post–AD 1300 occupation. Both DePratter (1983:207–210) and Anderson (1990:564–565) argue that this architectural shift from a community administered through the rituals of platform mound ceremonialism to one of council house meetings indicates the breakdown of a hierarchical authority structure and its replacement by one of greater horizontal integration and egalitarian status relations. Anderson further observes that the later (Irene period) mortuary complex exhibits increased indications of regional warfare and a general impoverishment of the occupants as evidenced by a decline in grave offerings.

All of these patterns could represent simply the effects of cyclical deterioration in centralized authority (i.e., cycling) associated with the instability of chiefdom organization. I suggest, however, that the long-term sequence is more indicative of the inability of simple chiefdoms along the coast to withstand megadrought, given their marginal agricultural systems and the accompanying decline of estuarine productivity. From this perspective, the Irene phase component at the Irene site chronicles the last stages of societal decline before the coastal zone was abandoned. In the next chapter, I present a method to identify late fifteenth- and sixteenth-century occupations, with the aim of demonstrating the abandonment of the Central Coast and other surrounding regions across this time horizon.

2

A Model of Ceramic Change for the Eastern Wing of South Appalachian Mississippian

All models are wrong; some models are useful.
—Box, Hunter, and Hunter, *Statistics for Experimenters: Design Innovation and Discovery*, 2005

The basic framework of a ceramic chronology model for the eastern wing of South Appalachian Mississippian has great general utility for organizing the historical sequences of individual regions, but my express purpose in presenting it here is much more specific: to demonstrate rigorously which regions experienced depopulation or abandonment during the fifteenth and sixteenth centuries, or what is culturally identified as the Lamar period. The discussion is divided into two parts. The first presents the details of a new sequence that I propose for the late prehistoric occupation of the Central South Carolina coast. Earlier, David Anderson (1982) and Michael Trinkley (1980, 1981a, 1981b, 1989) conducted important foundational studies of ceramic description and cultural chronology for the period of interest. Building on their findings, and with the support of Robert Morgan of the Francis Marion National Forest, I developed a rigorous program of ceramic type description for the Late Woodland and Mississippian periods and organized and calibrated it chronologically through the strategic selection of radiocarbon samples from deposits representing the full span of known occupational phases. Anchored by this finalized Central Coast sequence, the second part of the chapter extends the model to the other regions of the eastern wing of SAM where Lamar occupation has been investigated. Ensuing chapters apply the model to the survey data from the Central Coast and to the other coastal regions of the SAM culture area in order to define the extent and nature of the Lamar period depopulation.

The Central Coast Chronology

When archaeologists first confront the material culture of a new region, they are forced inevitably to devise ways of bringing chronological order to a messy and diverse collection of artifacts. Initial classifications are always incomplete and cursory because of observational confusion, classificatory fuzziness, and inadequate sample sizes. Once a relatively effective system is devised in the early stages of investigation, however, it tends to be perpetuated because it is tied so closely to the phase structure of a region (see O'Brien 2000). In the case of the Central Coast, this problem was aggravated by the onset of Cultural Resource Management in the late 1980s and 1990s, which ushered in an unprecedented influx of new researchers tasked with describing ceramic patterns, with which they were unfamiliar, across large tracts of land. Under these conditions, and with the added pressures of budget and time constraints, there was little motivation to improve on the original typologies. Moreover, the varied levels of expertise and training that this shift in funding entailed resulted in serious problems of classificatory consistency and replicability. This is a common problem in archaeology (see Fish 1978; Whittaker et al. 1998), but very little attention has ever been devoted to measuring intra-analyst error rates, let alone inter-analyst agreement.

One of the most debilitating problems in conducting regional analysis on the Central Coast is the failure of many analysts to partition plain sherds into culture-historic types. This shortcoming seriously impedes attempts to compare and seriate the proportion of surface treatment types, which is an especially important feature of Mississippian assemblages since plain surfaces are abundantly represented. As I have taken the opportunity to revisit and reanalyze some of these earlier collections, I have noted more subtle intra-analyst inconsistencies in the identification of rim and body decoration, in thresholds for analytical inclusion based on sherd size, in criteria for identifying complicated stamped motifs, and in criteria for distinguishing series and surface treatment classes. My own work is not immune to these problems.

After living with this issue for more than two decades, I concluded that the only way to achieve the necessary level of classificatory consistency in conducting regional analysis would be to reanalyze the collections that offered the easiest access and greatest strategic value for constructing a detailed chronology. Unfortunately, my early attempts were *ad hoc* and would require yet another "re-revisit" to complete. Ultimately, I wanted the analysis to record information not only on technology and design but also on function and formation processes, in other words, on a wide set of attributes that would inform on larger issues of cultural process *sensu* Binford (1965).

It was Binford's contention that our path to understanding culture change would best be served by the construction of multivariate classification systems in which attribute associations measuring formal variation (morphological and decorative) across different dimensions (technical and design) could be first isolated and then compared to yield a more detailed picture of both culture process and chronology. Of greatest importance was the notion that these dimensions should be independently measured and analytically separated so that the character and direction of their interaction could be rigorously observed. Many early culture-historic classifications conflated the attribute dimensions in question in ways that promoted confusion and, at times, distorted our ability to understand change by obscuring the possibility of certain associations that might be inconsistent with an analyst's *a priori* assumptions about temporal positions and homologous relationships (O'Brien and Lyman 2002). Maintaining analytical independence improves our ability to discriminate chronological trends in artifact classes at much finer levels of detail, because these dimensions are apt to change at different rates. A hybrid type that earlier interjected stress on the prevailing typlological system, then, fits comfortably within a framework of multidimensional analysis and provides new insights into the processes that transformed cultural systems.

Phase Construction

Because the archaeological record of the Francis Marion National Forest and adjacent properties has been constructed from a preponderance of survey level data and restricted test excavations, quality contexts containing discrete deposits are not numerous. Nevertheless, sufficiently representative contexts exist that can be enlisted to construct a basic and near complete ceramic phase chronology. Moreover, in coordination with the Forest Service, I have been able to obtain radiocarbon dates for several of these contexts (Table 2.1), providing a basis for calibrating the phases in time. Descriptions of the contexts are presented here.

Feature 1 at the Fogarty Creek site (38BK552) was excavated by Dan Elliott (1984) during a brief salvage operation in the summer of 1983. The pit was truncated during road construction, exposing an oval stain measuring 1.3-×-2.5 m. Fill from the pit consisted of dark brown sandy loam, faunal bone, some oyster and periwinkle shell, lithics, rock, and a number of large sherds. Differentiated soil coloration suggested to Elliott that several pits may have been responsible for the formation of the feature, but the narrow range of ceramic types (Late Woodland tradition) and large sherd sizes in fact indicate that most of the material contained in the pit was derived from one or several filling episodes closely proximate in time. Many of the large sherds exhibited edge and facial wear patterns, suggesting that they had been used as recycled tools, most likely in the operation of the pit. The organically rich content of the fill further implies that the feature represents an abandoned roasting pit or ceramic pit kiln (Cable et al. 2013:130–151). Fill was removed without vertical separation in two halves. The radiocarbon sample (Beta-331757) consisted of carbonized hickory nutshell combined from three soil samples (#710–712) taken from both halves.

An oyster midden at the Sewee Camp site (38CH426) was sampled with a 1-×-1 m test unit, in conjunction with a project Palmetto Research conducted for the Charleston Nature Conservancy (Cable et al. 2013:153–158). The unit was placed adjacent to a shovel test that yielded terminal Late Woodland ceramics. The midden was thin, extending to depths of only 20 to 30 cm. A sample of 18.5 gm of deer long bone from the associated shovel test was submitted for radiocarbon analysis (Beta-331759).

Espenshade (1989) excavated a 1-×-2 m unit into a thin oyster midden (20 to 30 cm) at 38CH1048 located near the Sewee Camp site. I (Cable et al. 2013: 223–224) revisited the collection from the excavation and confirmed Espenshade's observation that the ceramics represented a nearly pure, very early Mississippian occupation (i.e., Jeremy phase). A sample of charcoal from below plow zone in the lower portion of the shell midden (Prov. # 51, TU 2, Level 3) was submitted for radiocarbon analysis (Beta-386634).

During his salvage work at the Fogarty Creek site (38BK552), Elliott (1984: 3–4) also established a trench consisting of three 1-×-1 m contiguous test units arranged in a line along the freshly graded roadbed in an area that had exposed a section of oyster midden, a human femur, and a grooved axe. A portion of a wall-trench structure was identified at the base of the excavation in TU 3, at the south end of the excavation trench. Elliott reports that a prepared floor of periwinkle shells covered the base of the excavation south of the wall trench in TU 3, suggesting that the remaining portion of the house extended southward outside of the excavated exposure. After consulting with Elliott (personal communication, August 14 and 16, 2015), and armed with the ceramic evidence, I interpret the results from the excavation in the following manner: An earlier Mississippian house that may have contained a human burial was later superimposed by an expansion of the shell midden at the edge of the habitation area (see Figure 1.2). Two radiocarbon dates were obtained from the overlying shell midden (Feature 2). The first (Beta-283028) consisted of material that had originally been identified as maize kernels (Wagner 1997) recovered from Level 1, S1/2 (#701) and Level 2, NE Quad (#702) in TU 1. However, the material could not have been maize because the fractionation ratio (-26.2 ‰) identifies a C3-pathway plant

Table 2.1. Radiocarbon Dates from South Carolina Central Coast Contexts

Site	Sample No.[a]	Method	Provenience	Phase Association	Material
38BK552	331757	AMS	Feature 1 (Pit)	Sewee Camp	Charred Hickory Nut Hulls
38CH426	331759	AMS	ST S4 (Oyster Midden)	Sewee Camp	Deer Bone Collagen
38CH1048	386634	AMS	TU 1/2 (Oyster Midden)	Jeremy	Charred Plant Material
38BK552	283028	AMS	Feature 2 (Oyster Midden)	Fogarty Creek	Charred Plant Material
38BK552	331758	AMS	Feature 2 (Oyster Midden)	Fogarty Creek	Charred Hickory Nut Hulls
38CH44	282199	Radio-metric	Test Unit (Clam Midden)	Fogarty Creek	Knobbed Whelk

[a]Beta-Analytic lab numbers.

(cf. O'Leary 1988). Maize is a C4-pathway plant and returns considerably lower fractionation ratios in the neighborhood of -13.0 ‰. Wagner (personal communication, July 27, 2015) informs that the material was most likely pine resin, which can be confused with maize kernels or cupules when preserved in a bubbly state. An additional sample of carbonized hickory nutshell (Beta-331758) from the same two-level proveniences and a third provenience (#707: TU3, Level 2, SE Quad) was later submitted for radiocarbon analysis. The strong presence of rim strips and other appliqué decorations in the deposit indicated that the deposit was associated with the Lamar period.

The North Tibwin sample from 38CH1977 comes from superimposed shell middens that were tested by Palmetto Research during a survey of the tract (Cable 2002a:419–422). A shovel test exposed a clam midden of approximately 55 cm in thickness that overlaid a 20 cm thick oyster midden in the southern portion of the site. The shovel test was placed close to the bank of Tibwin Creek, where an associated assortment of sherds eroding out of the bank was recovered. This collection numbered 918 sherds, many of which were large enough to al-

Table 2.1. *continued*

Conventional ^{14}C Age B.P. (±1-σ)	$^{13}C/^{12}C$ Ratio	Calibrated Age[b] (±1-σ)	1-σ cal AD Age Ranges	Area under Probability Curve
790 ± 30	-25.3 ‰	AD 1240 ± 22		
780 ± 30	-21.9 ‰	AD 1245 ± 20		
660 ± 30	-22.9 ‰	AD 1335 ± 39	1285–1306 1363–1385	.333 .349
580 ± 40	-26.2 ‰	AD 1358 ± 36	1312–1358 1388–1410	.461 .221
560 ± 30	-24.7 ‰	AD 1368 ± 37	1322–1347	.339
1110 ± 40	-0.9 ‰	AD 1389 ± 61[c]	1393–1416	.343

[b]OxCal 4.2[93] Program (Bonk Ramsey 1995, 2001), terrestrial samples calibrated using IntCal13 Curve and Marine sample (knobbed whelk) calibrated using Marine 13 Curve (Reimer et al. 2013).
[c]Marine Reservoir Correction (ΔR) of 148 ± 70 for *Busycon carica* (knobbed whelk) collected in 1840 from Bulls Bay, South Carolina (Thomas 2008) applied to calibration.

low the identification of complicated stamped motifs and patterns of rim decoration. The absence of rim strips (a chronological marker of Lamar assemblages in the region), the dominance of undecorated rims, and a secondary contribution of Town Creek-Irene axis (see Reid 1967) rim decorations (e.g., notches, hollow reed punctations, rosettes, pellets, nodes) indicate that both of these contexts are associated with a regional pre–Lamar phase (Tibwin).

One other radiocarbon date (Beta-282199) deserves mention, although the associated ceramic data have yet to be analyzed. This particular assay was obtained on a knobbed whelk shell recovered in a test unit excavated at 38CH44 in the Salt Pond cluster (DePratter 2007). The midden consisted primarily of hard clam. Earlier surface collections from the site (J. Gardner 1992) yielded two rim strips and an incised sherd whose culture-historic affiliation is not known. Although incising appears at or after AD 1400 in Early Lamar assemblages across the SAM culture area, it also occurs in much earlier Thoms Creek and Refuge assemblages on the Central Coast. Direct inspection of the incised sherd from Salt Pond would be required to confirm its culture-historic association. Rely-

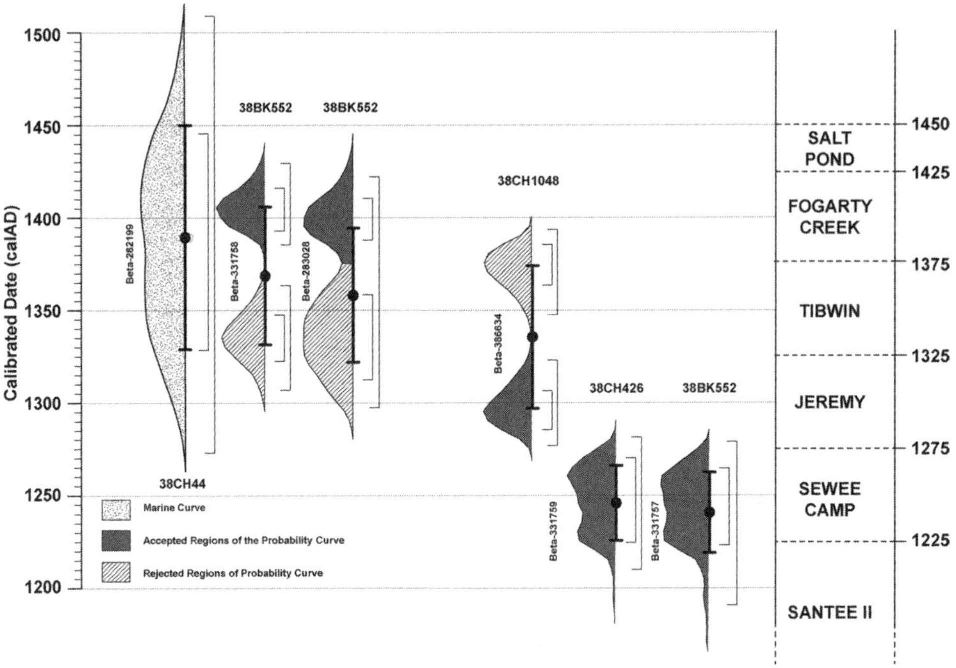

Figure 2.1. Central Coast culture chronology.

ing on marine shell samples in localities that do not yet have the benefit of a region-specific marine reservoir correction (ΔR) is not advised, but in this instance, a reasonable value can be used from an offset derived from a knobbed whelk shell collected in 1840 in Bulls Bay (Thomas 2008). The resulting calibration yields an extremely wide one-sigma range of AD 1328 to 1450, which is characteristic of calibrations based on the marine curve, but the mean falls at AD 1389 and is consistent with the age of the oyster midden (Feature 2) at the Fogarty Creek site.

The dating of these contexts, which appear to represent the entire chronological range of phases represented in the larger data base compiled from investigations conducted in the Francis Marion National Forest, produces a compressed late prehistoric chronology that covers only about two hundred years, extending from terminal Late Woodland to Early Lamar (Figure 2.1). There are multiple date options for the samples whose means fall into the fourteenth century. If the means for the three latest dates are accepted, then it could be concluded that the latest occupation on the Central Coast terminated prior to AD 1400. However, given what I know about the dating of Early Lamar phases in other regions, it seems much more likely that the later date options more ac-

Table 2.2. Comparison of Bayesian End Boundary Ages with Those of the Proposed Chronology

Phase	Proposed End Date (cal AD)	Bayesian End Date[a] (cal AD ± 1-σ)
Fogarty Creek	1425	1368 ± 40
Tibwin	1375	1340 ± 29
Jeremy	1325	1313 ± 23
Sewee Camp	1275	1273 ± 18
McClellanville	1225	1235 ± 30

[a]OxCal 4.3.2. program (Bronk Ramsey 2009a).

curately reflect the age of the Fogarty Creek phase occupation. As a working model, then, I initially proposed a sequence of four ceramic phases of 50-years duration each, ending at AD 1425.

Although the current model suffers from a general paucity of dates and a significant void in the Tibwin phase interval altogether, a Bayesian test of the sequence confirms its internal consistency (Bronk Ramsey 2009a, OxCal 4.3.2). A 6-phase model assuming no overlap (the earliest and latest phases were vacant boundary holders) yielded an Agreement Index (AI) of 116.3%. An AI of 60% approximates a chi-square probability of .05. However, the Gibbs Sampler disagrees with my proposed phase boundaries (Table 2.2). The differences are not great, but if the means of estimated end dates produced by the Gibbs Sampler are used to establish phase boundaries, the Mississippian phases would end slightly earlier than I propose. Nevertheless, if the time ranges that I infer are entered into the model, an AI of 125.1% is obtained. More dates are needed to fine-tune the absolute chronology, but the model indicates that my generalized estimate of phase boundaries constitutes an excellent fit with the available data.

Ceramic Seriations

I arranged ceramic data from the selected contexts in a set of comparisons that take on the formal appearance of the classic Ford (1962) frequency seriation but represent in fact a range of ordering methods as outlined by O'Brien and Lyman (1999). The first comparison uses the radiocarbon date associations to anchor and order paste type percentages. This method is not a true seriation but is often referred to as absolute seriation (O'Brien and Lyman 1999:132–

136). Braun (1985:509) employed it to construct a temporal gradient of sherd thickness based on associated radiocarbon dates, while Harrington (1954) applied it to plotting hole diameters of historic kaolin pipe stems. Others (Binford 1962; Plog and Hantman 1990) have reduced the results of absolute seriation to regression equations. When this technique is used to produce interval scale chrono-clines from straight-line regression models, however, it assumes that change occurs at a steady rate (see O'Brien and Lyman 1999:136), an assumption that is probably not warranted in many situations. The paste seriation described here is not enlisted for this purpose; the data are insufficiently detailed to achieve such an undertaking. Instead, the technique is used only to identify the general direction of change.

A second comparison conforms more closely to the classic structure of a frequency seriation. This technique orders aggregates of artifacts from specific kinds of recovery units (i.e., sites, deposits, surface collections) based on their intrinsic qualities independent of stratigraphy or any other technique of time ordering (see O'Brien and Lyman 1999:109–110). Here, the attribute of exterior surface treatment type is seriated independent of paste type. However, the collections are isolated by paste type so that the degree of temporal overlap between traditional ceramic types in the region (i.e., combinations of paste and exterior surface treatment attributes) can be monitored and interpreted.

Finally, a third comparison monitors change in rim decoration through an occurrence seriation. This kind of seriation measures only the presence/absence of an attribute or type (Dempsey and Baumhoff 1963; O'Brien and Lyman 1999: 119–121). The technique is particularly well suited for ordering attributes with low occurrence rates where relative frequencies within contexts are subject to high stochastic error.

All of these comparisons monitor change at the attribute level and as such conform to what LeBlanc (1975) has termed "microseriation." It is his opinion that the seriation of individual attributes will produce a finer temporal resolution than traditional culture-historic types. O'Brien and Lyman (1999:131) point out that there is no *a priori* reason for attributes to occupy tighter temporal spans than types (i.e., combinations of attributes) and Duff (1996) has demonstrated that ceramic types can produce fine-grained temporal seriations equal in quality to those produced from single attributes. However, I contend that the value of isolating attributes resides not in achieving a finer-scaled seriation as an end product but in improving our ability to trace the overall sequence of changes across technological and design dimensions so that the implications of these changes can be better understood within a social adaptational framework.

The seriations here rely primarily on sherd frequency data, although analysis has progressed far enough in several of these contexts to allow for the use of vessel membership data to avoid the pitfalls of biased proportions that arise when aggregates from a small number of vessels are overemphasized in a larger sherd assemblage. In an attempt to avoid this biasing effect, I have included in the larger analysis program a control variable that Orton and colleagues (1993:172) refer to as the "Nuclear Sherd Family." Sherd families are formed from cross mends and/or attribute matching and represent the total number of sherds that can be demonstrated or inferred to be members of a single pot. Nuclear sherd families represent a subset of this category, consisting of the sherds from a single pot found in a single recovery unit. Because my application of the concept relies almost exclusively on physical attribute correspondence, I refer to the control variable as "Inferred Nuclear Sherd Family" (INSF).

It is generally a simple procedure to identify INSFs in small and dispersed recovery units such as the interval shovel tests that comprise the bulk of the regional database. A repeated pattern in the shovel test data is that the total number of INSFs is nearly equal to the total sherd count. This characteristic is especially diagnostic of the Mississippian and Late Woodland occupations, which represent longer-term, sedentary settlements. In this context sherds from broken pots are scattered far and wide as a consequence of midden formation processes, size transformation from trampling, and storage and reuse of larger sherds from pot breaks (see Deal 1998; DeBoer and Lathrap 1979; LaMotta and Schiffer 2005; Schiffer 1987). Larger-sized recovery units (i.e., test units and block excavations) complicate the process and substantially increase the amount of time it takes to build INSF data sets (cf. Newell et al. 1949:75–77).

I produced INSF data for three block excavations of 2, 4, and 4 m^2 respectively on a small site with majority Middle Woodland components at Fort Bragg, North Carolina (Cable 2011a:320–364). The blocks were located in different areas of an arcuate refuse midden where INSFs with multiple sherd memberships would be expected. Of the 84 identified INSFs, 76% were composed of only 1 to 3 sherds each, while 13% had 4 to 11 sherds. Seven percent had memberships of 12 to 33 sherds and only one INSF contained numbers of sherds (n = 194) sufficient to constitute what is commonly viewed as a partial vessel.

The test unit (TU 4) excavated into the thin oyster midden at Sewee Camp that serves as one of the seriation contexts for this study showed an even greater degree of disorganization (Cable et al. 2013:158–164). Of the 58 INSFs, 78% were represented by only 1 sherd each and another 19% were characterized by memberships consisting of only 2 to 4 sherds. Just two INSFs had greater memberships, of 10 and 20 sherds respectively. The 1-×-2 m unit Espenshade (1989) excavated into the oyster midden at 38CH1048 produced 24 INSFs; all but one of these were represented by only one sherd (Cable et al. 2013:223–224). The remaining INSF in this case was composed of just two sherds. These data suggest that raw sherd counts will have negligible biasing effects for shovel test data and low bias for test units in midden deposits in the region. Furthermore, eliminating just the obvious sherd aggregates recovered from specific contexts, whether those contexts be shovel tests or test units, should lower this kind of bias to negligible levels.

Paste Seriation

Trinkley (1980:416–417, 1981a:19) used paste as the primary basis for distinguishing between the early (Jeremy) and late (Pee Dee) Mississippian ceramic series in the region; otherwise, he would not have been able to partition his abundant sherd collections from Jeremy Island and Awendaw Midden into only two large groupings. This is a fact not fully appreciated by later researchers who, at times, have conflated paste with design motifs and rim decoration, resulting in the creation of hybrid types of potentially dubious validity. Trinkley described Jeremy paste as containing moderate amounts of medium-to-coarse quartz sand grains, although he noted that some examples exhibited much finer sand particles. In contrast, Pee Dee paste was said to be composed of coarser quartz sand inclusions (coarse to very coarse) and was characterized by a "sugary texture," a term J. Jefferson Reid (1967:6) enlisted to describe the paste texture of Mississippian pottery at the Town Creek site. Trinkley found evidence that Jeremy was stratigraphically earlier and exhibited almost exclusively undecorated rims, while Pee Dee series ceramics were characterized by an approximate 10% representation of Town Creek-Irene axis (see Reid 1967:65) decorations (i.e., nodes, hollow reed punctations, punctated lips, rosettes, lip notching, pellets, and zoned punctations). The earlier McClellanville series, which Trinkley (1981b:11) had difficulty positioning chronologically but is now known to occur in the later half of the Late Woodland period (see Cable et al. 2013:169–180), was also defined on the basis of paste. Compact McClellanville paste was described as containing moderate amounts of medium quartz sand in a homogenized matrix, with some examples exhibiting even finer sand grains. Employing the Type-Variety system, Anderson (1982) arrived at similar categories, but, due to his classificatory method, the types conflated different dimensions of variation and his paste groupings are not as clearly differentiated.

Both Trinkley and Anderson described a clear trajectory of development, progressing from finer textured Late Woodland paste toward gradually coarsening pastes in the Mississippian sequence. That there is temporal directionality to changes in paste structure is not a new idea and researchers in the Mississippi Valley have been successful in documenting such changes over long spans of time (cf. Braun 1987; Feathers 1990; Hoard et al. 1995; Neff 1992; O'Brien and

Holland 1990; O'Brien et al. 1994; Tite et al. 2001). The observed technological shifts have been explained as products of selective forces aimed at improving the performance of pottery for intended uses reflective of an overall adaptation (e.g., stable storage, serving, sumptuary offering, cooking). I have yet to undertake scientific measurement of strength, toughness, and thermal shock fitness, as these researchers have, but based on their findings at least some evidence can be brought to bear on the nature of changes in the Central Coast paste sequence. Each of the broad paste classes that I have identified is briefly described here to provide an *a posteriori* foundation for the seriation.

Much confusion concerning the cultural and chronological relationship between Santee Simple Stamped as defined by Anderson (1982:302–308) and Trinkley's (1981b:13–15) McClellanville series has occurred over the years. Were these researchers describing virtually the same phenomenon, or were the entities sequential? Alternatively, do they represent synchronous regional variants? A survey conducted next to Sewee Camp below Awendaw, South Carolina (Cable et al. 2103:145–180), has provided the first significant and comprehensive view of this material since these early seminal works.

Within the 350-year span of the Late Woodland period, there is sufficient change in paste characteristics to merit a division into two formally distinct and sequential series. The earlier series is softer with greater oxidation, while the later paste is harder and exhibits evidence of being fired in a partially reducing atmosphere, as core colors range between brown and gray. Since it is evident that Trinkley was describing something very similar to the paste (i.e., well fired, sandy feel, sugary appearance, and gray cores) and surface treatment characteristics (i.e., simple stamped, plain, cord marked, and fabric impressed) of the later Late Woodland series at Sewee Camp, I have assigned it the name "McClellanville." By default, I have designated the earlier series by the name "Santee," which is similar but not precisely equivalent to Anderson's Santee Simple Stamped type.

The two series exhibit similar variation in sand composition classes, of which there are three. Class 1 consists of well-sorted and uniform silt and fine clear quartz sands of less than 0.5 mm diameter, with a mode of 0.1 to 0.3 mm. The paste in these specimens is virtually free of larger sand inclusions (Figure 2.2a). Class 2 exhibits an identical background matrix, but approximately 10 to 15% of the core surface contains larger rounded and subrounded clear quartz sand grains ranging between 0.75 and 1.50 mm in diameter (Figure 2.2b). Class 3 also contains a fine sand background but displays a much higher concentration of subrounded clear quartz sand ranging between 0.5 and 2.0 mm in diameter (Figure 2.2c). Approximately 25 to 40% of the area of core faces is composed of these larger inclusions. The reasons for this variability is not yet well understood.

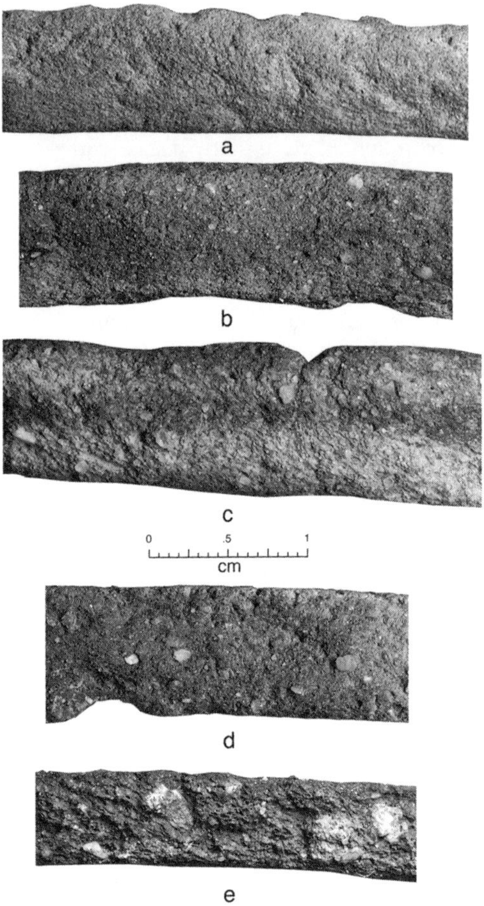

Figure 2.2. Typical sherd cores of McClellanville series paste variants (a–c) and Jeremy series paste variants (d–e).

Classes 2 and 3 may represent natural variation within a single technological system, but the intentional winnowing out of larger inclusions in the Class 1 paste suggests that it may have been used for a special range of functions.

The Jeremy series emerges at the end of the Late Woodland period and occurs with some simple stamped specimens in the middle of the thirteenth century (Cable et al. 2013:145–169). Unlike the Late Woodland types, which have densely packed fine sand throughout, the Jeremy series is characterized by a fine silty clay matrix. It is likely that the type as defined here contains only a subset of the ceramics Trinkley described as "Jeremy." From his description, it would appear that he included some examples that I have classified as either McClellanville or Pee Dee 1. A very high percentage of the Jeremy series exhibits gray and dark gray cores, the color of which extends to the exterior and interior wall surfaces (Figure 2.2d). This would suggest that most pots of this class were fired in a moderately well-controlled reducing atmosphere. Quartz sand inclusions generally range from 0.5 to 2.0 mm in size and are sparsely (10 to 15% of the matrix) to moderately (20 to 25% of the matrix) abundant. Variation in sand contribution has not yet been broken out into separate classes. Other inclusions found in Jeremy specimens include hematite nodules and fine and coarse limestone.

Limestone is particularly abundant in early Jeremy Simple Stamped, where it can occur in very coarse to granule sizes (Figure 2.2e). Hoard and colleagues (1995) argue that limestone temper can provide a threefold advantage over sand tempering at relatively low firing temperatures (i.e., below 600° or 700°C). First, mixing limestone with fine, sticky clay can release calcium ions through flocculation, rendering the composite grog stiffer and more easily workable without the added support of large amounts of temper to retard vessel wall collapse. Second, the addition of limestone potentially permits the production of thinner, globular vessels. Finally, due to ionization, limestone can make a fired clay body more resistant to mechanical and (probably) thermal shock. Hoard and colleagues (1995) also indicate that firing pots in a reducing atmosphere will inhibit lime spalling at temperatures up to about 700°C. The fact that Jeremy series paste is light in temper and exhibits clear evidence of being fired in a reducing atmosphere suggests that all of these factors may have been at play in a rapidly changing ceramic tradition. This interpretation is supported by the absence of lime blowouts in the Jeremy series pottery, which is a common feature of the earlier Wando series (Marcoux and Poplin 2012). Firing temperatures below 700°C would not have resulted in low-grade vitrification, which may explain the somewhat weak bond producing a "silty" feel of exposed and eroded surfaces on Jeremy series sherds. This practice appears not to have continued in the later stages of Jeremy series manufacture, but the addition of fine limestone might not be as evident from macroscopic examination.

Pee Dee 1 series paste is very hard and exhibits evidence of the initial stages of vitrification, which begins at firing temperatures between about 800° and 950°C. Although some examples are still "silty" to the feel, most feel "waxy." Gray cores are still abundant but clear and graded cores are common, suggesting that firing atmospheres were not fully reducing. This change may relate to changes in firing temperatures or to the time the clays were allowed to sinter before removal from the kiln. Quartz sand inclusions generally range from 0.5 to 2.0 mm in size and make up 15 to 20% of the clay body, as revealed by 2D examination of cores (Figure 2.3a). The mix of grain sizes is estimated at about 75% medium-to-coarse (0.25–1.0 mm) and 25% very coarse (1.0–2.0 mm). Kilikogulou and colleagues (1995; 1998) indicate that 20% quartz temper volume is optimal for producing pottery that is resistant to mechanical shock damage with only minimal

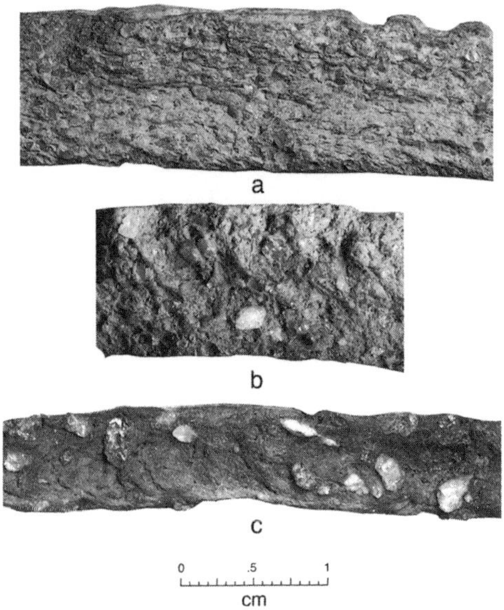

Figure 2.3. Typical sherd cores of Pee Dee series ceramic paste variants: (a) Pee Dee 1; (b) Pee Dee 2; (c) Pee Dee 2.5.

reductions in strength for ceramics that achieve low-grade vitrification, which seems to be applicable to the Pee Dee 1 series.

Pee Dee 2 series paste is also extremely hard and exhibits evidence of the initial stages of vitrification, suggesting the attainment of firing temperatures between 800° and 950°C. Unlike Pee Dee 1, however, all of the members of this series are "waxy" to the feel, which suggests that an optimal and consistent firing atmosphere for low-level vitrification had been achieved. Although gray cores are still abundant, clear or graded cores are most common. The principal difference between Pee Dee 1 and 2 paste types is the increased volume and coarseness of the quartz sand temper grains in the latter (see Figure 2.3b). Quartz sand particle size ranges are similar to Pee Dee 1, generally between 0.5 to 2.0 mm, but temper volume makes up between 35 and 45% of the paste when exposed cores are examined. The mix of grain sizes is estimated at about 50% medium-to-coarse (0.25–1.0 mm), 45% very coarse (1.0–2.0 mm), and about 5% gravel or fine granule (2.0–3.0 mm).

Kilikogulou and colleagues (1995; 1998) argue that quartz temper volumes exceeding 20% do not achieve any higher efficiency for toughness and mechanical fracture resistance; rather, higher volumes increasingly lower the strength of a vessel to withstand impact and thermal shock. Precisely why this modification was made in the Pee Dee 2 series is unknown, but the addition of greater amounts of very coarse and granule quartz sand might have improved survival during firing. Larger temper particles, especially platy types like shell or mica sheets, reduce crack propagation during firing (O'Brien et al. 1994:278). Quartz is not an ideal material for this purpose because its expansion coefficients differ significantly from clay, which weakens the bond of the matrix. However, Müller and colleagues (2015) note that increased volume of temper, which is characteristic of Pee Dee 2 paste, achieves higher fracture energy by impeding the progress of cracks. If higher firing temperatures increased manufacturing failures in pots constructed from Pee Dee 1 paste because of the lower volume of temper, then increasing quartz sand volume may have solved this problem. This factor would have also improved performance during subsequent cooking episodes.

The clay body of the Pee Dee 2.5 series is equal in hardness to Pee Dee 2 and exhibits a similar range of core and surface colors, suggesting that the firing technology was the same. However, the size and distribution of sand grains is distinctive. Average sand grain size is much larger (Figure 2.3c). Gravel-/granule-sized sand dominates, making up as much as 70% of observable sand grains. Sand volume, however, is vastly reduced, ranging only between 15 and 20%. Because sand mass is concentrated in larger grain sizes, large areas of the cores consist almost exclusively of clay. This configuration returns sand volume to the optimal range as described by Kilikogulou and colleagues (1995; 1998), which produces a sufficiently tough and strong paste body. However, the advantage delivered by greater temper volume to reduce fracture incidents during firing is lost. Pee Dee 2.5 paste, then, seems to have been designed to reduce impact fractures, a concern for pots that are moved and handled frequently, such as serving vessels.

Figure 2.4 displays an absolute seriation for the five-paste series. Each series distribution conforms relatively well to the foundational principles of the seriation model (see O'Brien and Lyman 1999:28–29, 126). That is, the seriated items must have a continuous and relatively short-lived occurrence in time (the principle of continuous occurrence), and their frequency distributions should approximate unimodal popularity curves of increasing, decreasing, or battleship shapes (the principle of normal distribution). Minor departures from these assumptions can be recognized, but because the contexts are primarily drawn from shell middens, such deviations are most likely explained by deposit mixing.

The thin lenses of shell midden that occur in Late Woodland and Mississippian occupations along the Central Coast contain artifacts with wide age ranges due to processes of secondary refuse accumulation in locales where food preparation activities such as cooking entailed significant deposit displacement through the operation of roasting pits, earth ovens, and other kinds of processing stations. Although each location within a laterally extensive shell midden tends to contain the residue from a very narrow time range, earlier material is also included in what would most commonly be considered a single stratum. Earlier material can enter a midden when the clearing or sweeping of an adjacent area pushes it

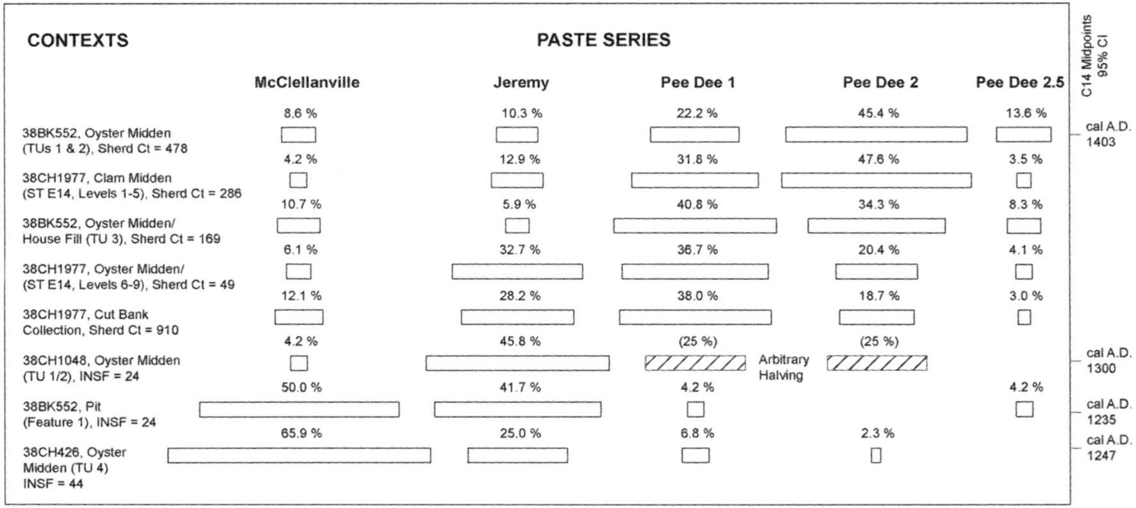

Figure 2.4. Central Coast Mississippian paste seriation. INSF = inferred nuclear sherd family.

up against the core of the midden, or it can be added by digging out and subsequently reusing features such as roasting pits. The other minor departures from the seriation model from the sweriation model are most probably a function of stochastic variation and under- and over-representation of certain deposits. For example, the Jeremy phase portion of the sample is extremely narrow in comparison to especially the Tibwin phase, even though the phases are thought to be of equal duration. That the series are sequential is not a revelation, as this trajectory was recognized by both Anderson and Trinkley. The Ford diagram, however, begins to provide an appreciation for the pace of change in paste configurations through this very narrow span of 200 years. Because the series appear to be linked to adaptive selection pressures operating on technological performance (i.e. crack propogation during firing, impact fracture resistance, vessel strength, etc.), it would not seem likely that entire manufacturing traditions would disappear abruptly even as newer production techniques gained in popularity. This suggests that there could be a relatively significant degree of temporal overlap in the life spans of these paste types.

I confronted this issue when I first examined the sherd assemblage from the Late Woodland pit (Feature 1) at the Fogarty Creek site (Cable et al. 2013:133–137). Here was an assemblage of large sherds that had been recycled for use as tools to perform functions related to the operation of the pit. It was remarkable, however, that the tools consisted of sherds with both McClellanville and Jeremy paste. The surface treatments on sherds from both series were of the Late Woodland tradition (i.e., simple stamped, check stamped, cord marked, and fabric impressed), leading me to propose that the temporal overlap was likely brief. I reasoned that the McClellanville sherds represented residuum that had rested in a provisional discard area (see Deal 1998:118–119) for a period of time before being recycled as tools. Obviously, we do not know the full range of conditions that bring artifacts that are seemingly offset chronologically into synchronic association. Overlapping use life histories can certainly contribute to this phenomenon, especially in the shortened time frame of interest here. Although household pots generally last fewer than 2 or 3 years in ethnographic situations, some types of larger, stationary vessels used principally for storage can last 10 to 20 years, or even 50 to 75 years in some situations (cf. P. Rice 1987:296–297; Shott 1996). In addition, some have suggested (Arnold 1999; Stark 2003:207; Stark and Longacre 1993) that inexperienced potters are more prone to adopting and creating innovations, which would, over time, result in overlapping stylistic change trajectories that may span one or two generations. All these factors could produce the subtle and short-lived overlap that I had earlier proposed for the Feature 1 assemblage. The results of the next seriation, however, suggest that paste series replacement may have been a relatively long and drawn out process in the region.

Frequency Seriation of Exterior Surfaces

If the paste series are chronologically ordered in the manner described by the previous seriation, then a complementary ordering should be evident in the patterns of exterior surface treatment. Moreover, these patterns should conform to the principles of a frequency seriation. That is, they should reflect continuous occurrence and unimodal distributions, at least in some of the surface treatment types. Table 2.3 displays percentage data for surface treatment types by context, organized by paste series. In some cases, collections from several contexts have been combined to produce larger sample sizes. An overview of the data reveals several general patterns. Namely, there is a shift over time from assemblages dominated by fabric impressed, cord marked, check stamped, and simple stamped toward collections dominated by plain and complicated stamped. If this shift were characterized by a brief period of transition, we would expect only minimal crossover associations of paste type and exterior surface treatment (i.e., hybrid types). However, a much more projected change trajectory emerges when inter- and intracontext associations are examined.

McClellanville and Jeremy series surface treatment types within individual contexts covary. That is, some contexts characterized by high percentages of McClellanville Simple Stamped also contain high percentages of Jeremy Simple Stamped (Cases 3 and 4; 7 and 8), while contexts with high percentages of McClellanville Complicated Stamped also contain high percentages of Jeremy Complicated Stamped (Cases 15 and 16; 10 and 11). Other contexts show gradations of this trajectory (Cases 20 and 21; 28 and 29), which appears to confirm the assumption of continuous change inherent in the paste seriation model. Covariance relationships of exterior surface treatment by context can also be seen in the later paste series. For instance, percentages of complicated stamped and simple stamped, although graded, are very similar for Pee Dee 1 (Case 27) and Jeremy (Case 28) in the trench excavation at the Fogarty Creek site. Other contexts (cf. Cases 19 and 20) display adjacent series with percentages that are different. This suggests that the time disparity between lineally related ceramic series will vary by context and that monitoring such incidences will provide a much richer appreciation of the pace of ceramic change and the nature of cultural phase transitions. Moreover, the Pee Dee 1 collection from the Tibwin Midden, Bank Cut (Case 14) is nearly identical to the Pee Dee 2 collection (Case 13) from this context.

The cases displayed in Table 2.3, however, represent only slices of the kinds of subassemblages that would be ideal to seriate. Ultimately, it would be desirable to group the slices into temporally discrete subassemblages, or what Rouse (1967:158) called "events." We know from the radiocarbon sample that the ceramics in this study were deposited over a span of not more than about 200 years.

The paste seriation indicates that there is a chronologically ordered and overlapping sequence of five paste types, which suggests that it is possible to analytically extract subassemblages that span durations of far less than 50 years each. Given the covariance of exterior surface treatments by paste type, we can expect rather significant proportions of different culture historic types (i.e., combinations of paste and exterior surface treatment) to coexist for variable periods and to be deposited in discrete and/or stratigraphically related contexts.

A major hurdle in extracting temporally organized subassemblages is presented by the problem of mixture. Shell midden deposits, which make up the bulk of the contexts, are horizontally accreting features that predominantly consist of deposits that accumulated over very short intervals of time (perhaps less than 10 years) at specific locations. Occasionally, locations within a midden are revisited, resulting in the deposition of additional layers that can sometimes be easily discerned by a change in molluscan fauna like the Tibwin clam-over-oyster midden at 38CH1977. At other times, the absence of variation in molluscan fauna in a particular deposit blurs stratigraphic breaks. Moreover, there is no guarantee that all of the sherds that occur in spatial coassociation within these midden layers are exclusively comprised of artifacts from the brief span of time when the target activities (events) were performed. These midden lenses were variably churned through cultural transformation processes such as house construction activities, episodic refill and clean-out of roasting pits, trampling during sheet accumulation, or localized midden mining for construction projects, etc. (Schiffer 1987). That such activities resulted in significant mixing is attested to by the routine occurrence of earlier Woodland and Archaic artifacts throughout the shell matrices.

Under such conditions, it is critical that analytical techniques be developed to distill mixed and ambiguously separated occupations into subassemblages of finer temporal resolution. A numerical index well suited for this purpose is the Brainerd-Robinson (BR) coefficient, also known as an index of agreement (Brainerd 1951; Robinson 1951). This measure was developed specifically to measure the similarity/dissimilarity between archaeological units based on relative frequencies (proportions) of associated artifact types or attributes. BR has figured prominently in studies searching for temporal sequencing of deposits when other evidence is not available (see Marquardt 1978), but it is equally useful for differentiating and identifying synchronous phenomena such as social groupings, political alliances, and interaction spheres (cf. DeBoer et al. 1996; Engelbrecht 1980; Hart 2012; Huntley 2008; Pugh 2010). This is because the BR coefficient is based on the general assumption that similar units will have similar proportions of artifact types or attributes whose structure of arrangement may result from a variety of causes other than the exclusive passage of time.

TABLE 2.3. CERAMIC TYPE PERCENTAGES FOR CONTEXTS USED TO CONSTRUCT THE SURFACE TREATMENT SERIATION

Case No.	Site	Provenience	Series	Fabric Impressed	Cord Marked	Check Stamped
29	38BK552	Combined	McClellanville	13.9	2.8	5.6
28	38BK552	Combined	Jeremy	0.0	0.0	1.9
27	38BK552	TU3	PD 1	0.0	1.6	1.6
26	38BK552	TUs 1 and 2	PD 1	0.0	1.0	0.0
25	38BK552	TU3	PD 2	0.0	0.0	0.0
24	38BK552	TUs 1 and 2	PD 2	0.0	0.0	0.0
23	38BK552	TU3	PD 2.5	0.0	0.0	0.0
22	38BK552	TUs 1 and 2	PD 2.5	0.0	0.0	1.7
	38BK552	Universe		0.9	0.5	0.9
21	38CH1977	E14, Combined	McClellanville	0.0	0.0	0.0
20	38CH1977	E14, Combined	Jeremy	0.0	0.0	2.3
19	38CH1977	E14, Combined	PD 1	0.0	0.0	0.0
18	38CH1977	E14, Combined.	PD 2	0.0	0.0	0.0
17	38CH1977	E14, Combined	PD 2.5	0.0	0.0	0.0
	38CH1977	Universe		0.0	0.0	0.3
16	38CH1977	Bank Cut	McClellanville	0.0	1.0	1.0
15	38CH1977	Bank Cut	Jeremy	0.0	0.0	0.4
14	38CH1977	Bank Cut	PD 1	0.0	0.0	0.0
13	38CH1977	Bank Cut	PD 2	0.0	0.0	0.0
12	38CH1977	Bank Cut	PD 2.5	0.0	0.0	0.0
	38CH1977	Universe		0.1	0.0	0.2
11	38CH1048	TU1/2	McClellanville	0.0	0.0	0.0
10	38CH1048	TU1/2	Jeremy	0.0	0.0	0.0

TABLE 2.3. *continued*

Simple Stamped	Plain	Burnished Plain	Complicated Stamped	Cob Impressed	Corrugated	Sherd Count
19.4	33.3	8.3	16.7	0.0	0.0	36
15.1	47.2	0.0	35.8	0.0	0.0	53
7.9	50.8	1.6	36.5	0.0	0.0	63
2.0	60.2	0.0	35.7	1.0	0.0	98
3.6	41.8	0.0	54.5	0.0	0.0	55
1.0	44.6	0.5	52.8	1.0	0.0	195
0.0	57.1	0.0	42.9	0.0	0.0	14
0.0	26.7	20.0	51.7	0.0	0.0	60
4.5	45.6	3.0	44.0	0.5	0.0	574
35.3	41.2	5.9	17.6	0.0	0.0	17
2.3	34.9	11.6	48.8	0.0	0.0	43
0.0	28.4	3.9	67.6	0.0	0.0	102
0.7	29.6	12.6	57.0	0.0	0.0	135
0.0	40.0	5.0	55.0	0.0	0.0	22
2.5	31.2	8.8	57.1	0.0	0.0	317
6.3	43.8	6.3	41.7	0.0	0.0	96
1.7	44.1	3.4	50.0	0.0	0.4	236
0.6	39.4	1.8	57.8	0.0	0.3	327
0.0	38.1	1.3	60.6	0.0	0.0	155
0.0	36.0	8.0	56.0	0.0	0.0	25
1.4	40.9	2.9	54.2	0.0	0.2	839
0.0	0.0	0.0	100.0	0.0	0.0	1
0.0	37.5	12.5	50.0	0.0	0.0	8

Continued on the next page

Table 2.3. *continued*

Case No.	Site	Provenience	Series	Fabric Impressed	Cord Marked	Check Stamped
9	38CH1048	TU1/2	PD1/PD2	0.0	16.7	0.0
8	38CH426	TU4/ST S4	McClellanville	0.0	21.7	4.3
7	38CH426	TU4/ST S4	Jeremy	0.0	0.0	0.0
6	38CH426	TU4/ST S4	PD 1	0.0	0.0	0.0
5	38CH426	TU4/ST S4	PD 2	0.0	0.0	0.0
4	38BK552	Feature 1, Pit	McClellanville	16.7	0.0	16.7
3	38BK552	Feature 1, Pit	Jeremy	0.0	0.0	0.0
2	38BK552	Feature 1, Pit	PD 1	0.0	0.0	0.0
1	38BK552	Feature 1, Pit	PD 2.5	0.0	0.0	0.0

PD = Pcc Dee.

Focusing on the apparent covariance of exterior surface treatment types, the objective of conducting BR analysis here is to group slices defined on the basis of paste classification into combinational units that can be inferred to represent approximate subassemblages or events. In the following analysis, comparisons were only computed for slices within, rather than between, individual contexts because it is assumed that each context reflects a slightly different occupation history. Computation of the pair-wise BR coefficients for the study were run using the resampleBRED script provided in Keith Kintigh's (2002) Tools for Quantitative Archaeology package. While there are several ways to express similarity between units, the method used here corresponds to the classic procedure. That is, the index of agreement (i.e., coefficient) between two analytical units is calculated as the sum of differences between each of the paired type proportions, subtracted from 200 (see Marquardt 1978:264). The value of 200 comes from the fact that the maximum possible agreement built into the calculation is 200%, shared equally by the samples compared. Accordingly, maximum similarity is signified by a value of 200, while maximum dissimilarity returns a value of zero. Based on the manner in which the index is calculated, types or categories with low abundances are de-emphasized because their pair-wise differences do not contribute significantly to the overall magnitude of the computed value. In some situations, this might pose a problem, but it is a desirable effect for the purposes

TABLE 2.3. *continued*

Simple Stamped	Plain	Burnished Plain	Complicated Stamped	Cob Impressed	Corrugated	Sherd Count
0.0	16.7	16.7	50.0	0.0	0.0	6
73.9	0.0	0.0	0.0	0.0	0.0	23
71.4	0.0	0.0	28.6	0.0	0.0	7
0.0	50.0	0.0	50.0	0.0	0.0	2
0.0	100.0	0.0	0.0	0.0	0.0	1
33.3	25.0	0.0	8.3	0.0	0.0	12
100.0	0.0	0.0	0.0	0.0	0.0	6
0.0	0.0	0.0	0.0	0.0	0.0	0
0.0	0.0	100.0	0.0	0.0	0.0	1

of this study because the explanatory power rests in comparing and contrasting the limited number of abundant surface treatment types.

In addition, the resampleBRED script estimates sampling errors based on respective sample sizes of the pair-wise comparisons and their degree of similarity. This is an important addition to the analysis presented here, because sample sizes vary greatly from one slice to the next and the relative strength of a particular BR coefficient can be more objectively evaluated when sample size effects are known. Calculation of probabilities relies on Monte Carlo simulations of a specified number of sample iterations (10,000 for this study) run with replacement for each comparison from an empirically derived population. The population from which the Monte Carlo samples are drawn consists of the pooled proportions or means of each pair-wise comparison of the slices and the sample size inputs correspond to the respective sample sizes of the slices within the pair-wise comparison. This procedure generates a sampling distribution of BR coefficients calculated from each paired Monte Carlo sample. Associated probabilities are then found by summing the cumulative probabilities of the sampling distribution of Monte Carlo BR coefficients that fall at or below the calculated BR coefficient of the actual pair-wise comparison. Low probabilities (i.e., < 0.05) suggest that the disparity between the compared units is unlikely to be the result of sampling error and that therefore the units are probably different, while higher probabilities (i.e., > 0.05) suggest that the compared units may have come from the same population. Three contexts (Fogarty Creek Site, Feature 2; Tibwin Midden, ST E14; Tibwin Midden, Cut Bank) in the study exhibit complex occupation histories in stratigraphically ambiguous deposits that will be better understood through an application of BR analysis.

Excavated data from Feature 2 (oyster midden) at the Fogarty Creek site consist substantively of a block trench measuring 1-×-3 m, segmented into three contiguous 1-×-1 m units. A fourth unit was extended from TU 3 but was not subjected to comparable artifact recovery methods. TUs 1 and 2 occupy the north end of the trench and represent deposits from an accumulating oyster midden. TU 3 also contains oyster midden, but there is an undifferentiated lower stratum containing the remnants of a wall-trench house that extends southward outside of the block excavation. This suggests that TU 3 contains a mixture of artifacts associated with two different activities (earlier house occupation and later oyster midden accumulation), a supposition that is borne out in the BR analysis to follow.

Tibwin Midden, ST E14 was excavated in arbitrary 10 cm levels in an enlarged shovel test (E14) measuring 0.4-×-0.4 m. The upper five levels sampled a spatially discrete clam midden overlying a laterally extensive oyster midden that was vertically positioned in the lower four levels of the shovel test. Although clear stratigraphic separation was evident, the lower oyster midden deposit contained very few sherds, which prevented its incorporation into the BR analysis. The cut bank collection from the Tibwin Midden represents a surface collection from the base of the eroded creek channel. ST E14 is nearby, but the compositions of the slices are somewhat different, supporting the contention that late prehistoric shell middens in the region were formed by lateral accretion and are composed of spatially and temporally discrete deposits that subtly vary from one place to the next within their boundaries.

My first attempt to discriminate subassemblages using BR analysis met with only partial success. Much of the problem arose from the bimodal distribution of plain sherds, which were popular in the early and late paste series but decreased in popularity in the intermediate series. This factor resulted in a convergence of BR values between the late and early portions of the seriation, once the popularity of simple stamping waned. This prompted a second BR analysis in which plain sherds were removed from the counts of each slice so that the behavior of stamped and burnished sherds could be more unambiguously observed. Again, this second analysis was only partially effective in grouping paste series into a coherent sequence, but it produced results different from those of the first BR analysis. This suggests that an additional problem in using BR analysis to discriminate subassemblages in these particular contexts arises from low variability in the late half of the seriation, which is characterized by the numerical dominance of just two types (plain and complicated stamped).

Curtis and Jackson (1962) observe that when faced with the inability to derive a single, unambiguous index to describe a relationship, it is often advisable to use multiple indicators. Heeding their advice, I reasoned that combining the results of the two BR analyses might yield a superior method of ordering and associating the various slices, given the limitations of a bimodal plain sherd distribution and the homogeneous nature of the surface treatment proportions in general. Table 2.4 presents the results of the two BR analyses in a single matrix. Both the pair-wise similarity coefficients and associated Monte Carlo probabilities are presented. The first value in each cell registers the results of the first BR analysis (which included all surface treatments), while the second value, separated from the first by a forward slash, represents the second BR analysis (which was run without plain sherd counts). With the exception of the McClellanville series, all of the slices exhibit relatively high similarity with one another. In most cases, the greatest similarity is shared by slices that are adjacent in time, based on the results of the paste seriation.

These patterns are not immediately obvious, but a simplified diagram is presented in Table 2.5, which takes the form of an occurrence seriation (Dempsey and Baumhoff 1963; O'Brien and Lyman 2002:119–121; Rowe 1961). In contrast to a frequency seriation, occurrence seriation tracks only the presence or absence of a trait or variable, not its proportional representation. In Table 2.5, cells in the BR matrix are classified by the various combinations of simulated probabilities for the two analyses that indicate "no difference" (> 0.05) in pair-wise comparisons. Two capital "Xs" indicate that a comparison yielded a statistical result of "no difference" (same) for both analyses; one small "x" indicates only one analysis produced a "no difference" probability; finally, a blank cell indicates that both analyses produced probabilities (< 0.05) indicating a difference in the pair-wise comparison.

Considering first the Feature 2 deposit at the Fogarty Creek site, the pattern of "no difference" occurrence yields a nearly perfect seriation when ordered by relative ages of paste series. Adjacent cells are the same for both analyses, while a singular presence extends to the next adjacent cell to the right (i.e., later slice). The continuity is broken by the PD 2.5, TU 3 slice, which is similar to all other slices due to its small sample size. As mentioned, both TU 3 slices consistently seriate earlier than the TUs 1 and 2 slices for all paste types, which confirms the suspicion that this test unit contains a mixture of older deposits associated with the house floor that underlies the later oyster midden. The overall pattern of association suggests strongly that the slices from Feature 2 represent a fine-grained sequence of continuity that would pose serious problems for constructing a sub-

Table 2.4. Brainerd-Robinson Coefficients and Sampling Error Probabilities for the Segregated Contextual Slices

Fogarty Creek Site Feature 2	Slice/Case	McClellanville Comb. (29)	Jeremy Comb. (28)	PD 1 TU 3 (27)
	McClellanville Combined (29)		0.03/0.02	0.02/0.01
	Jeremy Combined (28)	133/114		0.81/0.35
	Pee Dee 1, TU 3 (27)	126/102	184/174	
	Pee Dee 1, TU 1&2 (26)	106/65	167/151	177/164
	Pee Dee 2, TU 3 (25)	107/63	163/148	159/161
	Pee Dee 2, TU 1&2 (24)	103/56	160/139	164/154
	Pee Dee 2.5, TU 3 (23)	101/50	160/136	167/148
	Pee Dee 2.5, TU 1&2 (22)	106/80	128/140	125/152

Tibwin Midden (E14)	Slice/Case	McClellanville Comb. (21)	Jeremy Comb. (20)	PD 1 Comb. (19)
	McClellanville Combined (21)		0.02/0.00	0.00/0.00
	Jeremy Combined (20)	121/87		0.07/0.02
	Pee Dee 1, Combined (19)	100/71	162/161	
	Pee Dee 2, Combined (18)	108/82	181/188	179/173
	Pee Dee 2.5, Clam (17)	117/77	176/167	183/194

Tibwin Midden (Cut Bank)	Slice/Case	McClellanville (16)	Jeremy (15)	PD 1 (14)
	McClellanville (16)		0.16/0.03	0.01/0.00
	Jeremy (15)	179/168		0.13/0.09
	Pee Dee 1 (14)	167/156	184/188	
	Pee Dee 2 (13)	162/152/	179/183	193/195
	Pee Dee 2.5 (12)	164/170	173/187	178/195

Note: In each matrix, the values in the upper division, which extend to the right, record sampling error probabilities whereas values in the lower division, which extend to the left, record the Brainerd-Robinson Similarity Coefficient for each comparison.

Table 2.4. *continued*

PD 1 TU 1&2 (26)	PD 2 TU 3 (25)	PD 2 TU 1&2 (24)	PD 2.5 TU 3 (23)	PD 2.5 TU 1&2 (22)	Sample Size
0.00/0.01	0.00/0.01	0.00/0.01	0.04/0.00	0.00/0.00	36/24
0.17/0.03	0.17/0.01	0.04/0.00	0.41/0.05	0.00/0.01	53/28
0.40/0.11	0.04/0.04	0.02/0.01	0.78/0.11	0.02/0.01	63/31
	0.15/0.61	0.04/0.34	0.78/0.36	0.00/0.00	98/39
164/190		0.74/0.38	0.55/0.51	0.06/0.00	55/32
164/187	186/191		0.62/0.14	0.04/0.00	195/108
167/180	170/191	166/188		0.10/0.05	14/6
133/140	154/140	156/143	136/141		60/44

PD 2 Comb (18)	PD 2.5 Comb (17)	
0.01/0.00	0.01/0.04	17/7
0.52/0.71	0.61/0.27	43/15
0.17/0.01	0.56/0.77	102/73
	0.62/0.33	135/95
182/179		22/8

PD 2 (13)	PD 2.5 (12)	
0.01/0.00	0.19/0.30	96/54
0.08/0.01	0.34/0.58	236/132
0.75/0.28	0.46/0.82	327/198
	0.47/0.12	155/96
178/179		25/16

TABLE 2.5. SIMILARITY OCCURRENCE SERIATION OF THE SEGREGATED CONTEXTUAL SLICES

Fogarty Creek Site Feature 2	Case/Slice	McClellanville Comb (29)	Jeremy Comb (28)	PD 1 TU 3 (27)
	McClellanville Combined (29)			
	Jeremy Combined (28)			XX
	Pee Dee 1, TU 3 (27)			
	Pee Dee 1, TU 1&2 (26)			
	Pee Dee 2, TU 3 (25)			
	Pee Dee 2, TU 1&2 (24)			
	Pee Dee 2.5, TU 3 (23)			
	Pee Dee 2.5, TU 1&2 (22)			

Tibwin Midden (E14)	Case/Slice	McClellanville Comb (21)	Jeremy Comb (20)	PD 1 Comb (19)
	McClellanville Combined (21)			
	Jeremy Combined (20)			x
	Pee Dee 1, Combined (19)			
	Pee Dee 2, Combined (18)			
	Pee Dee 2.5, Combined (17)			

Tibwin Midden (Cut Bank)	Case/Slice	McClellanville (16)	Jeremy (15)	PD 1 (14)
	McClellanville (16)		x	
	Jeremy (15)			XX
	Pee Dee 1 (14)			
	Pee Dee 2 (13)			
	Pee Dee 2.5 (12)			

Note: "XX" defines comparisons that are similar for both "plain included" and "stamped/burnished only" in the Brainerd-Robinson Matrix; "x" defines comparisons that are similar for either "plain included" or "stamped/burnished only" in the Brainerd-Robinson Matrix, but not both; and empty cells indicate comparisons in the Brainerd-Robinson matrix that were different in both instances.

PD = Pee Dee.

TABLE 2.5. *continued*

PD 1 TU 1&2 (26)	PD 2 TU 3 (25)	PD 2 TU 1&2 (24)	PD 2.5 TU 3 (23)	PD 2.5 TU 1&2 (22)	Sample Size
			x		36/24
x	x		XX		53/28
XX			XX		63/31
	XX	x	XX		98/39
		XX	XX	x	55/32
			XX		195/108
				x	14/6
					60/44

PD 2 Comb (18)	PD 2.5 Comb (17)	
		17/7
XX	XX	43/15
x	XX	102/73
	XX	135/95
		22/8

PD 2 (13)	PD 2.5 (12)	
	XX	96/54
x	XX	236/132
XX	XX	327/198
	XX	155/96
		25/16

assemblage that could be argued to represent a distilled moment in time (or in Rouse's [1967] terminology, an event). However, a reasonably serviceable approximation of events can be accomplished by combining adjacent slices, beginning at the early end of the sequence. Thus, Case 28 can be paired with Case 27, Case 25 can be combined with Case 26, and Cases 23 and 24 can in turn be grouped together. This procedure would leave isolated Case 22 (PD 2.5, TU 1 and 2), which moreover appears to be dissimilar enough to all other slices to justify this partitioning.

The same pattern of progressive continuity characterizes the other two contexts. Ceramics from Tibwin Midden, ST E14 come from an extremely limited horizontal space within a clam deposit that may have been deposited over a shorter interval than either the Fogarty Creek or Tibwin Midden, Cut Bank contexts, which resulted in the harvesting of sherds from a much larger areal extent within the same midden. It is therefore not surprising that all of the slices are relatively similar, with the exception of the McClellanville slice. Variation is most evident in the Pee Dee 1 slice, which shares only one "no difference" probability with the Jeremy and Pee Dee 2 slices. It could be argued from this pattern that all four of the Mississippian slices recovered from the shovel test are contemporaneous, but this fails to explain the less-than-unanimous similarity of the Pee Dee 1 slice. If the clam deposit was laid down quickly, the contention that all the Mississippian slices are contemporaneous becomes less likely to be true. Given the probability that the deposit, although discrete, most probably contains a mixture of material contemporaneous with the targeted event and earlier material, a more reasonable approach to forming subassemblages would correspond to the adjacent cell procedure already described. The resulting subassemblages would consist of (1) McClellanville, (2) Jeremy/Pee Dee 1, and (3) Pee Dee 2 and Pee Dee 2.5. The Tibwin Midden, Cut Bank context exhibits a similar pattern and corresponds even more closely with the Fogarty Creek, Feature 2 context. It is argued here that the same combination of slices outlined above for the ST E14 context are appropriately applied.

None of the BR analyses of the three contexts perfectly captures the targeted events, but they can serve as accurate approximations once their limitations are recognized. For this reason, these combinations will be referred to as "Seriated Sub-Assemblages" (SSAs) in this study. They are choppy, artificial representations of the cultural sequence in which types may be under- or overrepresented due to the partitioning methodology, but they provide a more informed view of ceramic change trajectories than could be achieved from seriations of traditionally defined culture-historic types in situations where significant chronological overlap is in play. Indeed, they describe the general structure of this overlap. It is true that the method presents several problems. First, the partition rule ex-

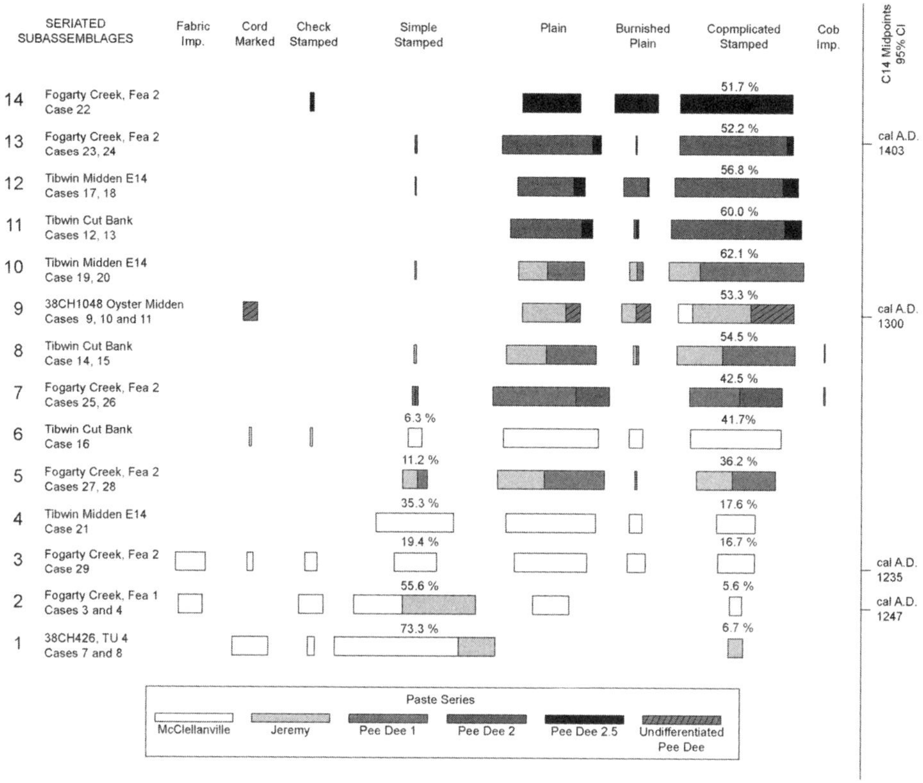

Figure 2.5. Central coast Mississippian subassemblage seriation. Cob imp = cob impressed.

cludes nonadjacent paste series (e.g., Jeremy versus Pee Dee 2) from membership in the same SSA. Second, one must concede that the paste trajectory is more continuous than this procedure allows. It is my impression, however, that the method provides a serviceable picture of the ceramic change trajectory.

Figure 2.5 presents a seriation of SSAs by exterior surface treatment constructed in the classic manner of the centered bar graph, informally known as the "Ford graph" (Ford 1962). Outwardly, the seriation is organized by percentages of complicated and simple stamped, which are arranged in nearly perfect unimodal popularity distributions. However, this ordering can only be accomplished with recourse to proportional paste composition, since the upper two-thirds of the seriation space is occupied by only two abundant surface treatment types, one of which (i.e., plain) is bimodal and hence confuses ordering decisions.

The proportional contribution of paste types is illustrated for each bar in the graph so that the decision criteria for placement can be readily seen. Once simple stamping virtually disappears from the seriation, there are no other surface treatment types or types with unimodal distributions available to reliably arrange complicated stamped percentages on either side of the peak in its popularity curve. At this point, the only recourse is to use the relative proportions of paste types to order cases. Ultimately, then, surface treatment, on its own, does not provide a strong basis for seriating deposits for this time range in the Central Coast sequence. However, both simple stamped and complicated stamped percentages appear to be potentially useful in broader, multivariate approaches to seriation because both of these curves are correlated with paste patterns that are demonstrated to be chronologically ordered.

The fact that the popularity curves of complicated stamped and simple stamped can be ordered in conformity with the principles of frequency seriation in such a way as to reflect the temporal ordering of the paste seriation strongly confirms the chronological sensitivity of these variables. There are several instances in which paste composition of a particular SSA appears slightly out of sequence. For instance, SSA 6 is exclusively composed of McClellanville series paste, but it is placed above SSAs 1, 2, and 4, which contain significant contributions of Jeremy and/or Pee Dee 1 paste. The homogenous paste composition of SSA 6 is clearly a result of the partitioning process. Had there been a methodological basis for further disaggregating the Jeremy slice (Case 20) from this context into early and late groupings, a heterogeneous SSA could have been constructed. Another slight violation is registered by SSA 5, which should be placed above SSA 6. Finally, the paste composition of SSA 7, which contains significant amounts of Pee Dee 2 paste, should occur later in the seriation. Given the nature of the partitioning method, though, we can assume that small proportions of Pee Dee 2 paste extend earlier than the chronological boundary depicted in Figure 2.5. Despite these contradictions, there is close agreement between the paste and surface treatment seriations, both of which can be arranged in unimodal distributions strongly suggestive of chronological ordering. Minor violations of this sort are easily accommodated within the seriation model, as long as what is being claimed is that, as Cowgill (1972:385) states, "units are in the right general position in the total sequence, not that the sequence is exactly right in detail."

The distributions of plain and burnished plain deserve some additional scrutiny. As mentioned, the plain curve is bimodal, which means that it exhibits two peaks. Thus, it would not provide a reliable basis to predict the unidimensional passage of time, given the temporal scale of the seriation illustrated in Figure 2.5. However, if we were to narrow the span of the seriation to include only the bottom half of the window or, alternatively, the top half, plain sherd percentages could clearly be used to help seriate subassemblages within a more contracted temporal span.

Burnished plain, by contrast, shows very little evidence of any discernible distributional pattern, as burnishing is a common, low-occurrence feature throughout the span under consideration. This strongly suggests that fluctuations in relative frequencies of this attribute are most likely the consequence of wide stochastic variation characteristic of small samples. SSA 14 (Case 22 in Table 2.3), the latest position in the seriation, may be an exception to this rule. Burnished plain makes up an unusually high proportion (20%) of this subassemblage and the sample size (n = 60) is large enough to suggest that this increase may reflect the beginning of another unimodal distribution sufficiently abundant to be seriated. Absolute date associations indicate that this material was deposited around the turn of the fifteenth century, a time in which burnished plain begins to increase in popularity in at least some other contemporaneous phases within the SAM culture area.

At the Mulberry site in the Wateree River valley at the Fall Zone, for instance, burnished plain popularity ranges between 25 and 38% of ceramic assemblages associated with the McDowell I and McDowell II phases, which date between approximately AD 1425 and 1525 (Cable et al. 1999; see also Caldwell 1974:89–95). In addition, Boudreaux's (2005:65, 76) Town Creek seriation (corrected here to exclude the category of unidentified surface treatments in his Table 2.11) shows low percentages of burnished plain (0.8 to 2.2%) in deposits associated with the Teal and Town Creek phases, while in Early Leak phase deposits plain burnished increases to around 12%.

The complicated stamped curve shown in Figure 2.5 is consistent with Early Mississippian through Early Lamar assemblages across the Piedmont (Hally and Rudolf 1986) and Valley and Ridge (Hally and Langford 1988) provinces of north Georgia and, also, in the Town Creek polity (Boudreaux 2005:61, 81). Based on the correlation of the surface treatment seriation and absolute dates, it can be estimated that complicated stamped appears around AD 1250 on the Central Coast and then rapidly gains in popularity until it peaks sometime around AD 1350, where its contribution to assemblages reaches 60 and 70%. Between AD 1350 and 1425, complicated stamped declines to levels of about 50 to 55%, as plain and plain burnished begin to increase in popularity. Boudreaux's (2005:81, Figure 2.22) Ford graph by phase for the Town Creek polity shows a similar trend, with complicated stamped percentages peaking in the Town Creek phase and then falling off in the succeeding Leak phase.

I have intentionally avoided drawing phase divisions in the paste and surface treatment seriations because, on their own, they are not definitive. One could certainly draw conclusions based on the clearly distinct distribution of paste types. SSAs dominated by McClellanville and Jeremy series ceramics could be inferred to represent the Sewee Camp phase, while SSAs exhibiting Jeremy and

Pee Dee 1 combinations could just as easily define the Jeremy phase. Accordingly, SSAs with Pee Dee 2 and Pee Dee 2.5 combinations could be classified as representative of the Tibwin phase; at the same time, it could be speculated that SSA 13 is the sole representative of the Fogarty Creek phase. As will be seen in the discussion to follow, however, these characteristics alone do not provide a rigorous basis for identifying phase boundaries. An occurrence seriation based on decorative elements is much more effective for this purpose.

Occurrence Seriation of Decorative Modes

Although SAM assemblages are characterized by an abundance of undecorated vessel rims and shoulders, there nevertheless exist a set of diagnostic decorative modes with restricted temporal popularity distributions that are duplicated across wide geographic space. In a sense, these modes represent what would traditionally be regarded as index types or horizon styles (O'Brien and Lyman 2002:45). In Willey's (1945:53) formulation, horizon styles could be represented by single attributes, a discrete kind of object, or a set of discrete object types. That these distinctive modes can be used to construct horizon styles in the larger culture area is well known, but the patterns have not yet been rigorously documented.

The first comprehensive attempt to trace the chronological span of decorative modes in Eastern SAM assemblages was undertaken by Chester DePratter and Chris Judge (1986), as an outgrowth of their investigations of Wateree Valley mound sites near Camden, South Carolina. DePratter's (1976, 1983, 1991) earlier work with Irene and related sites in Georgia also figured prominently in their chronology (Figure 2.6). The model was developed without the aid of either direct radiocarbon associations—at least for the Wateree sites—or modern calibration curves. Still, the time scale is remarkably accurate and the chronology they devised will require only minor adjustments when a larger regional database of assemblage contexts and absolute dates is obtained. As an outcome of a testing project conducted subsequently at the Wateree Valley mound sites of Belmont Neck, Adamson, and Mulberry (Cable et al. 1999), I have suggested that the time scale be shifted 50 years backward (right-hand side time scale in Figure 2.6).

The Wateree Valley decorative modes can be grouped into three horizon style associations. Around AD 1250, notched lips, punctated lips, zoned punctation on vessel exteriors and reed punctations incidental to lips emerge and continue to be popular until approximately AD 1400. An intermediate association consists of ticks at or below the lip, shoulder ticks, appliqués such as pellets and rosettes, riveted nodes, and punctated and segmented rim strips. The temporal span of this intermediate group overlaps with the other horizon styles as mod-

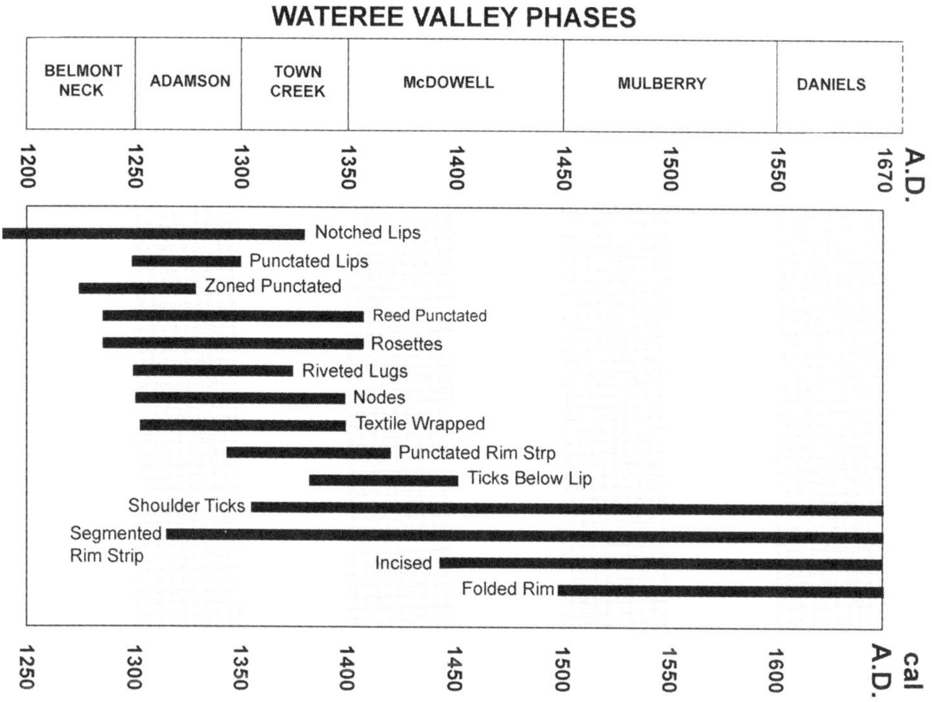

Figure 2.6. Wateree Valley decorative mode sequence (adapted from DePratter and Judge 1986).

eled. Recent information suggests that these attributes emerge around AD 1350 or 1375 and continue throughout the fifteenth century and later. Incising (after AD 1400) and folded rims (after AD 1450) most diagnostically represent the latest horizon style association.

In his ceramic seriation study of the Town Creek polity, Boudreaux (2005:64–68) traced the temporal associations of decorative modes and surface treatment types in selected subassemblages using the method of occurrence seriation. As discussed, this method allows for the ordering of infrequently occurring types or attributes through the construction of a presence/absence matrix (Cowgill 1972:388–389). The perfect seriation of this form brings all of the occurrences (marked by "Xs" in the matrix) for each column into contiguous vertical alignments in which the march of time is detected by offsetting horizontal positions. Because of stochastic variation, it is not likely that all of the boxes in a projected column span will be filled when sample sizes are small, however; in this case, the responsibility falls to the researcher to infer the actual pattern.

Boudreaux established four groups or segments based on the timed appearance of certain decorative modes, all of which correlated with his surface treatment seriations and radiocarbon associations. These included: Segment 1, top-thickened rims and small punctations; Segment 2, rosettes; Segment 3, punctated strips; and Segment 4, segmented/notched strips, large punctations, and folded rims. He defined three phases spanning the period from AD 1000 to 1500, two of which were further partitioned into early and late subphases. With the exception of top-thickened rims, the decorated modes he identified are widely distributed across the Eastern SAM culture area and the fact that they have offset and bounded chronological spans strongly suggests that they qualify as horizon style markers. Below I will describe the morphological structure of the Central Coast decorative modes, pointing out differences with Eastern SAM assemblages from nearby regions when appropriate. Once this descriptive framework is presented, it will be possible to construct an occurrence seriation for the Central Coast Mississippian sequence.

Even though I have characterized some kinds of rims in the Central Coast assemblage as "thickened," it is unclear to me how my types correlate with Boudreaux's. His top-thickened mode possesses chronological significance, as it occurs only in early assemblages, but this particular mode has not been recognized in the Central Coast region. Nevertheless, four other types of thickened rims are found in the Central Coast samples. Untrimmed rolled rims exhibit irregular, unshaped terminations of the folded over rim coil (Figure 2.7a). Rolled rims exhibit rim coils that are bent back to contact with the exterior vessel wall and the contact between the two elements is uniformly shaped (Figure 2.7b, c). Generally, the thickened portion of rolled rims display a rounded profile. Exterior thickened rims were formed by the addition of a final rim coil not fully blended along the exterior edge of the vessel, creating the appearance of an irregularly smoothed band (Figure 2.7d). Finally, interior thickened rims were formed by welding a coil to the interior portion the vessel above and adjacent to the final exterior coil (figure 2.7e). The interior coil was left unblended on the lip. One other modification seen in some necked jars in the Central Coast sample is a wide, recessed, and smoothed band (Figure 2.7f). The example illustrated here exhibits a sequence of two concave smoothed zones. Thinning of the zone below the lip in these specimens is similar, although not identical to, collard rims common to the Appalachian Summit region of North Carolina and points westward (see Dickens 1976:178–181). Thickened rims clearly have functional significance. Such features would have provided utility in handling vessels that were commonly moved around. The same may be true of recessed neckbands.

Lip notching is not recorded for the Town Creek seriation, although it plays a significant role in characterizing the early part of the Central Coast sequence.

Figure 2.7. Central Coast decorative rim modes I: (a–c) thickened rolled; (d) exterior thickened; (e) interior thickened; (f) thinned bands; (g–h) notched lips; (i–k) punctated lips.

I suggest that this is not a real difference in the two regions, however, because Reid's (1967:Table II) ceramic analysis indicates that notching is actually prevalent at Town Creek. It is especially abundant in the premound humus layer, confirming its association with early occupation at the site. Several lip and rim notching techniques are evident in Central Coast assemblages. These include (1) zigzag cutouts along the edge of thickened rims (Figure 2.7g); (2) deep incisions along the top of lips (Figure 2.7h); (3) fingernail incising along the top of lips; (4) impressions made with solid convex dowels that are visible as indentations along the rim edge (Figure 2.7i); and (5) rectangular notches along the rim edge (Figure 2.7j). Larger samples might reveal chronological patterning in the distribution of these various techniques, but for the purposes of this study all lip/rim notching is combined into a single mode.

Another decorative mode not identified in Boudreaux's seriation sample is punctated lips. This mode commonly occurs in the assemblages from the Camden mound sites and the Central Coast. Two variations were found in the Central Coast sample: (1) small punctations created by an oblong stylus with a convex termination (figure 2.7k) and (2) narrow, slightly curved linear punctations that may have been produced using fingernails or a section of a reed or bone. A feature of application that distinguishes punctated from notched lips is that the rim edge is not modified in the former.

In most other respects, Central Coast decorative modes follow closely the structure that Boudreaux describes for the early portion of his sequence. His mode categories differ slightly from those I have used, but these differences can be easily resolved to compare the seriations. Boudreaux's small punctated mode includes all hollow reed and solid punctations with diameters smaller than 10 mm. Following DePratter and Judge (1986), I distinguish hollow reed punctations from solid and/or small punctations occurring on the exteriors of vessels below the lip/rim edge. The latter are not present in the seriation sample but do occur in the region (Figure 2.8a). Generally, hollow reed punctations are distributed in one or two linear arrays immediately below the lips of vessels and run along the entire circumference of the rim (Figure 2.8b, c), although they occasionally occur along vessel shoulders (Figure 2.8d). All of the punctations in the Central Coast sample exhibit diameters less than 10 mm. Boudreaux's large punctated mode includes both hollow reed punctations and fingernail impressions, which are absent from the Central Coast sample.

Rosettes (Figure 2.8e) and pellets (Figure 2.8f) are closely related modes. Each was formed by attaching lines of small, rounded balls of clay below the rim edge or along vessel shoulder junctures. Rosettes are punctated with a hollow reed, distinguishing them from the unelaborated pellets, which sometimes occur as ovals or even subrectangular blocks. Pellets can be hollow, suggesting that one technique for their formation was to coat a perishable item such as a berry with clay.

Nodes are large, molded spheres of clay that are applied to the exterior surface of pots meant to serve as carrying lugs (Figure 2.8g). They are generally attached to vessel walls by rivets, although in rare cases they may simply be blended to the surface, which creates a much weaker bond. Often times the presence of a node is inferred by the presence of an unfilled rivet hole (Figure 2.8 d, h), which is distinguished from a drilled mend hole by the absence of abrasion. Generally,

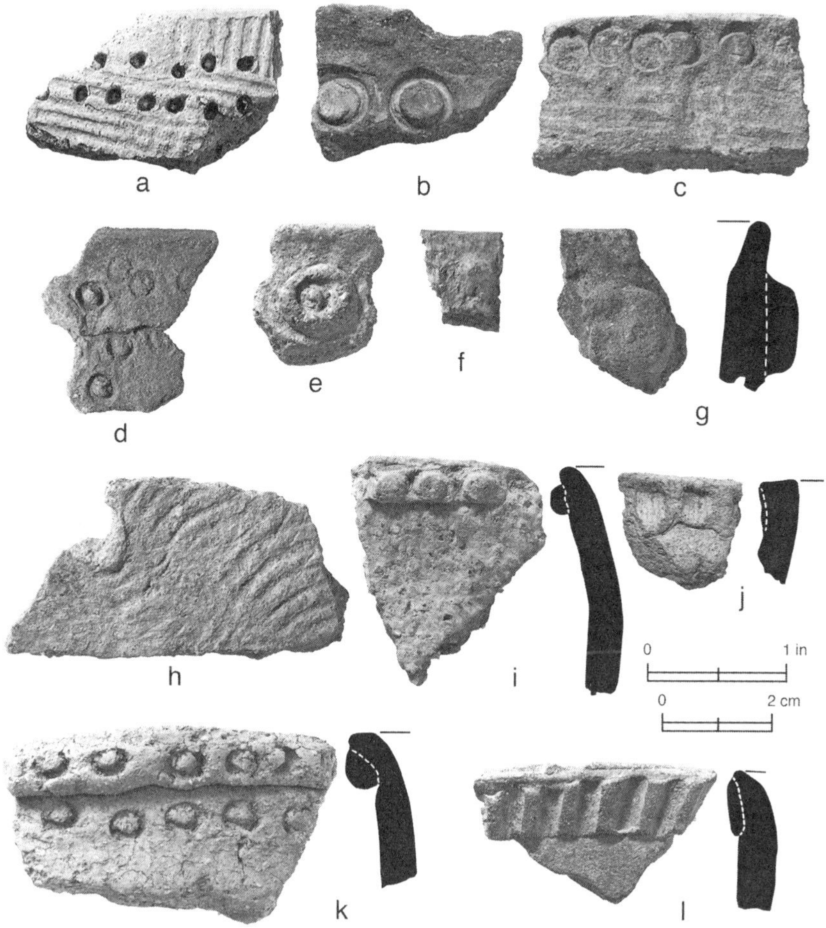

Figure 2.8. Central coast decorative rim modes II: (a–d) body punctations; (e) rosettes; (f) pellets; (g–h) nodes; (i–l) strips.

rivet holes are also characterized by untrimmed extrusions of clay resulting from molding the hole prior to attaching the node. Nodes commonly exhibit a hollow reed punctation in the center, affecting a similar appearance to rosettes. Most nodes are additionally accompanied by surrounding hollow reed punctations on the vessel exterior (Figure 2.8d), which are distributed around the perimeter of rims in sets of two or four lines.

Rim strips consist of flattened clay coils applied to the exterior of a vessel at or just below the lip. They are decorated with hollow reed punctations (Figures 2.8i, k), or various types of notched or fluted segmentation (Figures 2.8j, l), and

Table 2.6. Occurrence Seriations for Ceramic Decorative Modes from Central Coast Slices

SAA	Undecorated[a]	Punctated Lip	Recessed Band	Notched	Hollow Reed Punctated
14					
13	12			X(1)	X(2)
12	8			X(1)	X(1)
11	46		X(3)		X(5)
10	12		X(1)	X(1)	
9	1				
8	75	X(1)	X(2)	X(3)	X(4)
7	5				
6	18	X(1)	X(1)	X(2)	X(1)
5	8			X(1)	X(1)
4					
3	7			X(1)	
2	6				
1	2				
Total	(200)	2	7	10	14

Note: The presence of decorative element is indicated by "X" and the frequency of occurrences is recorded in parentheses.
[a]Undecorated category includes thickened and rolled rims.

rarely with solid punctations. In Central Coast assemblages, notches are commonly made with a solid rectangular stylus (Figure 2.8l). One example exhibits evidence of having been notched with a semicircular stylus with narrow ridges that may represent the exterior surface of a reed or shaped marine shell (Figure 2.8j). All of the rim strips in the seriation sample and all others that have been recovered across the Central Coast region originate at or within one or two mm of the lip. Strips that occur at distances of more than about five mm below the lip occur in many assemblages that postdate AD 1450. Folded rims have not been positively identified in the Central Coast region. They are distinguished from the visually similar rim strip by the fact that the last coil is folded back onto the vessel exterior. In Boudreaux's seriation, folded rims occur very late in the Town Creek sequence.

Table 2.6. *continued*

Rosette	Pellet	Node	Hollow Reed Rim Strip	Segmented Rim Strip	Total Decorated
	X(1)		X(2)	X(1)	4
X(1)	X(1)		X(3)	X(1)	9
					2
					8
					2
					0
					10
					0
					5
					2
					0
					1
					0
					0
1	2	0	5	2	43

If the SSAs defined in the previous section accurately reflect a time-ordered sequence, then the composition of decorative modes by SSAs should also exhibit a coherent ordering consistent with those described for nearby regions within the SAM culture area. Table 2.6 displays the frequencies of decorative modes for each SSA, ordered from latest (i.e., 14) to earliest (i.e., 1). Most of this sample consists of undecorated rims, but the smaller numbers of decorated modes exhibit patterns consistent with the DePratter and Judge (1986) chronology and the seriation performed by Boudreaux (2005) for Town Creek. Accompanying frequency data in the table are presence values (Xs) arranged in occurrence seriation format. This method of seriation is best applied to datasets that are comprised of rare attributes or in situations where the formation processes of deposits are uncertain and it is suspected that frequencies of types or modes may be subject to high stochastic error (see Cowgill 1972). Three clear divisions can be identified in the table. The latest SSAs (13 and 14) are characterized by the presence of appliqués (i.e., pellets, rosettes, and punctated and segmented rim strips), while the intermediate SSAs (5 through 11) are dominated by hollow reed punctations, punctated lips, recessed bands, and notches. Appliqué elements are conspicuously absent. Finally, the earliest SSAs (1 through 4) contain assemblages that are almost exclusively made up of undecorated rims, with the exception of a single notched application.

In conjunction with the paste and surface treatment seriations, this additional evidence provides the basis for assigning the SSAs to specific phases. SSAs 13 and 14 are composed primarily of Pee Dee 2 and Pee Dee 2.5 paste variants (see Figure 2.5) and appliqué decorative modes. This grouping defines the Fogarty Creek phase. The earliest end of the sequence (SSAs 1 through 4) represents the Sewee Camp phase, which is characterized by a predominance of McClellanville paste and lesser proportions of Jeremy paste, along with relatively high incidences of simple stamped, check stamped, cord marked, and fabric impressed, as well as low proportions of complicated stamped and primarily undecorated rims. Partitioning of the intermediate phases is not as straightforward. A distinction between the diagnostic decorative mode patterns is not obvious, and the only recourse at present is to divide the SSAs by paste composition. The later SSAs in this grouping (SSAs 11 and 12) are primarily composed of Pee Dee 2 paste with minor associations of Pee Dee 2.5, while the earlier ones (SSAs 5 through 10) consist of combinations of Pee Dee 1 and Jeremy paste. One exception is SSA 6, which consists exclusively of McClellanville paste. Although it appears slightly out of place with reference to the paste seriation, the decorative modes are consistent with the other SSAs assigned to the Jeremy phase, again suggesting that the technological and decorative dimensions of the assemblages have slightly different change trajectories (see Table 2.6). The later intermediate grouping, then, would coincide with what I have named the Tibwin phase, while the earlier one would correspond to the Jeremy phase.

Additional support for this partitioning is provided by inspection of the relative proportions of diagnostic decorative modes as displayed in Table 2.7. The Jeremy phase SSAs contain higher percentages of punctated lips and notching, while the Tibwin phase SSAs have significantly greater proportions of hollow reed punctations. A BR analysis was run on these two groupings to see if this statistical technique could effectively differentiate the two phases. The resultant BR coefficient was 125, which is relatively low, indicating an agreement of only 62.5%. The associated probability that the two samples are different, however, is only 0.13. If the sample sizes of the two groupings were increased by only a few more items, though, a significant difference could be established. Assuming the

Table 2.7. Decorative Mode Percentages by Phase, Central Coast

Phase	Punctated Lip	Recessed Band	Notched	Hollow Reed Punctated	Rosette	Pellet	Node	Hollow Reed Punctated Rim Strip	Segmented Rim Strip	Frequency
Fogarty Creek	0.0	0.0	7.7	15.4	7.7	15.4	0.0	38.5	15.4	13
Tibwin	0.0	30.0	10.0	60.0	0.0	0.0	0.0	0.0	0.0	10
Jeremy	10.5	21.1	36.8	31.6	0.0	0.0	0.0	0.0	0.0	19
Sewee Camp	0.0	0.0	100.0	0.0	0.0	0.0	0.0	0.0	0.0	1

same proportions as displayed in Table 2.7 and equal sample sizes of only 20 items, the resultant probability that the two groupings would be the same is only 0.044 and it could then be argued that they are probably different. Only larger sample sizes would allow the differentiation of the subtly different phases, but the differences in paste composition suggest strongly that Tibwin and Jeremy represent temporal divisions within an evolving ceramic tradition.

Regional Correlation of Eastern SAM Ceramics

With the local sequence described, it is possible to develop a broader perspective on inter-regional patterns of Eastern SAM ceramic change and to construct a more detailed system of horizon markers composed of decorative mode associations. I begin with the Central Coast chronology and show how occurrence patterns of decorative modes within it relate to occupation sequences in adjoining regions. Inter-regional integration is a necessary step in building the arguments presented in this study, because it provides a basis for evaluating where there is good evidence of fifteenth- and sixteenth-century occupation and where there is not.

A logical place to begin is a comparison of the Town Creek and Central Coast decorative mode seriations, since the data have been structured in a manner that allows effective comparability. Table 2.8 presents the occurrence seriations for the two regions. Several caveats concerning the construction of the matrices should be noted. First, the percentages of the various decorative modes are presented along with the more common presence-absence designations marked

Table 2.8. Comparison of the Town Creek and Central Coast Decorative Mode Occurrence Seriations

Regional Sequence	Phase	Age Range (cal AD)	Notched Lips	Small Punctated	Rosettes	Pellets
Town Creek						
	Late Leak	1400–1550?		2.4	9.5	9.5
	Hiatus					
	Early Leak	1300–1400				28.6
	Late Town Creek	1250–1300		18.2	36.4	45.5
	Early Town Creek	1150–1250		75.0	25.0	
	Teal	1000–1150		100.0		
	Late Leak	1400–1550?		—	x	x
	Hiatus					
	Early Leak	1300–1400				X
	Late Town Creek	1250–1300		X	X	X
	Early Town Creek	1150–1250		X	X	
	Teal	1000–1150		X		
Central Coast						
	Fogarty Creek	1375–1425	7.7	15.4[a]	7.7	15.4
	Tibwin	1325–1375	14.3	85.7[a]		
	Jeremy	1275–1325	46.7	53.3[b]		
	Sewee Camp*	1200–1275	100.0			
	Fogarty Creek	1375–1425	x	X	x	X
	Tibwin	1325–1375	X	X		
	Jeremy	1275–1325	X	X		
	Sewee Camp	1200–1275	X			

Note: Town Creek data from Boudreaux (2005:64–68). Occurrence seriation symbols represent the following percentage intervals: — = less than 3%, x = 3–10% percent, and X = > 10%.

TABLE 2.8. *continued*

Punctated Strips	Large Punctated	Folded/ Fluted	Nodes	Segmented/ Notched Strips	Folded/ Punctated	Frequency
14.3	4.8	2.4	9.5	14.3	33.3	42
50.0	7.1	7.1	7.1			14
						11
						8
						1
X	x	—	x	X	X	42
X	x	x	x			14
						11
						8
						1
38.5				15.4		13
						7
						15
						1
X				X		

[a]Hollow reed punctations
[b]Both hollow reed punctations and lip punctations.

by "Xs." It is useful to see just how the relative importance of certain modes might shift over time, in spite of the small sample sizes. Likewise, the percentages allow for another useful modification of the simple occurrence format by the ordination of presence-absence designations into substantial and minor contributors. I have chosen to segregate this information into three ordinal classes: less than 3% (-), 3 to 10% (x), and greater than 10% (X). Finally, some modification to the modes was required to bring the seriations into alignment. Data on thickened and recessed band rim types were eliminated because there are unresolved problems of definition and recognition between the two sequences in these instances. In addition, the "small punctated" mode in Boudreaux's seriation is equated with two modes in the Central Coast seriation (punctated lips and hollow reed punctations).

The earlier portions of both sequences are dominated by punctated modes, while the correlated Early Leak and Fogarty Creek assemblages record the first appearance of rim strips. Two major disconformities in the early part of the seriation require further elaboration. First, rosettes and pellets emerge much earlier in the Town Creek seriation. As modeled in the Central Coast sequence, these modes are not documented to occur earlier than the Fogarty Creek phase in association with rim strips; however, in the Town Creek seriation their first appearance precedes rim strips in the Early and Late Town Creek phases. This may suggest that rosettes and pellets in fact occur in the Tibwin phase of the Central Coast seriation but have not yet been positively identified.

A second disconformity concerns the relative ages of the early phases. In the Town Creek seriation, rosettes are said to emerge sometime between cal AD 1150 and 1250, during the Early Town Creek phase. Evidence of their first emergence in the Central Coast seriation is estimated to occur on or after cal AD 1375. Even conceding that this mode may appear first in the later stages of the Tibwin phase, the disparity in dating is probably too large to be explained by a lag in trait adoption alone. A problem with the Town Creek calibration is that only one of the contexts (Level A, Town Creek site) used to define the Early (Level A and Feature 16, Town Creek site) and Late (Level X, Town Creek site and Leak) Town Creek phases was directly dated (see Boudreaux 2005:83–85; Table 2.6). The other radiocarbon dates relied upon to date these phases were drawn from the Payne and Leak sites and their contexts and assemblages probably include a mixture of components. Level A is a midden zone situated below the mound at Town Creek and the associated radiocarbon date appears to have been derived from two samples (FSU-184/FSU-174) that may have been averaged (see Boudreaux 2005:Table 2.15). The assay yielded a conventional age of 745 ± 140 BP, a calibrated mean age of AD 1233 ± 119, and a most probable 1-σ range (53%) of cal AD 1154 to 1329 (OxCal 4.3.2). Problems with the wide sigma and a nonspecific sheet midden context aside, this particular date does not resolve the timing of the first appearance of rosettes and pellets because neither of these modes were present in the Level A context. Only top-thickened rims and small punctations were listed by Boudreaux, which is consistent with a Jeremy phase correlation.

Based on these observations, I argue that the Town Creek sequence is calibrated too early and that a better approximation of the age ranges of the phases could be achieved through alignment with the Central Coast sequence. The Early Leak phase is similar to the Fogarty Creek phase (i.e., pellets and punctated rim strips) and it should therefore date between cal AD 1375 and 1425. The Late Town Creek phase should correlate with the Tibwin phase and occur between cal AD 1325 and 1375. The Early Town Creek phase, therefore, should be correlated with the Jeremy phase, positioned between cal AD 1275 and 1325. Finally, the Teal phase appears to be constructed from an assemblage of long duration and I venture to suggest that the latest portion of the Teal phase correlates with the Sewee Camp phase.

Because assemblages dating to later than cal AD 1425 are missing from the Central Coast seriation, it is necessary to move beyond this region to construct a model of fifteenth- and sixteenth-century horizon markers and occupation sequences. Throughout the Eastern SAM culture area, such assemblages are identified by the presence of folded rims and incising, but the relative importance of these traits change over time and they occur in associations with other traits that provide the bases for defining more refined sequences. The construction of such a model is an undertaking that has not yet been systematically developed and one that will occupy the remainder of this discussion because it holds the key to understanding the latest occupation histories of different regions within the larger culture area.

Folded rims are of numerical importance in the latest of the phases (i.e., Late Leak) in the Town Creek seriation. I have inserted an empty row between the Early and Late Leak phases in Table 2.8, however, because there appears to be a significant time separation between them. Although Boudreaux (2007:60–62) characterizes the Late Leak occupation (cal AD 1400 to 1550) as sequentially continuous with Early Leak at Town Creek, there is not strong evidence to support this interpretation. The contexts he uses to define the Late Leak phase appear to be more in line with the Caraway occupation, which he dates between AD 1550 and 1700. One of these two Late Leak contexts (sq. 160–170L40/Pit) yielded charcoal with a conventional radiocarbon age of 300 ± 60 BP (Boudreaux 2005:Table 2.15, Beta-184061). Assays this close to the modern era are problematic, but the most probable (90.5%) 2-σ calibrated date option produces a range of AD 1447 to 1675, with a mean of AD 1587 (Oxcal 4.3.2). Although this date could fall between cal AD 1450 and 1550, it is also entirely consistent with a late sixteenth- or seventeenth-century age. Analysis of some 3,000 individual glass beads from various contexts at the site indicates an early- to mid-seventeenth-century time frame for at least a portion of the Caraway component (Boudreaux 2007:62). The other context, Feature 13, has not been directly dated, but the assemblage is very similar to that of 160–170L40/Pit, based on the extremely close position they share on Boudreaux's (2005:70) multidimensional scaling plot. If the Late Leak assemblage, as constructed, actually represents Caraway occupation, then the hiatus I have identified would span a period of as much as 150 to 175 years, an interval that could easily accommodate two or three additional phases.

An idea of what such intervening phases might look like, if they exist at all in the Town Creek polity, can be supplied by test excavations conducted at the Adamson and Mulberry sites as part of a project sponsored by the Wateree Archaeological Research Project (University of South Carolina) and the National Geographic Society (Cable et al. 1999). At the Adamson site, a block of three 1-×-1-m test units was excavated over a house pit of approximately 50 to 60 cm in depth. The floor of the structure contained two superimposed prepared clay hearths, separated by a thin lens of sandy silt wash. A laminated sand lens of approximately 10 cm in thickness also capped the upper house fill; this lens was in turn buried by 40 cm of later occupation debris. The upper portion of this late midden was plowed and segregated from the lower undisturbed loamy matrix during excavation. Four recovery contexts were differentiated: (1) Upper Late Occupation Matrix, (2) Lower Late Occupation Matrix and Flood Lens, (3) House Fill and Floor, and (4) Pre-House Occupation Matrix.

The Mulberry site test excavation consisted of a 2-×-2 m test unit (TU 1) excavated about 20 m distant from the base of Mound B. The fill from this excavation is what might be called a "kitchen midden" because of the high density of deer bone, carbonized corn, other ethnobotanical remains, dark organic staining, and large numbers of sherds. Refuse middens of this type are commonly found along the flanks of Mississippian platform mounds (M. Smith and Mark Williams 1994) and I speculated that this particular deposit might have been formed as a consequence of feasting in conjunction with elite ritual activities associated with the mound. Boudreaux (2005:55, 227, 396) has recognized a similar deposit, Level X, around the base of the mound at Town Creek that he referred to as a "flank midden." Other than several smudge pits that may have been instrumental in tanning deer hides in the Mound B midden at Mulberry, no other features were encountered during the excavation of TU 1. Cultural material extended to depths of around 50 cm below ground surface and the unit was excavated in arbitrary 10 cm levels. The upper two levels consisted of plow zone and were combined into a single provenience because vertical artifact associations in this matrix were jumbled. Also included in this context were artifacts from plow scars extending into the top of Level 3 in otherwise minimally disturbed midden. The levels below plow zone exhibited lighter soil coloration and declining frequencies of faunal bone and artifacts.

Table 2.9 displays an occurrence seriation of diagnostic decorative modes from the six Wateree Valley contexts. Again, although relative frequencies are presented, the principles of occurrence seriation were used to order the contexts. The Adamson site contexts seriate earlier, with the exception of the Upper Late Occupation Matrix. This late matrix consists of a mixture of redeposited sheet midden, intact sheet midden, and flood deposits containing ceramics dating between cal AD 1350 and 1450. The early portion of the seriation correlates well with the Town Creek and Central Coast sequences. Adamson phase deposits are dominated by notched lips and hollow reed punctations with minor contributions of punctated lips and rosettes, supporting the claim that rosettes appear before rim strips in contexts dating to the mid-fourteenth century. The Lower Late Occupation Matrix, which may also be mixed, correlates with the Early Leak and Fogarty Creek phases, principally characterized by the appearance of punctated (hollow reed) and segmented rim strips. High percentages of notched rims, however, suggest a degree of mixing with earlier deposits, primarily those of the Adamson phase.

A radiocarbon date (Beta-131285) was obtained from the Adamson phase context presented in Table 2.9. The sample consisted of wood charcoal from several different tree species recovered in the fill of the lower hearth and from the fill between the two floor plaster coatings (Cable et al. 1999). It returned a conventional age of 660 ± 50 BP, leaving us again with a bimodal probability distribution (Oxcal 4.3.2). The late option specifies a 1-σ range of AD 1354 to 1389, while the early 1-σ range is AD 1281 to 1317. Both options are equally probable. Since the material dated was wood, it is likely that there is an offset between the targeted event (i.e., the initial occupation of the house) and the age of the wood sample (see Schiffer 1986). Wood samples can yield radiocarbon ages that are easily 25 to 50 years older than the event of interest. Appending an offset of 25 years to the earlier date option would position the age of the house squarely within the range of cal AD 1300 to 1350, while a similar offset for the later option would yield a late fourteenth- or early fifteenth-century result, which would seem too late for the ceramic assemblage.

The latest contexts in the Wateree seriation contain elements that are either absent from the Town Creek and Central Coast seriations or occur in different proportional relationships. Shoulder and lip ticks appear for the first time (Figure 2.9a–d), rim strips ascend to greater dominance (Figure 2.9j–q), segmented rim strips increase in popularity relative to punctated strips, and nodes make their appearance in minor numbers. Nodes may occur slightly earlier than

Table 2.9. Wateree Valley Decorative Mode Seriations from Excavated Contexts

Site	Phase	Age Range (cal AD[a])	Context	Regional Correlation
Mulberry (38KE12)	McDowell II	1450–1500	Upper Midden (0–20 cm bd)	
Mulberry (38KE12)	McDowell I	1400–1450	Lower Midden (20–40 cm bd)	
Adamson (38KE11)	McDowell I / Wateree-Town Creek	1350–1450	Upper Late Occupation Matrix	
Adamson (38KE11)	Wateree-Town Creek	1350–1400	Lower Late Occupation Matrix	Early Leak/ Fogarty Creek
Adamson (38KE11)	Adamson	1300–1350	House Fill/Floor	Late Town Creek/ Tibwin
Adamson (38KE11)	pre-Adamson	pre-1300	Early Occupation Matrix	Early Town Creek/ Jeremy

Site	Phase	Age Range (cal AD[a])	Context	Regional Correlation
Mulberry (38KE12)	McDowell II	1450–1500	Upper Midden (0–20 cm bd)	
Mulberry (38KE12)	McDowell I	1400–1450	Lower Midden (20–40 cm bd)	
Adamson (38KE11)	McDowell I / Wateree-Town Creek	1350–1450	Upper Late Occupation Matrix	
Adamson (38KE11)	Wateree-Town Creek	1350–1400	Lower Late Occupation Matrix	Early Leak/ Fogarty Creek
Adamson (38KE11)	Adamson	1300–1350	House Fill/Floor	Late Town Creek/ Tibwin
Adamson (38KE11)	pre-Adamson	pre-1300	Early Occupation Matrix	Early Town Creek/ Jeremy

Note: Occurrence seriation symbols represent the following percentage intervals: — = less than 3%, x = 3–10%, and X = > 10% .

[a] Date ranges correspond to the right-hand time scale suggested by Cable et al. (1999) in Figure 2.6.

Table 2.9. *continued*

Notched Lip	Punctated Lip	Hollow Reed Punctated	Solid Punctations	Rosettes	Pellets	Nodes/Lugs	Hollow Reed Rim Strip	Segmented Rim Strip	Shoulder Ticks	Folded	Lip Ticks	Frequency[b]
1.0	1.0	1.0	1.9	1.0	1.0		2.9[c]	56.7	30.8			104 (3)
		11.1		8.9	13.3	2.2	9.9	21.2	31.1		2.2	45 (2)
13.3	6.7	6.7		6.7		6.7		46.6	13.3			15 (2)
37.5	6.3			25.0			6.3	25.0				16 (0)
42.9	7.1	28.6		7.1					14.3			14 (0)
100.0												3 (0)
—	—	—	—	—	—		—	X	X			104 (3)
		X		x	X	—	x	X	X		x	45 (2)
X	x	x		x		x		X	X			15 (2)
X	x			X			x	X				16 (0)
X	x	X		x					X			14 (0)
X												3 (0)

[b] Parentheses denote additional numbers of rim strips of indeterminate decoration that were used to prorate the relative contributions of rim strips with identifiable decoration.

[c] Includes two hollow reed punctated rim strips and one pelleted rim strip.

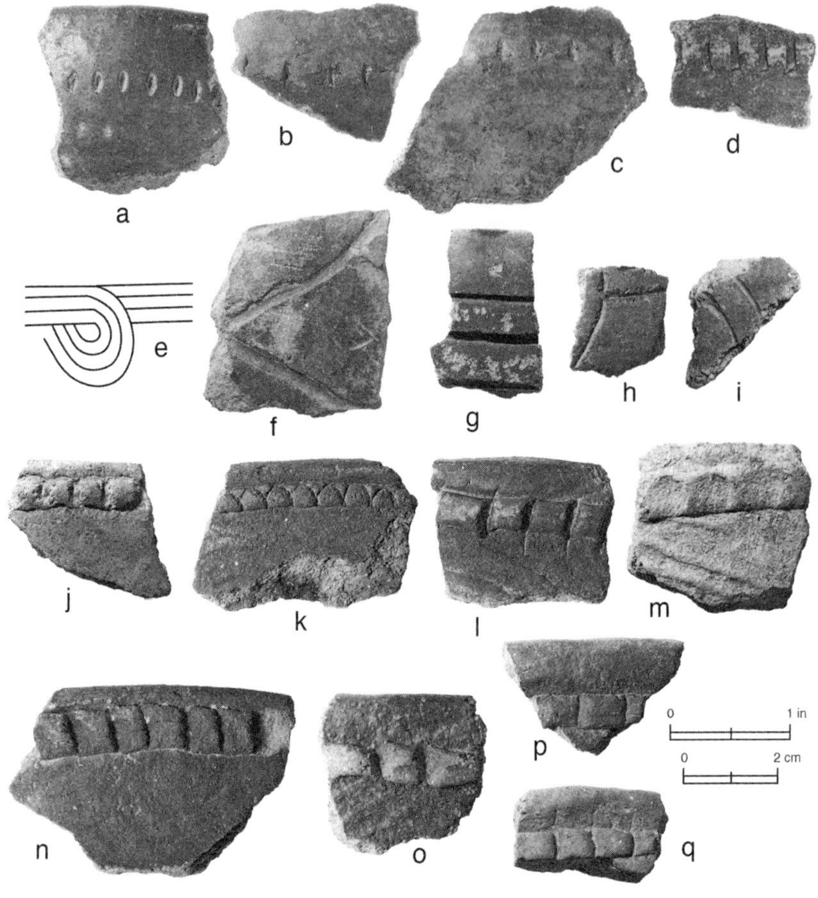

Figure 2.9. Decorative modes from McDowell I and II phase contexts at the Mulberry site: (a–d) shoulder ticks; (e–i) Lamar Incised; (j–q) punctuated and segmented strips.

this, however, as a single node is reported from the Early Leak contexts in the Town Creek seriation. Judging by its mixture of early and late modes, the Upper Late Occupation Matrix at Adamson is a mixed assemblage. The presence of shoulder ticks, though, suggests that occupation at the Adamson site extended into the middle fifteenth century.

The DePratter and Judge (1986) model in Figure 2.6 reveals that the decorative modes from TU 1 at the Mulberry site most closely correspond to the McDowell phase. It is in this interval that shoulder ticks, rim ticks, and incising first appear in substantial numbers. Incising has not been discussed up until now because it does not contribute to the assemblages of the Central Coast or

Town Creek sequences. Further, it is sometimes difficult to extract such information from previously compiled reports because of the radically different ways in which incising is handled by different researchers. While some prefer to categorize it as a decorative mode, it is more often considered a surface treatment. Here, I categorize it as a surface treatment type and hence separate it from the decorative mode seriation.

Incised sherds (Figure 2.9e–i), commonly referred to as Lamar Incised or Lamar Burnished Incised, make up 1.2% (n = 8) of the sherds with identifiable surface treatments (n = 682) in the McDowell I deposit (TU 1, Levels 3 and 4). In the McDowell II phase deposit (TU 1, Levels 1 and 2), incised representation increases to 3.4% (n = 58 of 1706). The absence of folded rims confirms a pre–Mulberry phase association for both contexts. About half of the incised sherds recovered from the unit are characterized by bold or wide incised lines (Figure 2.9f, g), while the remaining portion of the sample exhibits narrower incision widths (Figure 2.9h, i). Although narrow incised lines occur throughout the Lamar period, they gradually increase in popularity in the Middle and Late Lamar periods (see Hally 1994a:147). The most common Lamar Incised motif at the Mulberry site is a concentric curvilinear scroll connected by parallel straight lines (Figure 2.9e), which is repeated in a band below the lip of cazuelas and other bowl forms. The motif illustrated was traced from a photograph of a vessel fragment recovered from the Mulberry site published in Ferguson (1974:Figure 5, top). All of the incised sherd examples illustrated in Figure 2.9 are probably fragments of this standard motif.

Shoulder ticks and rim ticks are both absent from the Town Creek seriation, but Reid (1967:26, 30, 38; see also Coe 1995:178, Figure 9.21D) illustrates several sherds that would qualify as possessing shoulder ticks in his Town Creek Mound analysis. In addition, Boudreaux lists 21 incised sherds in his rim treatment and decoration tabulation (Boudreaux 2005:Table 2), which would amount to less than 0.2% of his overall sample of exterior surface treatments. It is likely, then, that a small McDowell phase-like component exists at Town Creek and is probably representative of terminal Mississippian occupation there. Therefore, while I would agree with Boudreaux (2005:87–89) that the Leak phase component at Town Creek probably ended around AD 1450, I suggest that a phase immediately succeeding Early Leak is missing from his seriation and is still in need of isolation from the larger set of contexts that comprise the Town Creek site.

Rim strips are abundant in the McDowell phase contexts at Mulberry. Segmented or notched strips increase in popularity at the expense of punctated strips, in line with the DePratter and Judge (1986) model. Proportions of punctated strips decrease from a high of 9.9% in McDowell I to 2.9% (including a pelleted strip) in McDowell II at Mulberry (see Table 2.9). The percentages of rim strips in the Wateree occurrence seriation are prorated to include the strips with indeterminate decoration so that the overall contribution of rim strips in each context is better characterized. This adjustment does not affect punctated and segmented strip representation.

A subtler change in rim strip data, however, is not systematically recorded in Eastern SAM ceramic analyses. Over time, a greater proportion of strips are attached farther down the vessel wall and away from the lip (see Figure 2.9, especially o–q). I recorded "strip position" as a continuous variable by measuring the distance between the lip and the origin of the fillet strip on the vessel wall. Strip position distances vary from 0.0 to 11.5 mm in the Mulberry test unit, with a mean of 5.11 ± 2.88 mm for measurable specimens (n = 62). However, there was no statistical difference between the McDowell I and II deposits. The sample (n = 26) from the Adamson site, made up of both test unit and shovel test specimens, should be somewhat older on average since it is made up of a mixture of Town Creek and (mainly) McDowell I phase deposits. As expected, it does yield a mean position closer to the lip ($\bar{x}$ = 4.35 ± 3.02 mm), but it is not statistically different from the Mulberry sample (t = 1.12, p = 0.132). If only the hollow reed punctated strips from Adamson are considered (n = 9), which should be associated primarily with the Town Creek phase, an even higher position relative to the lip is indicated ($\bar{x}$ = 3.51 ± 3.36 mm). Again, this refined sample is not significantly different from the Mulberry sample (t = 1.53, p = 0.065). However, the trend is clearly in the direction of higher positions of strips relative to the lip in earlier assemblages. Strip positions in the sample from the Fogarty Creek site, where it is relatively certain that the assemblage is discretely associated with the turn of the fifteenth century, are more closely aligned with the lip. This sample (n = 11) yields a mean of 0.61 ± 0.74 mm with a range of 0 to 1.9 mm, which is significantly different and closer to the lip than even the sample of hollow reed punctated strips in the Adamson collection (t = 3.27, p = 0.002). Although this variable was not measured in Boudreaux's sample from Town Creek, only one of eight examples that he illustrates exceeds a position appreciably greater than 3 mm (see Boudreaux 2005:Figure 2.10). Strip position might be expected to exhibit regional variation, but the Wateree data demonstrate a clear chronological trend toward greater separation in the fifteenth century between strip origin and vessel lips.

The Wateree seriation suggests continuity of the McDowell phases with the Wateree-Town Creek phase, which I estimate ended somewhere between AD 1400 and 1425. I have appended the suffix "Wateree" to the original phase name to avoid confusion with the phase of the same name in the Town Creek sequence. I favor a terminal date of around AD 1425 for the Wateree-Town Creek phase because of the dating of the Fogarty Creek phase in the Central Coast sequence. The earlier end date in Table 2.9 and Figure 2.6 reflects the chronology in the Wateree Valley as it has been understood since the late 1990s.

Just how far the McDowell phases extend into later prehistory has not been established, but, through a collaboration with the US Forest Service, I obtained two AMS dates from the McDowell II deposit in TU 1 that help to calibrate their chronological positions. Each sample consisted of white-tailed deer bone, one from Level 1 (Beta-440954) and one from Level 2 (Beta-440953). White-tailed deer bone, although not as ideal as annual plants for reducing the gap between the age of a targeted event and a radiocarbon sample (Schiffer 1986), still represents an effective alternative to undifferentiated wood charcoal because deer rarely survive to ages beyond eight years in situations where they are subject to heavy human predation. For instance, Kie and White (1985:115–116) conducted a study of white-tailed deer populations in the Welder Wildlife Refuge, Texas, in which they found that about 28% of females attained an age greater than 4.5 years, while only about 16% of the smaller male population survived to this age. Deer older than 10.5 years were not observed in their study.

Conventional ages for the two AMS samples were nearly identical, 365 ± 17 and 362 ± 24 BP, which, when calibrated, yield means between cal AD 1533 and 1540 (Figure 2.10). The probability densities, however, are bimodal and their distributions reveal that there is virtually no chance that the true ages of the samples coincide with their means, which fall in the gutter of the distribution between the two modes. Instead, we are presented with two date options, whose ranges are most clearly observed in the pooled conventional age, which demonstrates virtually no overlap. Accordingly, the true date range of the samples is either one or the other; it cannot be an average of the two. The early option, which has a 2-σ range of cal AD 1470 to 1514, is the most probable (60.9%), while the 2-σ range of the later, less probable option (34.5%) is cal AD 1575 to 1624. Although the earlier option is more probable, to favor it over the other solely on this basis would expose the decision to a relatively high risk of error. Fortunately, the associated ceramic assemblage provides corroboration. The late option identifies an interval of time that would predict an assemblage dominated by plain over stamped surface treatments and/or wide folded rims (e.g., the Barnett or Bell phases in Williams and Shapiro 1990). Neither is evident in the TU 1 assemblage, strongly arguing that the early option correctly dates the deposit. Thus, the estimated chronological position of the McDowell II phase should lay within the time range of AD 1475 and AD 1525.

According to the DePratter and Judge (1986) chronology, the early part of the sixteenth century should be occupied by the Mulberry phase, which is predicted to include substantial representations of folded rims, none of which were identified in the TU 1 excavation. The structure of the Mulberry phase was mod-

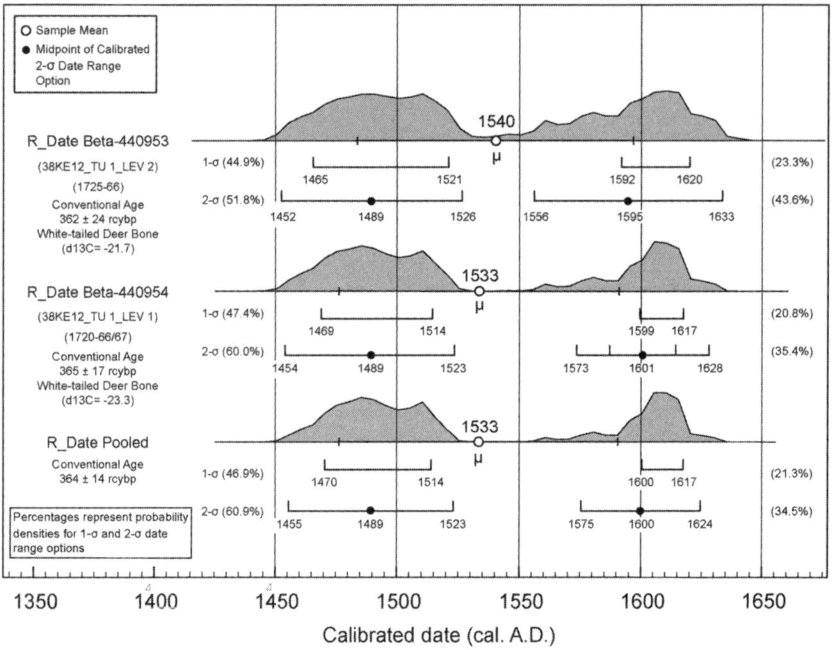

Figure 2.10. Probability densities of accelerator mass spectrometry (AMS) dates obtained on deer bone from McDowell II contexts in TU 1, Mulberry site (adapted from OxCal v4.3.2 program [Bronk Ramsey 2013]; r.5 IntCal 3 atmospheric curve [Reimer et al. 2013]).

eled on data recovered in a 1-×-5 m stratigraphic trench excavated into Mound B (DePratter 1985) at the Mulberry site, supplemented by large sherds collected from underwater locations along Big Pine Tree Creek and the Wateree River channel on the site's northern boundary (DePratter and Amer 1988; Judge 1986). In his master's thesis, Judge (1987:40–50) synthesized this information to more fully define two ceramic phases, Town Creek (or Wateree-Town Creek) and Mulberry. Although the McDowell phase was not recognized in these collections, it is likely that ceramics associated with it are included in the Mulberry phase definition. Only the presence of folded rims would have served to differentiate the two phases, as lip and shoulder ticks, segmented and punctated rim strips, and incising are shared modes. Even so, folded rims are not numerous in Judge's (1987:50) tabulation of Mulberry phase rim modes. They make up only 2 to 3% of all rims and only 5 to 7% of the combined strip/fold category, a fold-to-strip ratio of only 0.05/0.07:1. Incised percentages are low as well, making up only 2 to 3% of exterior surface treatments. After discussions with Judge (personal communication summer 2018), it appears that most of the material he identified as folded rims in his study are actually examples of thickened rims, which suggests that the Mulberry phase representation in his collection is even smaller than the folded-to-strip ratio generated above would indicate.

Had the contributions of the McDowell and Mulberry phases been segregated, which would have been virtually impossible to accomplish with the information Judge had at his disposal, it is likely that higher percentages of both folded rims and incised sherds would be found to characterize the Mulberry phase. This interpretation may affect the estimate of the ages of the earlier stages of Mound B construction, which at present are assigned exclusively to the Mulberry phase. The ramifications of inserting a McDowell phase do not include ruling out the possibility that the initial stages of mound construction began during the very early Mulberry phase. However, the presence of a pure McDowell phase midden near the flank of the mound in TU 1 strongly suggests that ritual/elite activities associated with the location of the mound were underway well before the initiation of the Mulberry phase and that at least the earlier stages of the mound were constructed during the McDowell phase. Further considerations of the Mulberry site occupation sequence discussed in chapter 7 of this book, in fact, strongly indicate that most of Mound B was constructed before the Mulberry phase.

A clean context representing the Mulberry phase in the Wateree sequence, then, has not yet been isolated, although archival visitations to earlier excavated collections may contain such a context. So, when it is claimed that Burke phase ceramics exhibit many more similarities with the Tugalo phase of the upper Savannah River than they do with any phase in the Wateree sequence (Beck 2013:52; Beck and Moore 2002:192, 195, 197; Moore 2002:168–169), one must step back and realize that apples and oranges are being compared. What Judge (1987) identified as the Mulberry phase in the Wateree sequence is more accurately characterized as a mixed assemblage dominated by the fifteenth-century McDowell phase and not the signature of sixteenth-century ceramic patterns. A final verdict on the relationship between the Burke and Mulberry phases awaits the moment when the Mulberry phase is properly defined.

If the original Mulberry phase spans nearly the entire breadth of the sixteenth century, then there is ample time for ceramic change of sufficient magnitude for us to distinguish two divisions, Mulberry I and II. A working model of what these phase divisions might look like can be constructed from a review of neighboring Middle Lamar ceramic assemblages (see Hally 1994a:147). Although there is not a plethora of defined phases from this period in the Eastern SAM culture area, three regions provide clearer glimpses of sixteenth-century ceramic assemblages. These are Port Royal Sound (Santa Elena), the upper Savannah

River valley (Tugalo phase), and the upper Catawba and Yadkin River valleys (Burke phase). Based on a review of sources describing the archaeology of these region, two trends in decorative modes distinguish sixteenth- from fifteenth-century assemblages: (1) the appearance and ascendency of folded rims and (2) peak incidences of incising.

Relevant information from Port Royal Sound comes from the Spanish occupation at Santa Elena, which spanned the period from AD 1566 to 1587. De-Pratter (2009) describes an assemblage of predominantly Irene series ceramics recovered from the back of one of two lots belonging to Governor Gutierre de Miranda, who lived at the fort between AD 1580 and 1587. DePratter argues that the ceramics that comprised this midden were probably obtained by the Spanish in trade with local Native Americans. Broken fragments from these Native American pots were likely dumped at the back of the lot, perhaps entirely during Miranda's occupancy. The presence of a small number of sherds belonging to the Altamaha series, which almost completely replaces Irene at the end of the sixteenth century in the Spanish mission settlements of the Georgia Sea Islands, would support this contention. Quantitative data on rim modes are not presented. However, visual inspection of illustrated rims with either folds or strips (n = 15) would imply the dominance of folded rims, at a 2:1 ratio (DePratter 2009: folded rims correspond to Figures 1.4 A, 1.8 A-C, E, G, 1.10 A, 1.14 D 1.15A, while rim strips include Figures 1.3 F, 1.8 F, 1.9 A, 1.14 E, and 1.15 B). Of course, this inference assumes that the photographs approximate an accurate representation of the entire assemblage. I viewed a sample of the rims from the Miranda Lot collection with DePratter and Heathly Johnson, an inspection that convinced me that this interpretation is accurate. Folded rims do appear to dominate the collection. Percentages of incising can be more objectively evaluated. Considering only the Irene series sherds and the indistinguishable stamped category in DePratter's (2009:38) Table 1.3, the incidence of incising in the Miranda Lot assemblage attains a level of 11.3% of identifiable surface treatments, a significant increase over the 3.4% representation in the McDowell II deposit recovered from TU 1 at the Mulberry site.

The currently accepted chronological span of the Tugalo phase in the upper Savannah River valley covers 150 years, stretching between AD 1450 and 1600 (Hally 1990:54). This duration is too long to represent a discrete ceramic phase and future work will no doubt refine the phase structure in this region. Hally's (1986) review of the approximately 7,000 sherds recovered from the flank midden situated on the northeast side of the mound at Tugalo (see Marshall Williams and Branch 1978) served as the type collection for the Tugalo phase. In that study, Hally reported an incised representation of 2%, but in a later revision based on additional input from the later occupations at Estatoe and Chauga he increased this figure to 8% (see Hally 1990:54). Quantitative data on rim forms are not available for the Tugalo phase, but Hally (1986:113) states that, with few exceptions, "all Estatoe phase vessel and rim forms and incised and stamped decorative motifs can be derived from those present in the Tugalo phase." Four rim forms characterize the Estatoe phase (Hally 1986:109): rolled, L-rims, rim strips, and folded rims. The ratio of folds to strips was 2.15:1. Although sketchy, it can be inferred that relatively substantial contributions of incising and a numerical dominance of folded rims relative to rim strips characterize the Tugalo phase. This inference is supported by the summary of the Tugalo phase by Anderson and colleagues (1986:42). They state that "folded-and-pinched rims increase in frequency to the point where they are the only rim form occurring on jars."

The Burke phase of the upper Catawba and Yadkin River valleys is characterized by soapstone-tempered Burke series pottery believed to primarily represent sixteenth-century occupation (Beck and Moore 2002). This region was visited by both the Soto and Pardo expeditions of the mid-sixteenth century and Spanish artifacts of this period have been recovered from several Burke phase sites. The burned remains of a large structure that is part of the documented Fort San Marcos outpost has recently been excavated at the Berry site, which is inferred to have been the chiefly village of Joara chronicled in Spanish documents (Beck et al. 2006, 2016). A radiocarbon date (Beta-97657) from a feature at the Broyhill-Dillard site in the nearby upper Yadkin valley confirms a sixteenth-century association for Burke series pottery with folded rims (Moore 2002:120–124, Table 23). The 2-σ calibrated range for the assay is AD 1438 to 1675 and the essentially unimodal density distribution (91.5%) indicates that the mean of AD 1576 (Oxcal 4.3.2) can be taken as the best estimate of the feature's age.

Moore (2002:74–89, 276–277) discusses three sites (31BK17, 31BK18, and the Berry site) in the upper Catawba River valley containing dominant Burke series assemblages. Four major rim forms are identified in these assemblages: folded, strip, thickened, and collard. The chronological associations of the latter two types are unclear and will not be considered here. At the Berry site, the fold-to-strip ratio is 6.83:1, while at 31BK17 it is 5.54:1 and at 31BK18 it is 4.13:1 (Moore 2002:279, Table 43). Small numbers of rims used to generate these ratios represent non-Burke series ceramics, but the precise impact on the proportional calculations cannot be determined because information on temper associations is not reported for 31BK17 and 31BK18. However, only 2 of the 49 rims with folds or strips were members of other series in the Berry site collection (Moore 2002:75, Table 5), suggesting that the conflation of temper does not significantly distort the general relationship.

Moore (2002:86) also reports on the incidence of Burke Incised at these three sites. Representation ranges from 2.4 to 6.4% of the total sherd inventories in his table, but it is unclear how these percentages were calculated. In addition, there are disagreements between the numbers of incised sherds reported in the table and those used in the text. It is likely that the sums of both Burke and non-Burke sherd counts were used to generate these percentages and it can be argued that had only Burke series sherds with identifiable surface treatments been used, Burke Incised percentages would be much higher. Indeed, although Moore (2002:Table 7) reports a Burke Incised representation for the Berry site of only 2.4%, an argument can be made that a truer reflection of its popularity would exceed 5% if only Burke series sherds are used in the calculation. Excluding sherds with indeterminate surface treatments, Burke Incised makes up 4.8% (119 of 2493) of this subsample. It is likely that the actual representation for the Middle Lamar occupation is higher, however, because the Burke series collection appears to be mixed with modes characteristic of earlier phases, including thickened and collard Pisgah-type rims, Pisgah and Etowah Complicated Stamped, and fabric and net impressed.

Based on what is known of sixteenth-century assemblages in neighboring regions, then, a generalized model for the two divisions of the Mulberry phase takes the following form. Shoulder ticks and rim strips should continue in Mulberry I from McDowell II, but folded rims will be added to the inventory. In addition, percentages of incising will increase slightly from a base level of about 4%, as represented by the McDowell II context at the Mulberry site. In Mulberry II, folded rims should ascend to dominance over rim strips and the incidence of incising should peak at 10% or greater. Shoulder ticks on jars and carinated bowls are common in McDowell I and II, and this feature may continue at the same level of popularity during at least Mulberry I.

Quantitative data on the incidence of shoulder ticks in other nearby regions are not readily extracted from reports, but evidence of their occurrence suggests both long-lived popularity and wide geographic distribution within the Eastern SAM culture area. Elaborations on shoulders are not always composed of short, notched ticks as is common in McDowell I and II. Linear arrays of circular, rectangular, and fingernail punctations also occur. Shoulder elaborations are present at Irene and Town Creek in what appear to be fifteenth-century contexts. Caldwell and McCann (1941:42, Figure 21) note the presence of shoulder ticks at Irene and include an illustration of a punctated array running along the shoulder break of an Irene Incised carinated bowl. At Town Creek, Reid (1967:Plates X, XIV, and XIX) illustrates ticking or notching on the shoulder of a carinated bowl. He also shows photographs of two sherds with arcuate punctations (fingernail?) aligned with the shoulder break. Middle Lamar associations of this mode are also documented in the Catawba and Upper Yadkin River valleys. Moore (2002:Plates 1, 25, and 27) illustrates two carinated bowls and one shouldered jar with punctated shoulder arrays on Burke and Cowans Ford series pots that probably represent sixteenth-century associations. Finally, Hally (1986:98–101) describes and illustrates three types of shoulder elaborations on Lamar Incised from eighteenth-century Estatoe phase contexts. These consisted of linear arrays of incised slashes (ticks), dot punctations, and cane punctations. It can be inferred that these shoulder elaborations extend much farther back in time, to probably as early as the Rembert phase in the middle and upper Savannah River valley.

Another decorative mode detail that appears to trend in the Middle and Late Lamar periods is strip position. During the fifteenth century, rim strips, as demonstrated in the McDowell contexts at the Mulberry site, are relatively narrow. Moreover, strips are generally positioned below the lip, and the contour of the strip is abruptly differentiated from the vessel wall. Mean strip width in the combined McDowell I and II contexts is 7.75 ± 1.94 mm (n = 84), with a range of 4 to 14 mm. Only 11% of rim strips in the McDowell phase sample are wider than 10 mm. During the sixteenth century, strip width increases and strip position returns closer to the lip, affecting the general appearance of a folded rim.

Although the construction technique of folded rims is generally thought to entail folding the uppermost fillet of a pot downward on the outside of the vessel wall to produce a continuous thickened collar around the orifice, this same visual effect can be achieved by merging and smoothing the junction between the lip and a closely abutting strip. Coe (1995:161–162) describes this latter method in relation to the construction of the Clarksville style rim, commonly regarded as an example of a folded rim type, which, he says, "was made with a strip of clay about 1 [inch] in width, applied to the jar such that the top edge was carefully blended with the lip." He contrasts this with what he describes as the Lamar style, "where the top and bottom edges of the applied strip were usually visible." This particular detail of construction on extant specimens classified as folded rims has never been rigorously investigated, which might explain why many researchers stress the equivalency of the two forms rather than their differences (see Hally 1994a:147; M. Williams and Shapiro 1990:6).

I contend that the shift in rim strip position between the fifteenth and sixteenth centuries represents a fundamental technological adaptation. Saunders (2001:86) has argued that folded rims are stronger than rim strips, presumably because they are more strongly bonded to vessel walls, which supplies a more secure gripping feature when moving pots. A possible stage in this technological adaptation may have brought rim strips close to the lip where the contact could be differentially smoothed or merged by blending the strip into the lip and folding the juncture outward. More attention to variation in rim construction techniques should provide a window into this changing dynamic.

Another feature of the general category of "thickened" rims (i.e., folds and strips, not to be confused with the Pisgah rim of the same name or the rim forms associated with earlier phases as discussed above) that may have some bearing on the composition of Mulberry phase assemblages is the trend documented in Lamar assemblages in Georgia for the width of strips and folds to increase through time (Hally 1978:153; Rudolph 1983:91; M. Smith 1994:178). In Georgia, Early Lamar "thickened" rims are narrow, averaging around 11 mm, while Middle Lamar widths increase to between 13 and 17 mm. Late Lamar/Historic widths can exceed 19 mm. Eastern SAM assemblages seem to exhibit this generalized trend, but the widths may be of a magnitude smaller. Mean strip width from the TU 1 midden at the Mulberry site was only 7.75 mm. However, this measurement may not be strictly comparable to the Early Lamar average of 11 mm for Georgia assemblages, because the Mulberry midden widths do not include the distance between the lip and the top of the strip. The standard method in Georgia incorporates this gap into width measurements (Hally 1978:151). A perusal of Hally's photographic plates of rims from the Little Egypt site (1978:Plate V g, h, l, n; Plate VI b, d; Plate VII d, e; Plate IX g) indicates that there may exist a number of rim sherds with applied strips that are positioned at significant distances below the lip. Without examining the actual sherds, I cannot flatly state that these examples represent rim strips rather than folded rims. However, if the widths of rim strips in the McDowell deposits from TU 1 at Mulberry are modified to include the lip-to-strip gap, a mean of 11.95 ± 3.42 mm (n = 69) is obtained, which is comparable to the Georgia Early Lamar data.

Since the Miranda Lot assemblage from Santa Elena is assumed to date to the last half of the sixteenth century, "thickened" rim widths should approach means of 13 to 17 mm, given trends observed for Middle Lamar assemblages in Georgia. To evaluate this expectation I measured the "strip" widths of the 15 sherds that I had earlier identified as representative of either rim strips or folded rims from an inspection of DePratter's (2009) artifact photographs using digital scaling techniques. It should be appreciated that because the rims were not directly measured, the resultant data are accurate but not necessarily precise. The overall mean for the sample is only 10.67 ± 2.82 mm, much lower than expected. If the rims definitively assigned to the Altamaha and Irene series are segregated, however, the Altamaha mean approaches the low end of the Georgia Middle Lamar range, 12.90 ± 2.37 mm (n = 5). The Irene sample, by contrast, yields a mean width of only 8.93 ± 2.76 mm (n = 6). These differences suggest that the Altamaha sample is younger and was deposited at the very end of the Miranda occupation, while the Irene sample may cover the total span of Spanish occupation at Santa Elena (1566–1587). Whether the Miranda Lot sample also includes contributions from earlier Irene occupation is a question that awaits an analysis of the entire Native American ceramic assemblage recovered from Santa Elena.

"Thickened" rims on Burke series sherds illustrated by Moore (2002) conform more closely to the expected range of widths for Middle Lamar in Georgia. Again, this information is derived from indirect digital measurements. The two reconstructed vessels (Moore 2002:Plates 25 and 26) from the Broyhill-Dillard site, one of which has an associated radiocarbon date mean of cal AD 1576, have notched/segmented folded rims with respective widths of 14.1 and 14.7 mm. Nine "thickened" rims are illustrated for the Berry site (Moore 2002:Plate 7a and b, Plate 12), including five with folds and four with strips. The former has a range of 12.0 to 17.0 mm in width and a mean of 14.5 ± 1.80 mm, while the strips, measured from the lip, range between 11.0 and 17.0 mm with a mean of 13.4 ± 2.8 mm.

Until actual measurements of complete collections of "thickened" rims are taken and reported, precisely how the width trajectory plays out in the Eastern SAM culture area during the sixteenth century will remain ambiguous. The Burke series illustrations suggest that increased fold/strip widths occur in the sixteenth century, but the Irene-series "thickened" rims from the Miranda Lot at Santa Elena display widths more compatible with Early Lamar assemblages in Georgia, even though this assemblage most likely has at least a partial mid-sixteenth-century association. This contrast may simply result from regional variation in the Lamar width trajectory of "thickened" rims, but it is also likely that assemblages from the first half of the sixteenth century will be shown to exhibit, on average, narrower fold/strip widths than those made in the last half of the century. Thus, mean widths of "thickened" rims may provide another variable useful in distinguishing between Mulberry I and II ceramic assemblages.

The scant glimpses that we now have of Late Lamar assemblages of the seventeenth and eighteenth centuries in the Eastern SAM culture area indicate continuity with the general trajectories of "thickened" rim characteristics and incising that are manifest in the sixteenth century. Folded rim widths appear to increase in the manner described for the Georgia Lamar sequence, where mean widths expand to 19 mm or greater. Digital estimations from rim sherd photos provide a basis to evaluate trends. Three lower Catawba River valley assemblages from this period have been reviewed in very general terms. These are the mid- to late-seventeenth-century material from the Belk Farm site and the early- to mid-eighteenth-century ceramics from the historic Catawba confederacy sites of Nassaw-Weyapee and Charraw Town. The ceramics from these closely related, although probably ethnically diverse, occupations belong to the Cowans Ford series, a sand-tempered tradition that appears to span the sixteenth through early eighteenth centuries (see Moore 2002:154–157).

Riggs (2010:Figure 3) provides photos of sherds belonging to three separate vessels with folded rims from the Belk Farm site. The folds were actually constructed from strips in the manner described by Coe (1995:161–162). Digital measurements return fold widths of 11.0, 16.5, and 17.5 mm, yielding a mean width of 15.0 mm. Folded rims from the later sites of Nassaw-Weyapee (Davis et al. 2015:Figure 2.10) and Charraw Town (Davis et al. 2015:Figure 2.13) are noticeably wider. Illustrated fold widths from Nassaw-Weyapee range between 15.9 and 27.1 mm with a mean of 18.8 ± 4.1 mm (n = 7), while those illustrated for Charraw Town range between 12.7 and 23.1 mm with a mean of 18.1 ± 4.1 mm (n = 7). These are compared with the narrower means that characterize the Middle Lamar fold widths for the Miranda Lot assemblage and the Burke phase, which range between 13 and 15 mm.

Hard quantitative data on the incidence of incising is not available from these lower Catawba Valley sites, but all contain incised sherds with brackets of six or seven lines each, a pattern most prevalent in the Middle and Late Lamar periods (Hally 1994a:147). In addition, the incised lines are generally fine, which tends to set Late Lamar styles apart from the bolder incised lines found in earlier Lamar contexts (see Riggs 2010:35). A review of Late Lamar assemblages in neighboring areas suggests that percentages of incised sherds decrease slightly from a peak of around 12 to 14% in Middle Lamar assemblages. The seventeenth-century Altamaha assemblage recovered from Mission Santa Catalina on St. Catherines Island, Georgia (AD 1602–1684) contains about 6% incised sherds (Saunders 2009:100–101). Once the mission was relocated to Amelia Island after 1684, incised sherd percentages decrease further in the excavated sample, to only 1.4% (Saunders 2009:102). Historic Creek assemblages of the seventeenth and eighteenth centuries in Georgia average about 7% incised, with a range of 0 to 13.5% (M. Smith 1992:65). Finally, early eighteenth-century deposits at the Little Egypt site, which Marvin Smith (1992:62–63) suggests may be ethnically Cherokee with an admixture of Upper Creek, contain between 4.4 and 9.2% incised.

This review of Middle and Late Lamar decorative modes makes it possible to construct a model of decorative mode style horizons encompassing the full reach of Eastern SAM sequences. Table 2.10 presents the model formatted as an occurrence seriation. Included in the table is an attempt to correlate the phases from the various regions of the Eastern SAM culture area. Bold, capitalized "X" characters indicate the chronological positions in which a mode is most abundant, while a lower case "x" reflects a contribution of only minor numerical importance.

The early portion of the seriation, which can be described as the Town Creek-Irene Axis horizon, is characterized by notched lips, punctated lips, small and solid punctations, and incidental hollow reed punctations. These modes continue to occur throughout most of the sequence, but there appears to be a chronological trend toward larger diameters through time as Boudreaux (2005:48, 86) discusses for the Town Creek seriation. His analysis distinguished an arbitrary division between large and small punctations at 10 mm, but with more and larger samples it will likely be shown that the trend is continuous and attention to punctated diameters from discrete contexts should identify modal breaks that will provide an even more sensitive chronological measure of assemblage change. While Boudreaux combined solid and hollow reed punctations in his small punctated category, I separate them. Rosettes, pellets, and nodes emerge at the end of the horizon and continue later.

The later horizons fit comfortably within divisions of the Lamar Complex as established by Hally (1994a). Early Lamar corresponds to the fifteenth century and marks the emergence of hollow reed punctated and notched rim strips, shoulder ticks/punctations, and incising on plain and burnished carinated vessels after around AD 1425. Notched strips are either segmented with a sharp stylus or notched with blunt, rectangular styli. Strip position occurs at varying distances below the lip. Middle Lamar marks the substantial occurrence of folded rims and peak percentages of incising. Fold widths increase from means of 11 to 12 mm to as much as 14 to 17 mm over time. Late Lamar is similarly marked by the dominance of folded rims and perhaps decreasing representations of incising. Mean widths of folds increase to greater than 18 mm, with some examples approaching 30 mm. Incised lines become narrower. A wide variety of notching techniques are displayed on folded rims during the Middle and Late Lamar horizons, including fingernail punctate, incision, solid rectangular and ovoid punctations, finger pinched flutes, "pie crust" scallops, and indentations at the base. Basal indentations probably have a later temporal distribution and attention to such variability will again increase the resolution of ceramic chronologies in the region.

I have not integrated the Late Leak phase of the Town Creek sequence into the model because there are associated anomalies in need of resolution. First, Boudreaux (2005:87) characterizes the component at Town Creek as terminating at around AD 1450. However, while the two contexts he ascribes to this component contain substantial representations of folded rims, they lack incised sherds, a pattern that would be unprecedented in Eastern SAM assemblages and clearly out of sync with the proposed early fifteenth-century chronological position. Substantial contributions of folded rims do not materialize until after about AD 1500 and are preceded for a lengthy duration by incising. The phase that immediately succeeds Early Leak should look similar to the McDowell phase deposits from the midden next to Mound B at the Mulberry site. Narrow rim strips positioned at variable distances below the lip should dominate decorative modes, along with substantial representations of shoulder ticks and/or punctations on the shoulders of plain and plain burnished carinated vessels. In addition, incised

Table 2.10. Occurrence Seriation Model for the Eastern Wing of South Appalachian Mississippian

Start Date (cal AD)	Horizon	Wateree	Upper Pee Dee	Central SC Coast	Upper Catawba and Yadkin	Lower Catawba
1700	Late Lamar					Nassaw-Weyapee
1625		Daniels	Caraway			Belk Farm
1575	Middle Lamar	Mulberry II			Burke	
1525		Mulberry I			Burke	
1475		McDowell II				
1425	Early Lamar	McDowell I	(Revised Late Leak)	(Salt Pond)		
1375		Wateree/Town Creek	Early Leak	Fogarty Creek		
1325		Adamson	Late Town Creek	Tibwin		
1275	Town Creek-Irene Axis	Belmont Neck	Early Town Creek	Jeremy		
1225		(Late Camden)	Teal	Sewee Camp		

Note: Capitalized "X" characters indicate the chronological positions in which a mode is most abundant, and a lowercase "x" reflects a contribution of only minor numerical importance.

percentages should approach 1 to 4% of the sherd inventory. Folded rims do not occur in the pre–AD 1450 contexts at the Adamson site and none were identified in the TU 1 midden assemblage at Mulberry.

The contexts that Boudreaux assigns to a pre–AD 1450 Late Leak phase component must surely date later and there should be a hiatus between it and the Early Leak phase. As discussed, the only dated context of the Late Leak phase (Sq 170L/40/Pit) in the Town Creek seriation yields a most probable 2-σ range of cal AD 1447 to 1675, which supplies a strong argument for a post–AD 1450 age. Moreover, the cal AD 1587 mean for the sample suggests the likelihood that

Table 2.10. *continued*

Notched Lips	Small/Solid Punctations	Hollow Reed Punctations	Rosettes and Pellets	Nodes	Hollow Reed Punctated Strips	Notched Strips	Shoulder Ticks	Incising	Folded Rims
								x	X
								x	X
		x		x	x	x	x	X	X
		x		x	x	X	x	X	x
		x		x	x	X	X	X	
		x	x	X	X	X	X	x	
	x	X	X	X	X	x			
x	X	X	X	x					
X	X	X	x						
X	X	x							

the true age of the pit falls much later, between AD 1550 and 1675, an age in better agreement with the substantial contribution of folded rims.

The absence of incising is more difficult to understand within the context of the regional seriation. Boudreaux (2005:94, endnote 4) states that incised sherds were not present in any of the samples he used in his study. The analysis sample included only sherds larger than 4 cm in diameter and smaller sherds were not systematically examined. However, he informs me that no obvious examples of incising were recognized in the culled sample of smaller sherds either (Tony Boudreaux, personal communication, October 22, 2015). Of the 53 incised or possibly incised sherds identified from the thousands of sherds that he has analyzed from the site, only 9 exhibit the bolder lines that he would definitely associate with Lamar Incised. He concedes, however, that the few identified contexts dating later than AD 1400 at Town Creek may not be representative of the entire range of phases that could populate the late end of the occupation sequence. Thus, it is likely that a larger sample of contexts or a closer examination of small sherds from the Late Leak contexts would resolve the apparent anomaly presented by the lack of association between folded rims and incising.

Clearly, Lamar Incised does exist at Town Creek and folded rims with widths that would qualify as members of either Middle or Late Lamar assemblages are definitely present. The four folded rim sherds illustrated in Coe (1995:Figure 9.9, examples A through D) were all recovered from Town Creek (R. P. Stephen Davis, personal communication, October 15, 2015), and fold widths range between 16.8 and 34.1 mm with a mean of 25.0 ± 7.1 mm, based on digital measurements. All but one of the sherds display a fold width of 24.5 mm or greater and almost certainly the wider widths are associated with seventeenth- or more likely eighteenth-century Late Lamar. The sherd with the narrower fold (Coe 1995:Figure 9.9, D) exhibits large (12.9 mm in diameter) hollow reed punctations, in keeping with Boudreaux's observation that such elements increase in diameter through time. The characteristics of this sherd are consistent with either Middle or Late Lamar. Thus, the folded rims and incised sherds that Boudreaux has identified at Town Creek are consistent with the documented but poorly defined Caraway phase, as is the radiocarbon date currently associated with the Late Leak phase.

That incising occurs in Caraway assemblages is little appreciated because of the standard practice of treating its occurrence in North Carolina as a decoration rather than a surface treatment. Ultimately, this practice results in embedding the incidence of incising within the larger exterior surface treatment classes of plain and plain burnished. The characterization of the ceramic assemblage from the Poole site, located on a tributary of the Yadkin River and the type-site for Caraway, is a case in point. Ward and Davis (1999:137; see also Wilson 1983:326–342) note the dominance of smoothed and burnished surfaces with secondary contributions of complicated stamped and simple stamped along with minor representations of net impressed, corn cob impressed, and brushed. Wilson (1983:327) observed that Coe's original description of the Caraway series could be read to imply that vessels were very seldom decorated, but he found that elaborations were relatively abundant when he sorted the collection. The inventory of decoration that Wilson discusses is consistent with the characteristics of the Middle and Late Lamar horizon styles as modeled here. These include a dominance of folded rims, the presence of incising between the shoulder and lip on some plain and burnished surfaces, and ticks below the lip, the latter of which was described by Wilson as "a band of short lines incised perpendicular to the lip" (Wilson 1983:330). Wilson also identified the occasional narrow pinched or notched rim strip placed adjacent to the lip. Folded rims predominated numerically, however, which again is consistent with Middle and Late Lamar assemblages. Many of the folded rims represent true folds, in which the effects of thickening are not manifest at the lip but taper outward below it with maximum thicknesses achieved at the base of the folds. Coe (1995:161) likened these specimens to the Clarksville style rim (Evans 1955:49–53), a seventeenth- and eighteenth-century trait found in Piedmont North Carolina and Virginia (see South 2005:30).

The Poole site represents the presumed palisaded settlement of Keyauwee visited by John Lawson in 1701 (Coe 1995:159–161) and it is generally believed that the Lamar occupation there covers much of the seventeenth century (Wilson 1983:324; 345). The only feature containing European trade items in the small area exposed during excavations in the 1930s was a large, shallow pit feature (Feature 4). Approximately 135 small blue and white glass trade beads and a bead refashioned from a section of a Kaolin pipe stem were recovered from the pit, all of which possess a long period of occurrence spanning AD 1600 to 1836. Caraway series ceramics from the pit include incised sherds, all of which are identified as Caraway Plain (BX9282-BX9339, BX9469-BX9494, RLA, University of North Carolina, Chapel Hill). Also pictured are a number of other incised Caraway Plain sherds from the plow zone, indicating that incising was abundantly represented at the site. Thus, it would appear that the modeled association of folded rims with incising holds for the seventeenth-century occupation in the Town Creek region, which suggests strongly that this relationship was also extant in the sixteenth century.

The paucity of incised sherds at Town Creek provides strong evidence that the Caraway occupation was brief or in some other manner light, but it does not establish that incising is absent from the ceramic inventory of this period. Finer-lined incising is not necessarily evidence of non-Pee Dee or non-Caraway culture associations. As noted, the trajectory of Lamar Incised stylistic change moves from bolder lines and simpler designs in Early Lamar to increasingly narrower lines, more complex designs, and greater numbers of lines in Middle and Late Lamar (Hally 1994a:147). The incised sherds from the Hardins site, Belk Farm, Nassaw-Weyapee, and Charraw Town illustrated by Riggs (2010) and Davis and colleagues (2015) demonstrate this trajectory from the sixteenth into the middle eighteenth centuries in the lower Catawba River valley. Very narrow incised lines appear to dominate by the second half of the seventeenth century.

The presence at Town Creek, in a Caraway burial, of a circular brass gorget that postdates AD 1630 and the recovery of numerous glass beads of the types that occur in Contact period Siouan sites in the Carolina Piedmont support the conclusion that most of the Caraway occupation at Town Creek dates to the latter half of the seventeenth century. Thus, there is a strong probability that many of the fine-lined incised sherds that Boudreaux believes derive from earlier, pre-Mississippian phases at Town Creek are actually affiliated with the Caraway occupation. Given what is known of the chronological structure of Lamar ceramics in the Eastern SAM culture area, then, a phase with the specific characteristics that Boudreaux ascribes to the Late Leak phase has yet to be identified in neighboring regions. Whether such a phase exists at Town Creek, however, requires further investigation.

Another issue that may contribute to the anomalous appearance of the assemblage is the possibility of mixing. Features positioned in the later stages of heavily occupied sites will have an increasingly greater probability of containing material from earlier phases in their fill. An examination of decorative mode composition in Table 2.8 indicates that an extremely high degree of diversity characterizes the Late Leak portion of the decorative mode seriation. The generous occurrence of earlier Town Creek-Irene Axis modes (i.e., small punctations, rosettes, pellets, and possibly nodes) with substantial contributions of folded rims suggests the strong possibility that the Late Leak contexts are mixed. Only more complete descriptions of discrete Middle and Late Lamar contexts would reveal whether the modes with early associations extend into the sixteenth and seventeenth centuries in substantial proportions. It would seem more likely, however, that their presence in late deposits at the site is the result of deposit mixing.

Discerning the character of ceramic assemblages dated to the late fifteenth and sixteenth centuries is critical to dating and understanding the events and relationships that transpired in the late prehistory of the Eastern SAM culture area. It provides a window into how to measure the occupation histories in each region and ultimately supplies the necessary proof for identifying regions that were depopulated versus regions that may have experienced influxes of population due to larger demographic processes of migration. Armed with this background, it is now possible to quantitatively demonstrate that the Mississippian occupation of the Central Coast of South Carolina ended by the middle of the fifteenth century and to consider the larger implications that this fact holds for understanding similar processes in neighboring regions.

3

The Fifteenth-Century Depopulation of the Central South Carolina Coast

I've come loaded with statistics, for I've noticed that a man can't prove anything without statistics.
—Mark Twain, Hartford Opera House, October 26, 1880

The Central Coast seriation is based on a restricted set of dated contexts that together describe the major trends in ceramic change in the region from about AD 1250 to 1425. This claim is confirmed by the ease with which the occurrence seriation of decorative modes correlates with the sequences from the Wateree Valley and Town Creek for this particular span of time. In this chapter, I argue that this seriation encapsulates the full range of Mississippian occupation periods in the Central Coast region and that there is a lack of evidence for continued settlement in the region after the first half of the fifteenth century. Proof of depopulation at this time comes from demonstrating that the structure of the larger regional sample is not statistically different from the seriation sample and then showing that decorative modes diagnostic of Middle and Late Lamar occupation extant in other Eastern SAM regions are conspicuously absent in the Central Coast record. I then consider factors that might have led to the abandonment, drawing on evidence from nearby coastal regions that appear to have been abandoned at the same time.

The Larger Regional Sequence

In many regions of the Southeast, developing a systematic regional ceramic database of sufficient reach to recover a representative sample of occupational components is financially unfeasible. However, because of the fortunate circumstance that the Central Coast coincides closely with the boundaries of the Francis Marion National Forest, an extensive record of Mississippian occupation has been identified and sampled systematically with interval scale shovel testing and limited test excavations because of federal mandates on the treatment of cultural resources. Consequently, most of the major tidal creek habitations that are the hallmarks of the Mississippian adaptation in the region have been extensively sampled with relatively comparable subsurface testing methodologies. In most situations, the minimal sampling unit was a 15 m-interval shovel test systematically dispersed across defined site areas, although some of these sites have been sampled at even greater intensity (i.e., 5 m- or 10 m-intervals). In some cases, dispersed test units and surface collections have been deployed to substitute for, or augment, shovel test data.

Table 3.1 presents decorative mode occurrences for the major Mississippian tidal creek habitation clusters, whose locations are illustrated in Figure 1.1. Several inland locations with aggregations of resource extraction sites that probably served as seasonal farmsteads are also included. The data are presented as percentages of total sherds exhibiting decorative modes recovered from each multisite cluster. This method was selected over frequency counts so that comparability across widely divergent sample sizes could be achieved. Thickened, thinned, rolled, and plain rims comprise the collapsed undecorated rim category. Between-cluster percentages of individual modes vary across the sample, indicating differences in occupation histories. Those clusters that display heavy occurrences of Town Creek-Irene Axis modes are also characterized by lower overall percentages of decorated rims, while those clusters with higher representations of Lamar modes (i.e., rim strips) display higher proportions of the former. Only a handful of rims (n = 7) referred to as "folded" have been identified in the corresponding survey reports for these assemblages. Examination of photographs and actual specimens resulted in the reclassification of five of these as strips (n = 3), thickened rims (n = 1), or rolled rims (n = 1). The remaining two specimens from the South Tibwin cluster (Brockington & Associates 2011) have not yet been verified, but it is likely that they will eventually be shown to represent some other type of "thickened" rim with an earlier chronological association. Based on these observations, the folded rim category in Table 3.1 has been assigned the status of unconfirmed and is not used in the following comparison.

To determine whether the seriation sample is representative of the larger occupation sequence, it was necessary to transform the data further to achieve comparability. The first step in this process combined the individual records for each data set into two grand averages. In the case of the seriation sample, this entailed combining the decorative mode data from the four phases identified in Table 2.7. For the regional sample, this procedure required collapsing the data from each cluster in Table 3.1 into a single occupational signature. Because sample sizes within the phases and clusters vary a great deal, a simple arithmetic average using frequency data would be misleading as it would bias the signatures in the direction of the units with the largest sample sizes. The information from each unit should have equal importance in characterizing occupational history; an effective means of achieving this end would be to calculate averages of percentage representation. Grand percentage averages for the two samples are presented at the bottom of Table 3.1. As can be seen, these modal averages are very similar, with the greatest disparity occurring in less common categories. The most popular modes (i.e., notches, hollow reed punctations, and combined rim strips) display nearly identical grand percentage averages. The greatest disparity, then, is evident in rarely occurring modes, which can most likely be attributed to stochastic variation attendant with small sample sizes.

A statistical test to determine just how similar the samples might be can be generated using the Brainerd-Robinson coefficient in the manner used in chapter 2 to group surface treatment percentages. Not only can this coefficient provide a relative measure of similarity between samples, but Kintigh's (2002) TFQA package also provides a basis for evaluating statistically the probability that two samples are derived from the same or different populations. To calculate BR statistics, however, it is necessary that cells within each unit of comparison sum to 100%. In the case at hand, the grand percentage averages for decorative modes sum only to 17.8% for the regional sample and 15.0% for the seriation sample, because undecorated rims claim the lion's share of representation in each sample. Although a BR coefficient could be calculated that would include the undecorated percentage averages, this procedure would obscure the contribution of the diagnostic decorative modes, which is the relationship of greatest interest. As a work-around, grand percentage averages for the decorative modes can be multiplied by the inverse proportion of decorated rims (1/p) for each corresponding sample (i.e., 0.178 for the regional sample and 0.150 for the seriation sample) to derive transformed percentages that sum to unity (100%).

The results of this procedure are displayed in Table 3.2. The calculated BR index between the samples is 179.2, which is relatively high. The greatest percentage differences are found in the rarest modal groupings (rosettes/pellets and nodes), which together make up two-thirds of the difference in the two signatures. If these two groupings were to be combined, the differences between the two signatures would be extremely negligible, producing a BR of 190.6. Accepting the lower BR coefficient, however, still supplies a strong case that the two samples are statistically the same. Kintigh's (2002) resampleBRED script produces a cumulative probability of 0.7274 in association with a BR of 179.2 given

Table 3.1. Decorative Mode Distributions for the Central Coast Regional Sample

Site Cluster	Site Number	Source	Recovery Units
Jeremy Island	38CH2	Trinkley (1980)	Surface Collection
North Tibwin	38CH1977, 1979, 1980,	Cable (2002)	15m Shovel Tests
South Tibwin	38CH1531	Brockington & Associates (2011)	15m Shovel Tests
Buck Hall	38CH644	Cable et al. (1991), Poplin (1993)	15m Shovel Tests, Test Units
Walnut Grove	38CH259	Wheaton et al. (1993)	15m Shovel Tests, Test Units
Walnut Grove	38CH300	Trinkley (1981a)	Test Units
Awendaw	38CH1116, 1039/1061, 138/533	Cable (2010)	15m Shovel Tests
Salt Pond	38CH44, 134, 503, 1440, 1446, 1447, 1450, 1709	Gardner (1992, 1993), Brock. & Assoc. (2011)	5, 15 and 30 m Shovel Tests
Sewee Camp	38CH146/1048, 38CH426	Cable et al. (2013)	15m Shovel Tests, Test Units
Maxville	38CH454	Trinkley (1980)	Test Units
Le Grand	38BK692,570, 704	Cable (2002)	15m Shovel Tests
Halfway Creek	38BK769,1176, 38CH1045, 1189, 2100, 2103	Cable (1991, 2008)	15m Shovel Tests, Test Units
Fogarty Creek[b]	38BK552	Cable (n.d.); Agha and Phillips 2006)	5 and 15m Shovel Tests
Royall Site	38BK2503	Cable (2013)	5 and 15m Shovel Tests
Combined[a]			
Percentage Averages	Regional Sample		
Percentage Averages	Seriation Sample		

Note: Data are presented as percentages of total sherds exhibiting decorative modes recovered from each multisite cluster.

[a]Combined = merged Halfway Creek, Fogarty Creek, and Royall site samples

TABLE 3.1. *continued*

Notched Lips	Punctated Lips	Other Punctations	Reed Punctations	Rosettes	Pellets	Nodes	Hollow Reed Punctated Rim Strips	Notched Rim Strips	Plain Rim Strip	Folded[c]	Total Rims	Total Decorated	Percent Decorated
1.3	0.0	0.0	1.3	0.0	0.0	0.0	0.0	0.0	0.0	0.0	77	2	2.6
8.0	0.0	0.0	8.0	0.0	0.0	0.0	0.0	0.0	0.0	0.0	75	12	16.0
1.7	0.0	0.6	14.9	0.0	0.0	5.1	0.0	0.6	0.0	1.1	175	41	23.4
6.3	0.0	0.0	15.6	0.0	0.0	0.0	1.6	0.0	0.0	0.0	64	15	23.4
4.8	0.0	2.4	0.0	0.0	0.0	0.0	2.4	0.0	0.0	0.0	42	4	9.5
2.5	0.0	0.0	9.5	0.0	0.0	2.7	0.0	0.0	0.0	0.0	325	48	14.8
2.9	0.0	0.0	5.7	2.9	0.0	0.0	0.0	0.0	0.0	0.0	35	4	11.4
5.9	0.0	2.9	17.6	0.0	0.0	2.9	2.9	2.9	0.0	0.0	34	12	35.3
0.0	0.0	0.0	4.3	0.0	0.0	0.0	0.0	0.0	0.0	0.0	23	1	4.3
2.1	0.0	0.0	11.7	0.0	0.0	3.3	0.0	0.0	0.0	0.0	583	105	18.0
8.3	0.0	0.0	0.0	0.0	0.0	0.0	0.0	4.2	0.0	0.0	24	3	12.5
10.0	0.0	10.0	10.0	0.0	0.0	0.0	10.0	20.0	0.0	0.0	10	6	60.0
2.6	0.0	2.6	7.9	0.0	0.0	5.3	10.5	5.3	2.6	0.0	38	14	36.8
20.0	0.0	0.0	0.0	20.0	0.0	0.0	0.0	0.0	0.0	0.0	5	2	40.0
5.7	0.0	3.8	7.5	1.9	0.0	3.8	9.4	7.5	1.9	0.0	53	22	41.5
4.1	0.0	0.8	8.0	0.4	0.0	1.5	1.4	1.3	0.2	0.1.	1510	269	17.8
3.8	0.4	0.0	6.7	0.4	0.8	0.0	2.1	0.8	0.0	0.0	239	36	15.0

[b]Elliott's (1983) test excavation data not included.

[c]Not verified

Table 3.2. Brainerd-Robinson Test of Similarity between the Central Coast Regional and Seriated Samples

Comparison	Notches	Small Punctations	Reed Punctations	Rosettes/Pellets	Nodes	Rim Strips	Frequency	Sum of Differences	Brainerd-Robinson Coefficient	Probability
Regional Sample	23.0	4.5	44.9	2.3	8.4	16.9	239			
Seriation Sample	25.3	2.7	44.7	8.0	0.0	19.3	36			
Percentage Difference	2.3	1.8	0.2	5.7	8.4	2.4		20.8	179.2	
Pooled Means	24.2	3.6	44.8	5.2	4.2	18.1	275			0.7274

the specified sample sizes, which means that a BR result of higher than 179.2 will occur in 28% of the samples drawn from the same population. The statistical decision from this comparison, then, is that the samples are not different, which strongly suggests that the range of phases represented in the seriation is identical to the ones contained in the regional sample. Therefore, a convincing case can be made that the Central Coast was abandoned, or at least severely depopulated, at the end of the Fogarty Creek phase, around AD 1425.

The issue of whether limited occupation continued in the first half of the fifteenth century is unresolved. If it did, the Eastern SAM seriation model predicts that it would approximate the structure of the McDowell I phase of the Wateree sequence. The principal diagnostics differentiating such a phase from Fogarty Creek are incised body sherds and shoulder ticks. There are several incised sherds in the regional database that might represent the Lamar Incised type, but these are accompanied by classificatory problems. Incising also occurs in Thoms Creek, Refuge, and Deptford series ceramics from the region, and it is seldom that such distinctions are made in survey reports. J. Gardner (1992) recovered five incised sherds from sites in the Salt Pond cluster. Only two were assigned culture historic identifications, one Thoms Creek and one Mississippian. The latter exhibited a notched lip and parallel incisions oblique to the rim, which could just as easily describe Refuge Incised, an Early Woodland type. Also, six incised sherds have been reported from the South Tibwin cluster (Brockington & Associates 2011). Four of these are unclassified and contain a fine-to-medium sand paste that would be consistent with Thoms Creek rather than the coarse pastes of later Mississippian ceramics in the region. The other two are classified as Mississippian rim sherds from the same vessel. None of these incised sherds are adequately described and in each instance verification awaits. Another incised Mississippian sherd has been reported from the Le Grand site (Cable 2002a). This sherd represents a shoulder fragment from a vessel with a strongly everted rim. Incising here consists of three very faint, parallel incisions running perpendicular to the plane of the rim, which appear to have been scratched onto the surface of a dried but unfired pot. Such a technique is inconsistent with the bold incised lines associated with early examples of Lamar Incised and it is not certain whether the sherd dates to the fifteenth century or not. Finally, a single, confirmed Lamar Incised sherd has been recovered from test excavations conducted at the Fogarty Creek site. The sherd is small, but the partial motif consists of a semicircular festoon element pendant to a straight line, which is entirely consistent with Lamar style incising. Thus, the status of a McDowell-like phase in the Central Coast sequence is confirmed but, by any measure, must be viewed as a very subdued interval of occupation. Only 5 of 11,367 sherds (0.04%) in the regional sample appear to represent good candidates for Lamar horizon incised sherds. This suggests that the presumed Salt Pond phase is likely to have been very brief and probably ended before AD 1450.

Contributing Factors

The timing of the abandonment of the Bulls Bay region, sometime between AD 1400 and 1450, corresponds precisely with the final stages of abandonment on the lower and middle reaches of the Savannah River valley. Anderson (1996; Anderson et al. 1995) attributes abandonment there to two interdependent factors, the innate organizational instability of chiefdom-level societies (i.e., chiefly cycling) and a series of multidecadal droughts that stretched agricultural production beyond the technological and political capabilities of the chiefdoms to distribute surpluses adequately enough to maintain the status quo. Ultimately, he argues, extended crop failures led to an erosion of elite authority and the collapse of the social system. Anderson's inferences concerning megadrought were derived from bald cypress tree-ring chronologies extracted from locations along the South Atlantic Slope, which reflected annual spring rainfall levels from AD 900 to 1985 (Stahle and Cleaveland 1992). Yearly variation in rainfall was high, but the application of cubic smoothing splines at 10- and 30-year intervals led to the successful identification of alternating decade-scale regimes of lower (droughts) and higher (wet periods) than average spring rainfall. Anderson and colleagues (1995) drilled down further into this reconstruction, transforming the rainfall data to reflect proxies of potential agricultural surpluses and shortfalls, assuming the ability of Mississippian societies to maintain food reserves of one year. This new curve enabled Anderson to more easily characterize the amplitudes of multidecadal droughts and wet periods that correlated with abandonments, migrations, and coalescences of Savannah River mound centers spanning a 300-year interval from AD 1150 to 1450. Droughts of particularly long duration and intensity occurred in the last half of the fourteenth and early part of the fifteenth centuries, and these intervals correlate with the construction of defensive fortifications, the widespread collapse of local chiefdoms, and, finally, the abandonment of the entire lower and middle Savannah River valleys.

Similar forcing events and processes probably befell the Central Coast population, but the social context was different. Anderson and colleagues (1995:259–260) argue that the Savannah River chiefdoms were supported by intensive agricultural economies fueled by the authority of the elite to levee tributes of crop surpluses that enabled them to manage shortfalls and successfully compete with neighboring polities. Lower-intensity agricultural production probably inhibited the development of a tribute-based redistribution system in the Central Coast region. The mainstay of Mississippian populations along the coast appears to have been estuarine resources, a subsistence base that would not have been as successfully stored across one or more years. This systemic difference is most strikingly manifest in the presence/absence of platform mounds. The Savannah River chiefdoms were characterized by multiple platform mound sites, while the Central Coast settlement system appears to have been devoid of any type of monumental architecture that could have served as the seat of chiefly authority. The nearest known mounds are located upstream on the Santee River at Scott's Lake (Ferguson 1973a), in addition to the Wee Tee mounds in Williamsburg County on the north side of the Santee River, about which little is known (Chester DePratter and Robert Morgan, personal communication, February 13, 2012). The former is 120 km distant from the mouth of the Santee, while the latter is situated about 34 km from the northernmost Central Coast Mississippian settlements along Jeremy Creek. Southward, only three Mississippian mound sites are known: one at Runnymeade Plantation on the Ashley River in Charleston County (Lansdell et al. 2005) and two in Beaufort County visited by C. B. Moore in 1897 (Larson 1998:54). All of these sites are situated well away from Bulls Bay, at respective distances of 32 and 110 km from the southernmost coastal clusters.

Certainly, platform mounds cannot be construed as a requisite feature of chiefdoms, but their absence in the Mississippian world suggests a less powerful hierarchical organization and perhaps one in which the ascriptive authority of a chiefly lineage had not yet been established. South's (2002) characterization of the palisaded thirteenth- and fourteenth-century Mississippian cemetery complex at Albemarle Point as a "moundless ceremonial center" captures the transegalitarian (see Hayden 1995) nature of the coastal settlements along much of the southern coast of South Carolina. In keeping with this interpretation, he concludes (South 2002:241) that the "ceremonial center cannot be classified as a large center, but rather a smaller one serving a limited number of nearby villages." In contrast to chiefdoms, such societies are probably better understood as emergent chiefdoms or big-man societies (Redmond 1998). Such societies contain within their metastructure both the capacity for temporary centralization of authority over autonomous villages by leaders with achieved status in response to situational crises and also the necessary apparatus to dissolve back into relatively unranked and independent relationships when crises subside. Carneiro (1998:36–37) contends that, historically, chiefdoms became permanently established at flashpoints when the need for stability provided by hereditary chiefs transcended the importance of village autonomy. In Mississippian society, platform mounds might be the principal indicator of just such flashpoints.

In this light, the contrasts between the Savannah River chiefdoms and the transegalitarian groups of the Central Coast Mississippian were not extreme. However, the latter would have lacked the characteristically entrenched redistributive and sociopolitical organization of chiefdoms, which Anderson argues was fatally destabilizing over the long run and ultimately led to collapse and regional abandonments in Mississippian societies. Yet, even without such internal structural stresses the Central Coast system underwent a collapse/abandonment event nearly identical to that registered for the Savannah River chiefdoms in the early fifteenth century. Chiefdom cycling is sufficient to explain localized collapses and abandonments of single mound centers, but no one has explained how chiefdom organizational instability could serve as the primary cause for the abandonment of entire regions. Clearly, a more sweeping external force must have been at work in the fifteenth century to bring about these synchronously profound events and I argue that they were brought about by megadrought and its attendant extreme climate-related events.

A likely drought-related event process that I think most succinctly explains the Central Coast abandonment is a severe collapse in estuarine productivity, the signature of which can be seen in the species succession of molluscan fauna in the shell middens of the coastal clusters. Trinkley (1980:407–409, 433) first noted this pattern during test excavations conducted at site 38CH454 in the Maxville cluster as part of his doctoral dissertation. There he found a nearly pure deposit of hard clam (quahog) overlying oyster shell in two separate test units and surmised that this may have been either a consequence of a shift in the season of reoccupation or a direct result of freshwater extinction of the local oyster population due to increased stream discharge. In later communications with Trinkley, Claassen (1986:124) reported that this successional pattern was widely spread across the Mississippian shell middens of the Central Coast region. Subsequent investigations have revealed that this process was not a simple substitution of one shellfish staple for another, however. Instead, the nearly pure hard clam deposits that invariably overlie sprawling oyster middens are by comparison significantly reduced in areal extent, suggesting the possibility that a severe shortfall in resource capture had ocurred in the estuaries.

When this successional pattern first came to my attention, I assumed that all of the clam middens would have been deposited in the very latest part of the occupation sequence. Surprisingly, however, they were not. In fact, they occur throughout the sequence and appear to correspond with the timing of abandonment, or reoccupation, at individual residential clusters. There are four clam middens with sufficient information to illustrate this pattern. Tibwin phase (AD 1325–1375) hard clam middens have been documented at 38CH1977 in the North Tibwin Tract (Cable 2002a) and at 38CH454 in the Maxville cluster (Trinkley 1980). In both cases, these deposits occur late in the site occupation sequence and mark the time of abandonment. The latest hard clam midden is located at 38CH44 (DePratter 2007; J. Gardner 1992) in the Salt Pond tract, where recovered ceramics indicate an association with either the later Fogarty Creek phase (AD 1375–1425) or the early stages of the hypothesized Salt Pond phase in the first half of the fifteenth century. A radiocarbon assay (see Table 2.1) on a knobbed whelk shell of cal AD 1389 ± 61 (1σ) supports an early fifteenth-century age assignment.

The Salt Pond tract also contains what appears to be the earliest clam midden in the sample. J. Gardner (1993:47–54) excavated a 1-×-2 m test unit in the shell midden at site 38CH134, recovering a collection of Mississippian complicated stamped and burnished sherds, which he characterized as Early Mississippian. A radiocarbon assay was run on a hard clam specimen from an adjacent shovel test in the midden, supplying additional perspective. The date was not included in Table 2.1 because of confusion concerning its actual conventional age. The reported age was 810 ± 50 BP (UGa-6694), and Gardner (1993:47) indicated that the sample had been corrected for fractionation, which would make the reported assay a conventional age. If so, the age of the shell would calibrate to a 1σ-range of AD 1559 ± 70 (Oxcal 4.3.2) applying a ΔR of 47 ± 60 for hard clam from Kiawah Island, South Carolina (Thomas 2008). However, the report (Kalin 1993) from the lab lists the calibrated 1σ-range as AD 1177–1264, which could only occur if fractionation had not been applied. This clearly demonstrates that the reported date must represent only the measured age of the sample and that a standard fractionation of 400 years for marine samples would need to be added to approximate the conventional age. The resulting conventional age would be 1210 ± 50 BP with fractionation. Calibration using the ΔR for the Kiawah Island hard clam, then, would return a modern 1σ calibration of AD 1224 ± 76, which is consistent with either the Sewee Camp or Jeremy phases.

Clam middens, then, have a relatively long span of temporal occurrence in the coastal clusters, but they always rest stratigraphically above oyster midden and are areally restricted. In each of the instances examined here, the clam midden was deposited at the end of the occupation sequence at a particular location. These coinciding patterns suggest that expansion along the coast was tied to estuarine exploitation and that early abandonments of clusters were directly linked to decreases in localized estuarine production. These early abandonments, then, could have resulted from overexploitation of fragile estuarine localities by the resident human population. Size measurements and age cohort data for oysters and clams, however, are not yet available to allow further evaluation of this hypothesis. Retraction of the settlement system begins at the end of the Tibwin phase and the differential geographic distribution of rim strips indicates that the latest occupations were confined to restricted areas. This pattern is accompanied by a decrease in population and, finally, abandonment of the region in the first half of the fifteenth century.

Claassen (1986:132–133) has identified two factors that may be responsible for the type of species succession seen in the Central Coast shell middens. First, decreased salinity should favor oysters over hard clams. Based on available data she suggests that the optimal salinity tolerance for oyster growth ranges between 14.1 and 22.2 ppt, while the optimal range for hard clam occurs between 24 and 28 ppt, with virtually no adult growth or larvae metamorphosis below 17.5 ppt. She further notes that inhibited growth and gonad formation will occur in either species when optimal salinity ranges are substantially exceeded in either direction. Freshwater floods can decimate both populations when the living environment is already characterized by low salinity. By contrast, both species are subjected to increased predator exploitation and disease when salinities exceed the upper limits of the optimal ranges. Importantly, hard clams are less susceptible to fluctuations in salinity because of their ability to remain closed and protected for longer periods. In this regard, Claassen cites a study by Burrell (1977) that examined a multiple week runoff event in the Santee River estuary in which hard clam populations were reduced by only 5%, while the oyster population saw a 66% decline.

The second factor Claassen identifies is levels of siltation. Oysters grow in beds or bars and are characterized by highly clumped concentrations and low mobility, while hard clams, although occurring in amorphous beds, are solitary and much more mobile. Increased siltation occurs when discharge from the creeks is low and water velocities decrease. Under these conditions, the reproductive success and growth of an oyster colony is severely reduced because silt begins to replace nutrients as the dominant constituent captured through filtration. In some cases, the progression of siltation can overwhelm and smother an entire oyster colony. Larval survival in the hard clam is not as adversely affected by low water exchange, and the ability of adults to move over short distances provides them with an opportunity to escape localized silt accumulations. In addition, hard clams have a greater degree of tolerance for silty or muddy substrates than do oysters, which generally build beds in sandy and hard substrates.

Claassen (1986:133) believes that species successions of molluscan fauna found in shell middens is a consequence primarily of transgressive and regressive events associated with changes in worldwide eustatic sea levels. In her model, transgressive regimes would foster conditions more favorable to oysters, while regressive regimes would induce hard clam expansion. The initial stages of the Little Ice Age correspond closely (ca. AD 1350 or 1400) with the timing of the species succession in the Central Coast shell middens and this event appears to have been accompanied by a slight drop in mean sea level along the South Carolina coast (Mark Brooks, personal communication November 11, 2015). Brook's conclusion is in agreement with the high-resolution reconstruction developed by Kemp and colleauges, based on salt marsh sediments on the United States Atlantic coast over the last 2,100 years (2011). They identify a relatively rapid period of sea level rise between AD 950 and 1400, followed by a stable to slightly negative regime from around AD 1400 into the nineteenth century, which they correlate with cooler temperatures. Yet, as minor as this sea level drop might have been, it could have had serious impacts on the estuary because the watershed of Bulls Bay is small and is characterized by an extremely shallow floor (Barbara Taylor, personal communication, November 9, 2015). The NOAA Navigational Chart (11518) for Bulls Bay reveals that nearly half of the area consists of near shelf that is less than one foot below low mean sea level at the time of this writing. Brooks informs me that the actual base of the low stand, as much as a meter lower than it was as of 2018, may very well have occurred between AD 1350 and 1450. Over a short period, then, the water levels in the tidal creeks might have lowered enough to overexpose established oyster bars to the air, resulting in significant die-offs.

Claassen does not elaborate on just how these sea level changes might favor one molluscan species over another. In fact, in her own study she finds that hard clam emerged as the dominant molluscan species in Winyah Bay just to the north of Bulls Bay much earlier than AD 1350 and before the onset of the Little Ice Age (Claassen 1986:126). In Winyah Bay, then, nearly pure hard clam middens were formed during the earlier transgressive event identified by Kemp and colleagues (2011) and not during the slight regression accompanying the Little Ice Age. Not only that, but the hard clam middens in Winyah Bay are extensive and thick, indicating a healthy and productive estuarine environment. In a study of nearly pure clam middens along Litchfield Beach near Winyah Bay, D. Jones and colleagues (2012) document a long historical dominance of hard clam dating back as early as cal AD 260 through cal AD 1700. Based on survivorship curves from dated contexts, these authors argue that the hard clam populations of the late prehistoric middens exhibited evidence of intensive overexploitation. Interestingly, one of the data points used in their study was a sample from 38CH44, the latest hard clam midden in the Central Coast Mississippian sequence. Its survivorship profile is composed of relatively older specimens, indicating that the Bulls Bay hard clam population may not have been subjected to the same degree of predation pressure as the Litchfield and Murrels Inlet samples. Hard clams were never a major constituent of the earlier oyster-dominated middens and it is likely that this older age profile is a function of the very late timing of intensive use of hard clams in the Bulls Bay region.

Both overexploitation and slight sea level regression may have played a part in the apparent collapse of molluscan productivity in the estuary, but neither factor would seem to sufficiently explain the successional pattern under investigation here. The necessary and sufficient factor appears to be supplied by a series of megadrought events that occurred after AD 1350. Anderson and colleagues (1995) identify a period of severe drought between AD 1359 and 1377 that has special relevance for the Bulls Bay estuaries. It was around this time that the human population began to decline and the incidence of hard clam middens increased in frequency. A proxy population curve for the Central Coast Mississippian based on ceramic representation of the phases in the regional sample illustrates the sharp decline beginning just after about AD 1350 (Figure 3.1). Stepping back into the drought data that Anderson and colleagues present, spring rainfall patterns indicate a period from AD 1154 to 1358 that was relatively favorable for agricultural production and also, by extension, for the maintenance of optimal levels of salinity to foster high oyster productivity in the estuaries. Sea level was transgressing and rainfall was sufficient to supply needed freshwater runoff to the system to maintain salinity levels within or near the optimal range for both oysters and hard clams. Predictably, this is also the period of near-exponential human population growth and settlement expansion around Bulls Bay. After AD 1358, however, there were two periods of severe drought, the one identified above and another one that closely followed between AD 1406 and 1432. Because the Bulls Bay estuary has an extremely limited freshwater catchment lacking a major river source, estuary circulation has always been poor and dominated by ocean water, creating a natural state of relatively high salinity that would have been exacerbated by the effects of long-term droughts. In both of these intervals, sufficiently long durations of consecutive years of low spring dis-

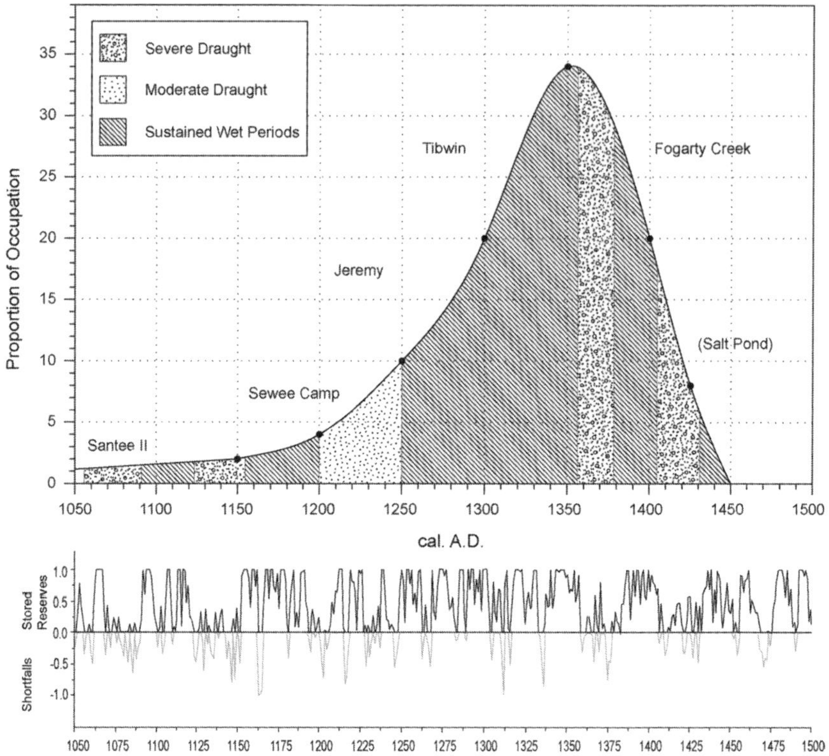

Figure 3.1. Comparison of proxy population curve for the Central Coast Mississippian settlement according to relative proportions of ceramic types with potential agricultural food reserves (A.D. 1050 to 1500) (modified from Anderson et al. 1995:Figure 3).

charge would have increased salinity even further and exerted intense stress on the oyster population, driving it to much lower levels of recruitment, lowering its resistance to disease, exposing it to increased predation of spat, and causing higher adult mortality rates.

Modern studies point to reduced runoff as a primary cause of dramatic declines in oyster productivity and reef destruction (Newell 1988). Seavey and colleagues (2012), for instance, have reported the results of a 29-year study of oyster declines along the Big Bend shoreline of Florida, a period during which oyster populations decreased by a net 66%. They found that decreased productivity was independent of potential overexploitation, but that the periods of greatest decline were correlated with diminished freshwater discharge into the estuary, which raised salinity. Oysters can tolerate fluctuations of salinity for short periods, but extended exposure to high salinity correlates with higher rates of disease and lower survivability of spat larvae (see also Claassen 1986). The researchers concluded that extended periods of high salinity in the estuary increased the mortality of older individuals and reduced recruitment of young oysters, which, in turn, increased the pace at which the reefs collapsed. Extended low discharge in this study was the result of modern water usage (i.e., water table depletion and poor flow retention), but these conditions are similar in impact to protracted droughts. Numerous studies have modeled the response of oyster populations to variation in freshwater input (both low and high discharge levels) and have reached similar conclusions (Buzan et al. 2009; Harding et al. 2008; Livingston et al. 2000; Petes et al. 2012; Power et al. 2006). That is, oyster declines are commonly linked to periods of lower-than-normal discharge from feeder creeks.

Hypothetically, the replacement of oyster with spatially limited clam middens at the approximate time of abandonment in each of the clusters examined here suggests that declines in oyster productivity played a significant role in both settlement shifts and localized abandonment processes. Not only does this decline signal the reduced contribution of an important resource niche for humans, it also suggests that the overall health of the estuary was degenerating. This is because oysters provide a vital service to the ecological balance of estuaries by filtering and removing sediments and plankton, impeding eutrophication and the effects of associated conditions of hypoxia and anoxia (Coen and Grizzle 2007; Jackson 2001; Kemp et al. 2005). Newell (1988) estimates that the pristine eastern oyster population of Chesapeake Bay was responsible for filtering the equivalent of the entire water column every three days. In addition, oyster reefs provide critical habitat for a wide range of organisms, including sessile plants and animals, microorganisms, and finfishes, which feed, breed, and inhabit the reefs (Hargis and Haven 1999). As oyster productivity declines, the healthy "veneer" of living oysters that encase the reefs also disappears and exposes the reef structure to increased erosion from waves and storms.

Concomitant reductions in filtering capacity from declining populations of suspension feeders would have ushered in eutrophication and increased hypoxia, which would have, in turn, adversely affected the productivity of other estuarine species, including, most importantly for the Bulls Bay Mississippian resource niche, finfish. The pace of eutrophication may have worsened due to a tendency for summer and spring rainfall to occur out of phase under drought conditions, which Stahle and Cleaveland (1992:1957–1958) attribute to long-term shifts in the migration and expansion of the Bermuda High (a semipermanent, subtropical zone of high pressure in the North Atlantic Ocean off the east coast of North America). This inverse relationship is strongest in the wettest and driest spring rainfall years, which suggests that summer dominant rainfall patterns may be a common feature of severe drought periods. Since summer rainfall is fueled by heat convection, the resultant thunderstorms would have increased the incidence of erosive flash flooding. Although increased freshwater discharge would have reduced estuarine salinity in the later part of the year, it would also have introduced higher concentrations of silt and nutrients into the estuaries, increasing eutrophication and further stressing the estuarine econiche. Moreover, during periods of decreased precipitation, the residence time of freshwater runoff in estuaries increases, which retards the flushing of phytoplankton and exposes the system to even higher levels of eutrophication (Lake 2003:1163; Scavia et al. 2002:156).

Given these factors, a strong case can be made that the oyster population in the Bulls Bay estuary collapsed during the latter half of the fourteenth and early part of the fifteenth centuries, primarily because of decreased freshwater inputs from tidal creeks that raised salinity to levels beyond the optimal limits of oyster growth and reproduction. In addition, summer-dominant rainfall patterns under drought conditions would have increased terrestrial erosion, resulting in increased siltation in the creek mouths and greater concentrations of silt in the suspended loads of the water column. Both factors would have adversely affected oyster and hard clam populations, but hard clams would have been able to withstand these perturbations longer. That the succession was not simply a replacement of one molluscan staple with another is born out by the restricted nature of the clam deposits and the fact that the deteriorating estuary was insufficiently productive to support a climaxing Mississippian population that had expanded its reach over a period of 200 years. Had the effects of drought not been so severe, it is likely that the local human inhabitants could have better managed the tendency to overexploit estuarine locations by the operation of a system of settlement shifts along the estuarine interface that they had begun to practice in less stressful times. Pluckhahn and Thompson's (2018) Crystal River study explores just such strategies for managing overexploitation in an estuarine ecosystem during the Woodland period in Florida, where the severity of drought conditions was not as extreme as the one I describe here for Bulls Bay at a later time.

4

The Cultural and Natural Geography of Megadrought

Before we can completely explain a phenomenon, we require not only to find its true cause, its chief relations to other causes, and all the conditions which determine how the cause operates, and what its effect and amount of effect are, but also all the coincidences.
—George Gore, *The Art of Scientific Discovery*, 1878

Strong evidence indicates that the Central Coast Mississippian population was severely impacted by multidecadal droughts in the late fourteenth and early fifteenth centuries. These severe droughts placed great stress on a resilient subsistence system that had been fine-tuned over thousands of years of development in the Eastern Woodlands. The two niches most adversely affected were the agricultural system, for which little is known but probably consisted of shifting corn-beans-squash farming and starchy seed cultivation, and the estuaries. The latter niche appears to have represented the most important component of the subsistence economy. It is likely, though, that all of the other resource niches experienced decreased productivity as well, including arboreal nuts and mast. As mentioned previously, Anderson (1996; Anderson et al. 1995) attributes abandonment events in the Southeast to the unstable organization of chiefdoms and their inability to manage long-term severe environmental deterioration. He further speculates that internecine and perhaps inter-regional warfare posed yet another factor, weakening the authority of the elite to conscript labor and to demand tribute. The Central Coast case, though, is instructive, because it suggests that regional collapse can occur at even lower levels of social organization, given relatively high population density fueled by subsistence intensification. The common denominator in these abandonments, then, might not be the instability of chiefdoms but rather the pressures placed on human systems of varying organizational type by the global processes of megadrought.

Population Dynamics in Late Prehistory

A phenomenon requiring much more attention than it has received so far in the study of abandonments is the Neolithic Demographic Transition (Bocquet-Appel 2002; Bocquet-Appel and Bar Yosef 2008; Bocquet-Appel and Naji 2006; Guerrero et al. 2008; Kohler et al. 2008), or the Agricultural Demographic Transition (ADT), as Bocquet-Appel (2011) has renamed it. The ADT describes the common shift in human fertility that accompanies the adoption and intensification of agriculture, an event that emerges at different times in different places across the Holocene. The model stipulates that the interaction during this era of increased food energy and female physiology with agricultural production led to earlier menarche and shorter birth intervals, increasing female fertility and enabling larger populations.

In North America, population profiles of early agriculturalists are relatively flat, with only slow population increases detected. However, around AD 500 or 600 population growth begins to accelerate, peaking in many areas between AD 1000 and 1300. Subsequently, population levels decline irregularly but consistently across many regions into the seventeenth century (Dean et al. 1994:73–76; Hill et al. 2004; Kohler et al. 2008). Summed probability curves for radiocarbon dates from different regions in the Vacant Quarter presented by Meeks and Anderson (2013) conform to this general model, as does the proxy population curve constructed for the Central Coast Mississippian (see Figure 3.1).

Much emphasis has been placed on the upswing portion of the population curve, its causes, and its relationship to the development of complex social organization. Less discussion has been devoted to the downswing in the late prehistoric period (see Hill et al. 2004). Societies positioned on the downswing tend to reproduce at below replacement (Muller 1997). When burial populations from this horizon are examined, there is a noticeable increase in infant mortality as well as declines in health (Armelagos et al. 1991; Kohler 2000; Lallo and Rose 1979; Larsen 2002, 2006; Milner 1986, 1991; Pechenkina et al. 2002; Pietrusewsky and Douglas 2001; Sheridan 2003; VanGerven et al. 1994). These trends are commonly viewed as density dependent outcomes, but intervals of climate change may have also influenced them. Between AD 1350 and 1500 the identifiable prehistoric cultures occupying southern Arizona and southern New Mexico vanished (Hill et al. 2004), and the area occupied in the northern Southwest shrank to include only the upper Rio Grande valley and several western outliers (i.e., Hopi, Zuni, Acoma, and Laguna). In the Southeast, vast areas of the American Bottom known as the Vacant Quarter were depopulated (Cobb and Butler 2002; Meeks and Anderson 2013; S. Williams 1990). On the Southeast Atlantic Slope, the middle and lower Savannah River valley (Anderson 1990), the central coast of South Carolina, and probably other locations along the Pee Dee (Boudreaux 2007) and the Santee-Congaree-Broad drainage systems (DePratter 1994:208) were abandoned or reorganized after AD 1350.

There is reason to believe that these continent-wide demographic processes were linked by global climatic forces expressed in coincident geographic patterns of prolonged drought and population declines and displacements. The climatic event of interest is what paleoclimatologists have named the Medieval Climatic Anomaly (MCA), a term that describes a period of recurring multidecadal droughts associated with increased global temperatures occurring between AD 1000 and 1500. On the South Atlantic Slope, the effects of this event were felt most intensely across the Coastal Plain in a vast area that I call the "Greater Desert of Ocute." This name represents an expansion of the term the sixteenth-century Spanish explorers coined to describe a small part of this deserted land, which they had occasion to pass through on their way to the interior regions of the Southeast. Below I describe the contours of this "desert" and the patterns of drought intensity that appear to coincide with it.

The Greater Desert of Ocute

During the Soto entrada of the early 1540s, the Spanish encountered a stretch of undeveloped and apparently uninhabited land between the chiefdoms of Ocute on the Oconee River and Cofitachequi on the middle Wateree River valley. Elvas, one of the chroniclers of the entrada, called it the "Desert of Ocute," a term shared by the members of the expedition (Hudson et al. 1984:72). Of course it was not really a desert, but the Spanish applied the term *desierto* to such places to convey the image of an uninhabited or deserted expanse (see DePratter 1983:33–36).

In their interpretation of the historic accounts, Hudson and colleagues (1984: 70–72) provide a vivid picture of the conditions experienced by Soto and his army as they moved through this vacant zone. The subjects of the Ocute chiefdom knew of no trails that would take the Spaniards to Cofitachequi, and Soto set out with dire warnings that his men would starve before reaching their destination. Once the party split off from the well-maintained path leading to the powerful Coosa chiefdom in northern Georgia (the historic Hightower trail), the Spaniards headed east, picking their way through a gradually deteriorating maze of trails and undeveloped forest. After crossing the Savannah River near the future site of Fort Moore, they travelled in a more southeasterly direction at first, apparently finding indistinct remnants of trails leading down to the chiefdom of Orista in Port Royal Sound (which was later contacted by the Spanish when the forts at Santa Elena were erected in the 1560s). Once reaching the North Fork of the Edisto River, they turned decidedly northward and, after much difficulty and without the guidance of trails, reached the confluence of the Broad and Saluda Rivers near present-day Columbia, South Carolina. The expedition pushed northward and crossed both rivers but could still find no evidence of a trail or town. Scouting parties on horseback were sent up the Broad River and down the Congaree River to search for settlements. After two days, one of the parties returned to camp accompanied by several natives from Hymahi, a village situated near the juncture of the Congaree and Wateree rivers. The Wateree valley was densely populated with villages on both sides of the capital town of Cofitachequi near Camden, South Carolina (Worth 1994:6).

Once Soto found the trail to Cofitachequi, the Desert of Ocute was left far behind and never revisited by himself or by Pardo a quarter of a century later. The firsthand inspection of the Desert of Ocute by the Spanish, then, was quite cursory and its boundaries were left undefined. From their accounts, it appears that no towns were situated within the area of their navigation through the wilderness surrounding the Savannah River, nor were any found in their scouting forays along the Congaree and Broad Rivers. Beyond this brief reconnaissance, however, very little of the potential area that may have lain uninhabited at this historical juncture was verified by the Spanish.

Hudson and colleagues (1984:72) found this wilderness, which covered a linear distance of 209 km between the two chiefdoms, to be extraordinary and in need of explanation. In their view, it represented a buffer zone between two powerful chiefdoms (see also DePratter 1983:33–37). Implied by this interpretation is that the size and power of the chiefdoms and their degree of animosity required an unprecedentedly large zone of separation to reduce hostilities. At the same time, an interview with Patofa, the captain of the native contingent that accompanied Soto from Ocute, revealed that, in his memory, there had been very little contact at all between the people of the two provinces (DePratter 1983:31–32). They were not engaged in any kind of pitched warfare and neither Patofa nor his contingent even knew how to get to Cofitachequi; nor did they know how to get out of the wilderness in which they found themselves. Skirmishes between the two groups occurred only when hunting parties unintentionally ran across one another. Patofa further confided that the people of Ocute were reluctant to expand or advance beyond their own boundaries because they were intimidated by the superiority of Cofitachequi. It is ironic that the Spanish appear to have enabled the most bellicose episode of military conflict between the two provinces in many years, or perhaps ever. According to Garcilaso (Varner and Varner 1951:292–293), Patofa and his war party secretly attacked Hymahi just before Soto's army arrived, pillaging the temple of its burial offerings, killing a good number of the inhabitants, and taking their scalps. When the slaughter was discovered, Soto ordered the warriors and porters from Ocute to return home so that their presence would not sabotage his attempt to form amiable relations with Cofitachequi. They gladly withdrew with scalps and gifts from the Spanish to present to their chief.

On the face of the ethnohistoric record, then, there is support for the idea that the deserted region represented a buffer zone, at least by default. There was animosity between the provinces and the potential that unrestrained violence could break out during episodes of contact did exist. However, as we begin to know more about the configuration of this desert, its extent appears rivaled in eastern North America only by the Vacant Quarter of the American Bottom. Certain regions of it are better understood than others, but a vast area from the Ogeechee River to the Cape Fear Valley can be tentatively included in this deserted zone, covering an area of roughly 140,000 km^2 or 34.5 million acres (Figure 4.1). This, then, is the potential boundary of what I call the "Greater Desert of Ocute" or GDO, a small fraction of which Soto penetrated in his search for Cofitachequi. If substantiated through continued archaeological investigation, this zone of depopulation would encompass an area much larger than is needed to buffer two warring polities and much larger than any of the buffer regions that appear to have been purposefully maintained in the fourteenth and early fifteenth centuries in the Eastern SAM culture area. It would rival the size of any of the vacated zones of North America or Europe and would require an explanation far more sweeping than inter-chiefdom animosities.

The Footprint of Megadroughts

Regional droughts of short, multiyear duration are a common feature of climatic patterns in North America. Examples of such droughts over the past century or so include the 1890s drought in the western United States, the Dust Bowl of the 1930s, the 1950s drought wave in the southwestern United States, and the post-1998, turn-of-the-century droughts across portions of the southern latitudes of the United States (Seager et al. 2009). These relatively brief interludes, however, pale in comparison to the multidecadal droughts that occurred during the Medieval Climatic Anomaly between AD 1000 and 1500 (B. Cook et al. 2014). The most extreme of these were the two multicentennial megadroughts in California (see E. Cook et al. 2010; Stine 1994). Although megadroughts are usually transcontinental in scope, the various regions of North America

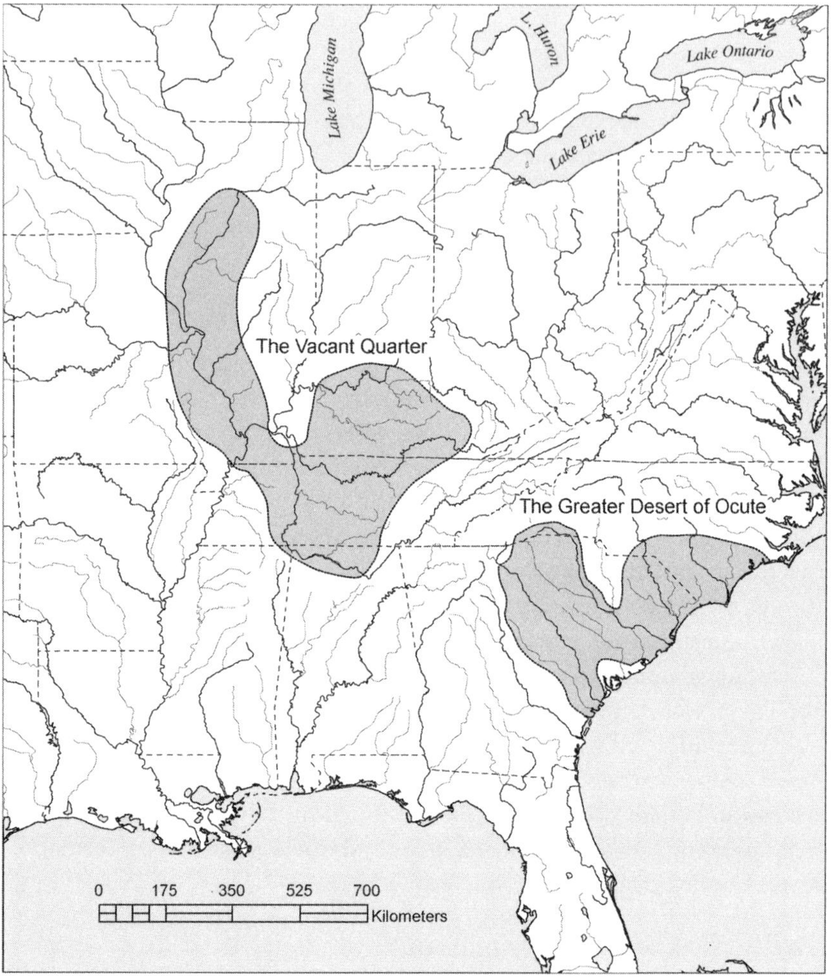

Figure 4.1. Late prehistoric abandonment regions in eastern North America.

have been shown to exhibit complexes of teleconnected drought conditions that are expressed as patterns of associations varying in time and space. Benjamin Cook and colleagues (2014:Figure 6) have identified five transcontinental megadrought patterns in the paleorecord of the MCA based on an arbitrary quartering of the United States into analytical regions. The Northeastern quarter was not considered in their study. The three most frequent patterns of regional megadrought associations are shown to be strongly teleconnected with averaged, detrended time series data for sea surface temperatures (SST) in the Pacific and Atlantic oceans expressed as indices of the Pacific Decadal Oscillation (PDO) and Atlantic Multidecadal Oscillation (AMO). Two of these patterns involve the southeastern United States (see Enfield et al. 2001; Kerr 2000). One identifies synchronous megadroughts in the Southeast, Central Plains, and Southwest, while the other associates the Southeast with megadroughts occurring in the Pacific Northwest and Central Plains.

The large-scale prehistoric abandonments that occurred during the MCA express very similar asynchronous, subcontinental associations. Benson and colleagues (2007) argue that there is strong evidence for the occurrence of three megadroughts between the early eleventh and late thirteenth centuries, and that these appear to have served as the catalyst for broad-scale abandonments in portions of the Southwest and Midwest. By AD 1300, the Anasazi and Fremont cultures of the Four Corners region of the Southwest had collapsed and the former resident populations were gone. The mid-eleventh-century abandonment of Chaco Canyon was the leading edge of this response. At the end of the thirteenth century, Cahokia collapsed and the subsequent depopulation of much of the surrounding central Mississippi valley occurred before AD 1350 (Benson et al. 2009). By this time, most of southern New Mexico had also been abandoned (Hill et al. 2004:Figure 3). All of these cultures were characterized by an economic dependence on rain-fed agriculture that had allowed populations to increase during favorable climatic conditions. Based on tree-ring reconstructions, Benson and colleagues (2007) have demonstrated that these megadrought events, especially the mid-twelfth and late thirteenth-century waves, were associated with a negative (colder detrended SST) PDO and a positive (hotter detrended SST) AMO climatic regime.

A second couplet of megadrought and abandonment sequences is documented in the late fourteenth and early fifteenth centuries. This time the entire geographic span of southern Arizona was depopulated (Hill et al. 2004). So were the Ohio, Tennessee, and Cumberland valleys of the Mid-Continent (Cobb and Butler 2002; Comstock and R. Cook 2018; Krus and Cobb 2018), as well as the area on the South Atlantic coast that I have identified as the GDO. The very different regional associations between the two waves of megadrought in the MCA suggest that while the earlier one was a function of a particular climatic regime forced in large part by a negative PDO and a positive AMO, the fifteenth-century wave probably owed its existence to a different set of climate forcing conditions. Since a positive AMO appears to be necessary for drought occurrence at such a large scale (B. Cook et al. 2014:390–392), it seems plausible that these contrasting regional patterns were the result of different temperature regimes in the Pacific Ocean (i.e., positive versus negative PDO). Accordingly, because the two patterns overlap geographically in the larger Southwest and Southeast regions, it is likely that the megadrought patterns Benjamin Cook and colleagues (2014) have identified will require redefinition before the influence of shifts in the sign of the PDO can be effectively addressed and understood within the broader context of the recurring geographic patterns of megadrought.

McCabe and colleagues (2004, 2008) have generated drought frequency distributions for the continental United States that provide a basis for observing the varying effects of climatic regimes constructed from the paired associations of positive and negative PDO and AMO. They obtained twentieth-century precipitation data from the National Climatic Data Center for each of the 344 climate divisions that comprise the coterminous United States. These data were then grouped into 20-year moving windows to facilitate analysis. In addition, data on PDO, AMO, and mean annual Northern Hemisphere temperature were gathered from repositories housed at the University of Washington, the National Oceanographic and Atmospheric Administration, and the Climate Research Unit, East Anglia, United Kingdom. Drought conditions were determined to exist in a particular climate division when annual precipitation fell within the lowest quartile (25%) of the 100-year record. Various alternative measures of drought frequency were also considered, including the Palmer Drought Severity Index (PDSI), and were found to yield similar spatio-temporal results. The 20-year moving windows, or drought frequencies, were subjected to rotated, principal components analysis. Fifty-two percent of the spatio-temporal variability in multidecadal drought frequency was explained by either PDO or AMO signs and an additional 22% was ascribed to a complex pattern of positive and negative trends in drought occurrence associated with Northern Hemisphere temperatures.

As a major finding of their research, McCabe and colleagues (2004:Figure 5) produced maps of the coterminous United States showing the density contours of drought frequency for the four basic scenarios of positive and negative combinations of PDO and AMO; these maps are redrafted here as Figure 4.2. It should be appreciated that drought frequency does not measure drought intensity, which can be expected to have been much higher in the MCA than in the twentieth century. Although it would be preferable to work with empirically generated drought frequency maps from the MCA, it appears likely that the general spatial patterns that McCabe and colleagues illustrate represent an accurate portrayal of the footprints of megadroughts that extend throughout a good portion of the late Holocene. Light fill in Figure 4.2 demarcates pluvial conditions with drought frequencies falling above the lowest quartile, while darker fill identifies increasing drought frequency within the lowest quartile of years.

Extensive droughts, as expected, occur when AMO is positive (Figure 4.2c, d), but the sign of PDO has a dramatic effect on their spatial occurrence. When PDO is negative (Figure 4.2d), droughts occur in the Midwest, and, in particu-

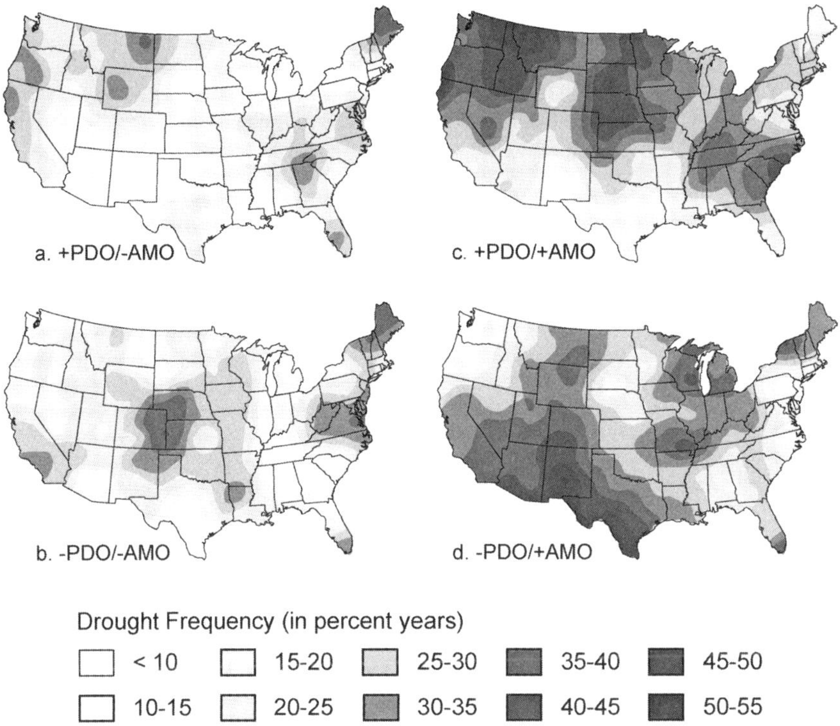

Figure 4.2. Geographic patterns of twentieth-century drought frequency for the continental United States (adapted from McCabe et al. 2004).

lar, the central Mississippi Valley, and across most of the Southwest. When PDO is positive (Figure 4.2c), however, major droughts are situated on the South Atlantic coast, the Ohio River catchment in the Mid-Continent, the Central Plains, and the Pacific Northwest. There are clear correlations between these drought patterns and the patterns of human system collapse and abandonment during the MCA. The -PDO/+AMO scenario, as argued by Benson and colleagues (2007), correlates with the eleventh- through thirteenth-century abandonment sequences at Cahokia and the surrounding central Mississippi Valley, as well as the Four Corners region of the Southwest. This abandonment sequence corresponds with the Southeast-Central Plains-Southwest pattern of megadrought identified by Benjamin Cook and colleagues (2014). By contrast, the +PDO/+AMO scenario correlates most closely with the fifteenth century abandonments in the GDO and Ohio River catchment zone and would seem to take the form of the pattern they (2014) identified for the Southeast-Central Plains-Northwest. The Hohokam of southern Arizona were the victims of an equally dramatic collapse and abandonment event, but this occurred during a wet period. In chapter 8 I will show how this seeming contradiction is actually understandable when one takes into consideration the larger implications of extreme weather events associated with global warming and megadrought.

Obviously, the intricacies of short-term climatic regimes cannot be fully explained by the generalities of SST functions, but there are clear conformances between regional abandonments and the larger spatial patterns of droughts associated with the PDO/AMO scenarios that McCabe and colleagues (2004, 2008) have mapped. It is of interest that the Northeast is never as affected by megadrought as most other regions in the coterminous United States. The extreme Northeast is clearly subject to recurrent droughts, but New York, Pennsylvania, and the Tidewater states are generally immune to harsh drought conditions in any of the scenarios discussed here. An exception is Chesapeake Bay, which appears to be subject to drought during colder periods. The drought of 1607–1612 that afflicted the Jamestown colony (see Blanton 2000; Stahle et al. 1998) may indicate that a regime of extreme negative PDO and AMO was at play at the turn of the seventeenth century (see Figure 4.2b). Certainly, this prediction is consistent with the colder climate of the Little Ice Age, which was fully expressed by AD 1500 (B. Cook et al. 2014).

Onward to the Cultural Footprint

With this basic outline of the geographic patterns of megadrought described, it is time to examine how it might be expressed archaeologically. Chapter 5 addresses the evidence for abandonment and depopulation in the GDO. Chapter 6 then lays out the evidence for the effects of demographic shifts associated with megadrought as these are expressed in the Ring of Drought Resilience, to which it is inferred that the displaced populations from the GDO migrated in the fifteenth century.

5

Regions of the Greater Desert of Ocute

No one can ever know for sure what a deserted area looks like.
—George Carlin, *Brain Droppings*, 1997

Comedian George Carlin was being facetious when he plucked this particular dropping from his brain, but its literal implications for studies of archaeological abandonment are profound. What does an abandoned area look like? Is it truly deserted or is it simply repurposed? Are there still remnants of societies hiding in the dark recesses of the forests or even in plain sight if we just had the tools to identify them? Such questions will need to be entertained if we want ultimately to understand the demographic processes of migration and displacement in the fifteenth century. I only attempt in this chapter to address the circumstantial evidence for depopulation within the regions of the GDO, leaving the more detailed questions of population dynamics to later endeavors.

A close inspection of the positive PDO/AMO scenario of drought frequency (see Figure 4.2c) indicates that there is a nearly perfect fit between the highest frequencies of drought years and the general outlines of the GDO. Coupled with the evidence of local megadrought revealed in tree-ring data covering the time range of the MCA (see Stahle and Cleveland 1992; Stahle et al. 2007), there is strong reason to believe that the adaptive systems of humans inhabiting the GDO were significantly affected by prolonged droughts in the fifteenth century. This chapter reviews the archaeological data upon which my mapping of the GDO's contours is based. Figure 5.1 shows a map of the GDO along with the locations of archaeological sites that are discussed in the text. In what follows, I review the late prehistoric archaeological records of the various regions encompassed by the GDO and discuss the basis on which each is evaluated to have been depopulated.

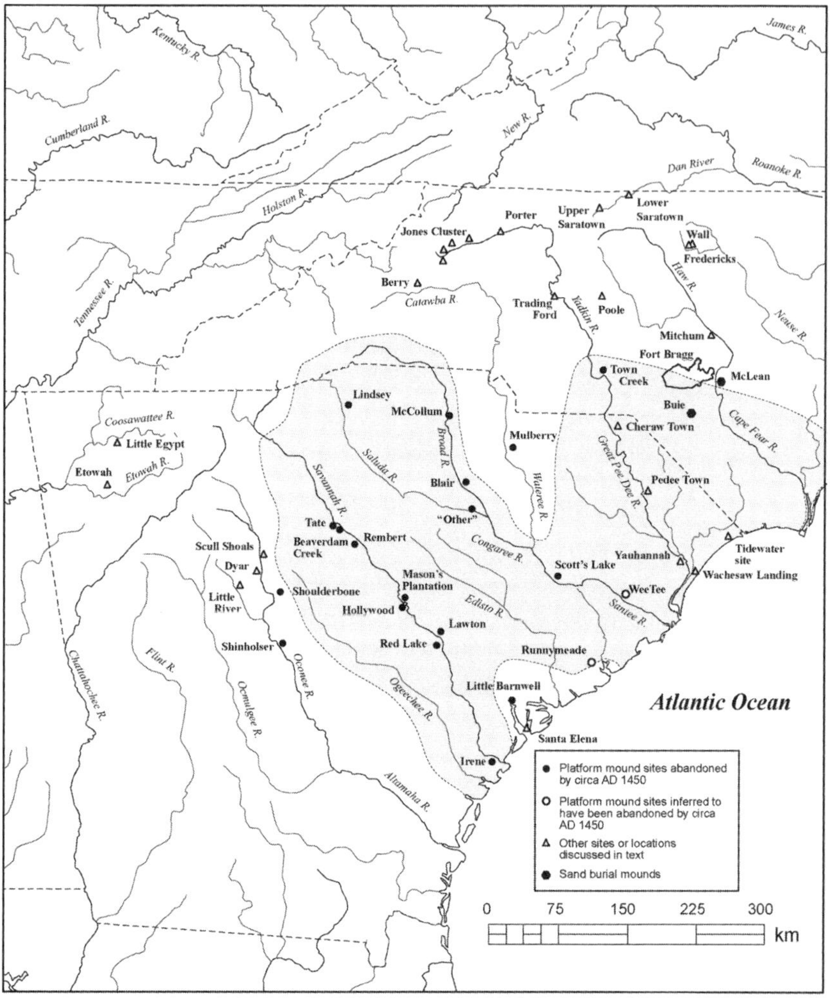

Figure 5.1. Greater Desert of Ocute in relation to major Mississippian sites discussed in the text.

NORTHEAST GEORGIA

Anderson's (1990a) study of the Savannah River valley convincingly documents a sequence of mound center abandonments across the entire lower and middle portions of the valley, beginning in earnest by AD 1300 to 1350 and ending with virtually complete abandonment around AD 1450. The fact that no Middle Lamar habitation sites have been found in this stretch of the river supplies additional evidence for major depopulation as opposed to a dispersed reorganization of resident populations. Indeed, the Ogeechee River valley to the south has long been suspected to have been abandoned at this same time (Mark Williams, personal communication, May 8, 2014).

Pluckhahn and McKivergan (2002:153–154) have shown that the coastal zones of the Ogeechee and Savannah rivers were densely populated with Savannah period sites, including conical burial mounds (n = 7) and one platform mound (Irene), but none of these appear to have been used or modified much later than around AD 1450. In an earlier study, Thomas (1993:41–42) proposed a large-scale abandonment of the coastal zone from Port Royal Sound, South Carolina, to at least as far south as St. Catherines Island, Georgia, during the Early Lamar period. He noted that the only confirmed evidence of sixteenth-century occupation there was represented by the assemblage of protohistoric pottery found at the Spanish colony of Santa Elena. I carved a wedge out of the GDO around Port Royal and Charleston, South Carolina, because the Pardo accounts of the 1560s (Hudson 1990) document a number of villages in this region. However, it is of interest that these villages were located considerably inland from the coastal fringe (see Hoffman 1990:69; Thomas 1993:45–46), suggesting that even where concentrations of population survived there was significant settlement relocation to the inland reaches of the drainages. Movement away from the coastal marshes seems to confirm that the collapse of the estuary in Bulls Bay may have been part of a much larger pattern of estuarine degradation during the fifteenth century.

Broad, Slauda, and Upper Santee

DePratter (1994:208–211) indicates that a similar pattern of abandonment characterized the Saluda, Broad, Congaree, and Santee rivers around AD 1450 (see Figure 5.1). Unlike the Ogeechee and Savannah Rivers, which were not traversed by the Soto expedition to any great degree, portions of the Broad and Congaree rivers were combed rather thoroughly by scouting parties when Soto found himself lost near the confluence of the Broad and Saluda rivers. Characterizations of the composition and number of scouting parties vary between the four chroniclers, and there are problems in assessing what represents firsthand testimony as opposed to secondhand factual liberties and embellishment (see Galloway 1997). What is consistently repeated, however, is that search parties were sent up the Broad River and down the Congaree River (Biedma in Bourne 1904b:12–13; Elvas in Bourne 1904a:61–62; Garcilaso in Varner and Varner 1962:287–295; Ranjel in Bourne 1904b:94–97).

Elvas and Ranjel respectively state that the parties consisted of a captain leading eight or ten cavalry each. Captain Añasco, who was sent down the Congaree, returned after two days and reported that he had found a small town (Hymahi or

Aymay) about twelve to thirteen leagues from Soto's camp (Elvas). Hudson and colleagues (1984:72) place this village near the confluence of the Congaree and Wateree rivers. Details diverge concerning other aspects of the search. Elvas indicates that two other parties of cavalry were sent "out into the country," suggesting that they may have gone overland. Johan Ruiz Lobillo, who Ranjel said went north, led one of the overland parties. Garcilaso reported that the other scouts found roads but no towns during their three-day reconnoiter. From Ranjel comes the only detailed accounting of the results of the other searches. Gallegos, who went up the Broad River, returned with an "Indian woman and news of a populated region." Lobillo, who went north, found only trails, while Romo, whose direction of search was never defined, encountered "four or five Indians, and not one would show any knowledge of his Lord's village or discover it."

It is possible that each party covered at least 65 to 70 km before returning to camp, based on Añasco's progress to Hymahi (*legua comuna* = 5.56 km). Thus, Gallegos would have passed the area of the Blair Mound on his trek up the northeast bank of the Broad River without discovering a village near it. It is unclear where the populated region might have been that Ranjel referred to and, without a translator (Perico went with Añasco to Hymahi), it is difficult to know what information was actually gathered beyond rumor or supposition. If Lobillo went overland to the north, it is likely that 65 to 70 km would have placed him very near the settlements of Cofitachequi, but he reported only finding trails. This suggests that he moved in a more northwesterly direction.

Garcilaso (Varner and Varner 1962:287) offers one additional detail about the captains' search patterns, which simultaneously explains Lobillo's route and provides added confidence that the scouting parties would have run up against towns had they existed in their search areas. An unnamed participant in the expedition told Garcilaso years later that the parties were paired. He relates that two "of the groups were to proceed upstream and two down, and they were so arranged and counseled that while one was adhering to the bank of the river, the other should pursue the same course a league inland, and each was to return within five days with an account of what he had found." Lobillo, then, was probably paralleling Gallegos about a league inland from the Broad River bottoms. There are troubling errors of fact in Garcilaso's account, including his description of Añasco's route, which he said was upstream instead of down, but it may very well be that on the larger point of the search strategy he was correct. If Romo was paralleling Añasco, then the four or five natives he brought back were most probably members of a hunting party from Hymahi.

As I reconstruct it, then, the 130 to 140 km stretch of the Broad and Congaree rivers combed by the captains was uninhabited in the 1540s, from just north of the Blair Mound to the confluence of the Congaree and Wateree rivers. That the McCollum Mound farther up the Broad River and the Lindsey Mound near Greenville on the Saluda River were apparently abandoned by AD 1450 furnishes additional support for the claim that these river valleys were virtually uninhabited during this time horizon. A possible mound is denoted by DePratter (1994) as "other" and is said to have been flooded during the construction of Lake Murray, near the mouth of the Saluda River. Its chronological history is unknown, but if the mound ever existed it probably corresponds to this same abandonment schedule. Other nearby regions to the east, which were not visited by Soto but reflect this same horizon of mound site abandonment and regional depopulation, provide evidence for a significant expansion of what is normally considered to have been the "Desert of Ocute." These include the Santee River valley represented by the Scott's Lake mound center (DePratter 1994:208–211), the Bulls Bay Mississippian of the central South Carolina coast (this study), and the mound at Town Creek on the upper Great Pee Dee River.

Upper Pee Dee and Yadkin

Boudreaux (2007:109–110) suggests that in its later stages of use Town Creek was transformed into a nonresidential mausoleum and ceremonial center, similar (although on a much-reduced scale) to the functional changes recorded at Cahokia and Moundville. Given this scenario, it would be reasonable to expect evidence of a contemporary population influx into the adjacent hinterland around Town Creek. Taryn Ricciardelli's (2014) regional analysis of Mississippian site distributions surrounding the mound seems to support this expectation. Based on the decorative mode seriation developed by Boudreaux for the Town Creek site, she was able to demonstrate that associated sites were clustered within an 18 km radius and that site frequency in this catchment increased after AD 1300 as the function of the mound site shifted to nonresidential use. Overall, it was concluded that the occupation history of the hinterland sites corresponded closely with that of Town Creek, which implies that the region was abandoned during the Late Leak phase at the same time that the use of Town Creek was discontinued.

Inspection of the decorative mode data Ricciardelli presents for Late Leak phase components reveals that eight hinterland sites have yielded folded rims (Ricciardelli 2014:32). Although this is consistent with Boudreaux's (2005) phase definition for the Late Leak phase, recall that I argued earlier that this phase, as constructed, is not continuous with the Early Leak phase. There should be at least an intervening McDowell-like phase between them, which is not represented in the Town Creek seriation. Folded rims do not seem to emerge in substantial representation until the sixteenth century and their presence in the survey data from the Town Creek polity indicates a significant Middle and/or Late

Lamar presence in the region. Whether all of the sites with folded rims are associated with the seventeenth-century Caraway component at Town Creek or whether some of it extends backward into the sixteenth century is a question that cannot yet be answered.

According to the accounts of the Pardo expeditions of the late 1560s, a powerful chiefdom was seated at a town identified as Guatari, located near present-day Salisbury, North Carolina, on the lower Yadkin River (DePratter et al. 1983: 151–156). Hudson (1990:92–93) speculated that the town's precise location was likely in the neighborhood of Trading Ford, the place where the Occaneechi Trading Path crossed the Yadkin River (Rights 1931). A portion of a site, Trading Ford (31Dv2), which Hudson felt might be a good match for Guatari, was excavated on the west bank of the Yadkin in 1938 (Coe 1995:160–165) during a drawdown of High Rock Lake. (It should be noted that Coe believed this site was Saponi, a town Lawson visited in 1701.) Noting that the Spanish accounts describe Guatari as a palisaded village with circular houses laid out like a Piedmont Village Tradition town (see Ward and Davis 1999:78–80), Hudson argued that the inhabitants were closely affiliated with the neighboring Siouan Hill Tribes. He further asserted that the pottery was comparable to what had been recovered from Upper Saratown on the Dan River.

Although I have not examined the ceramics from the Trading Ford excavations, Steve Davis (personal communication, June 30, 2014) took a second look at the Research Laboratories of Anthropology (RLA) collections in response to discussions we had on the subject. It is his opinion that much of the pottery could be classified as Caraway but was probably earlier than the type assemblage from the Poole site. He did not find examples of rim strips or folded rims, but some sherds exhibited large reed punctations below the lip and several complicated stamped sherds were of the "exploded" application typical of the Middle and Late Lamar Cowan's Ford series of the middle Catawba valley (Moore 2002:147, 158). It is likely, however, that the collection at one time did contain what Coe (1995:161–163) referred to as "Lamar style" rims, because he specifically stated that such rims were recovered from Trading Ford. This style consisted of rims with narrow appliqué strips positioned well below the lip (see Coe 1995:Figure 9.9G from Town Creek), in the typical manner earlier described for the late fifteenth-century McDowell phase rim strips at the Mulberry site. Coe contrasted the "Lamar style" rim with what he called the "Clarksville style," which characterized the Caraway series ceramics from the Poole site. These rims were thickened with a clay strip of about an inch in width, and the top edge was carefully blended with the lip. The "Clarkesville style" rim, then, corresponds to what is generally considered a type of folded rim and should have a later temporal occurrence (Middle to Late Lamar horizon). Although rim strips represent the nearly exclusive style of vessel orifice thickening in the fifteenth century, they have a diminished importance in sixteenth-century SAM assemblages (e.g., the Miranda Lot collection from Santa Elena [DePratter 2009] and the Burke phase sites on the upper Catawba [Moore 2002:79]), in which folded rims ascend to dominance.

Thus, there is nothing known about the ceramics from the excavations that would disqualify the Trading Ford site from being the presumed town of Guatari visited by Pardo in the 1560s and later chosen as the site of Fort Santiago (DePratter et al. 1983:154). Nor does it preclude the possibility that Trading Ford also contains a substantial McDowell-like component. If indeed Guatari represents the re-establishment of a regional center by the former inhabitants of Town Creek, where only a very subdued McDowell-like occupation is represented, then the presence of this phase at the Trading Ford site would be expected. Such a sequence, in fact, would agree with Robin Beck's (2013:52) suggestion that there is a direct historical link between Town Creek and Caraway. The available ceramic evidence strongly suggests that Guatari was participating in the Lamar horizon interaction sphere of the Eastern SAM tradition, rather than in that of the protohistoric Piedmont Hill Tribes.

At the same time, it is undeniable that significant interaction between Eastern SAM and Dan River groups was taking place, and there are hints that ceramic traits were flowing in both directions (see Wilson 1983:499). Jane Eastman's (1999:81–132) ceramic sequence shows an unusually heavy overlay of Lamar traits in the late prehistoric and protohistoric features at the Hairston site in the Upper Saratown site cluster. Complicated stamping, which was never an important component of the Dan River series, ascends to the second most frequent exterior surface treatment (15%) in protohistoric deposits. In keeping with its Dan River origins, net impressed (50%) still dominates the assemblage. Moreover, about 40% of the rims are folded. Two pit features from the protohistoric Nifong site on the Dan River yielded radiocarbon dates with sixteenth- and seventeenth-century date options and assemblages of large rim sherds, 20% of which were folded (Eastman et al. 1997:86, 110). Sherds from one partial vessel were shared between the features, suggesting to Eastman that they were closely contemporary. The two assays (Beta-105449 and Beta-105450) are statistically the same ($t = 0.878$, $\chi^2_{.05} = 3.84$, $df = 1$), and a modern calibration based on the pooled mean (293 ± 31 BP conventional age) would most probably place the features within the sixteenth century. The pooled calibrated mean is AD 1577 ± 52, with a most probable 2σ error range of 1492–1603 AD (Oxcal 4.3.2).

Although folded rims are seen to be diagnostic of Middle and Late Lamar Eastern SAM assemblages, there is evidence from the Hairston site that the trait may have occurred somewhat earlier in late prehistoric Dan River occupations. Eastman identified a large storage pit (Feature 22) at Hairston that contained a ceramic assemblage with a mix of paste groups transitional between the Dan River and Oldtown ceramic series. Consistent with Dan River, net impressed overwhelmingly dominated identifiable exterior surfaces (82%) and complicated stamped sherds were absent. However, fully 54% of the rims were folded. A radiocarbon date was obtained on wood charcoal from the feature (Beta-105651), which yielded a conventional age of 570 ± 50 BP. A modern calibration of the date produces two options at 1σ, AD 1311–1359 and 1387–1417 (Oxcal 4.3.2). Given the transitional nature of the assemblage, Eastman favored the late range for the assay, although even at 2σ the maximum youngest limit is only AD 1431. Since wood charcoal can be significantly older than the targeted event (see Schiffer 1986), however, it is feasible that the actual age of the pit extends into the last half of the fifteenth century.

If this sample accurately dates the assemblage, then a strong case can be made for an earlier appearance of folded rims in the Piedmont Village tradition than in the Lamar horizon of the Eastern SAM culture area, at least in North and South Carolina. Building on this assumption, it can be further suggested that folded rims diffused southward during the time of greatest interaction between the two areas, which appears to have been in the protohistoric/Early Contact periods (AD 1450–1620). After AD 1620, the occurrence of complicated stamping drops off to levels of not more than 1% and folded rims disappear altogether in the Upper Saratown assemblages of the Middle and Late Contact periods. As an aside, the Hairston site findings may represent a key to a better understanding of the apparently anomalous association of folded rims with earlier decorative modes in Boudreaux's Late Leak phase reconstruction. Perhaps we will find that folded rims do occur much earlier at Town Creek than in the rest of the Eastern SAM culture area.

This is not the case for Caraway, which maintains relatively high percentages of complicated stamping and folded rims throughout the seventeenth century. Wilson's (1983:326–341) description of the seventeenth-century Caraway ceramics from the Poole site, for instance, indicates that about 7% of the exterior surface treatments were complicated stamped, while folds occurred on 26% of the rims. It is probably not a coincidence that the greatest period of interaction materializes at the time Town Creek was abandoned and Guatari emerged at a location about 90 km closer to Upper Saratown. Indeed, this may represent a shift in the direction of regional interaction brought about by abandonment events transpiring in the GDO. Greater geographic separation, in fact, may have weakened communication with more southerly situated Eastern SAM polities. There are, of course, other pathways via which folded rims may have diffused into the northern branch of the Eastern SAM culture area, the most likely of which would have been through the Appalachian Mountains or the Ohio Valley (see Dickens 1976:200; P. Gardner 1980:52). Only dated contexts and better ceramic assemblage description will allow us to successfully map out the timing and diffusionary course of folded rims in the region.

Although major settlement changes and relocations characterize the post-AD 1450 occupation history of the upper Great Pee Dee valley, the region was not abandoned. Instead, there was a northward shift of the power center from Town Creek to Guatari on the lower Yadkin. Pardo was informed that the chiefdom of Guatari held sway over 39 lesser chiefs, which led Hudson (1990: 91) to speculate that these subject towns most likely were situated along the Yadkin and its tributaries and possibly on the upper Dan and Cape Fear rivers. Ricciardelli's survey data strongly suggest that the area around Town Creek was still occupied by Caraway groups into the sixteenth and seventeenth centuries; these settlements could very well have been occupied by subjects of Guatari. Coe (1995:Figure 9.11B) illustrates a typical multiline, concentric scroll Lamar Incised rim sherd with a notched strip positioned below the lip that came from a site about 20 km south of Trading Ford near Pooletown, North Carolina (R. P. S. Davis, personal communication, October 15, 2015), which evidences late fifteenth- or sixteenth-century occupation. Moreover, the southern edge of this polity may have reached as far south as present-day Cheraw, South Carolina. Both the Soto and Pardo accounts note a village respectively referred to as Ilapi and Ylasi that DePratter and colleagues (1983:155) locate near Cheraw based on measured trip distances in the Pardo accounts. Triangulation of the distances could place this town somewhat farther north, over the present state line, but it is clear from the direction and distance information in the accounts that the settlement was located in the upper Great Pee Dee valley.

The status of settlement upstream from Trading Ford on the Yadkin is still somewhat unclear, but archaeological investigation suggests that this region was densely populated. Surveys and excavations by Wake Forest University (Woodall 1984, 1990, 1999, 2009) and studies by others (see E. Jones and Ellis 2016; Moore 2002:100–124) shed light on this settlement system. The late prehistoric underpinnings correspond to Dan River (sand- and quartz-tempered) occupations from the Great Bend southward and to Smyth (steatite-tempered) components to the west on the upper Yadkin. The Smyth occupations may have given rise to the Lamar settlements of Jones Mound, the Nelson Mound and Triangle, the R. T. Lenoir Burial Pit, the Jones site, Broyhill-Dillard, and an assortment of Happy Valley phase sites, all of which are represented by steatite-tempered Burke series pottery. Alternatively, the Smyth occupation could have continued without acculturation in the upper valley, being replaced in the far upper reaches of the valley by Burke phase populations extending their range northward from the upper Catawba valley (see Rogers 1993).

The Porter site (Woodall 1999), situated 30 km east of the Burke phase occupation on the Yadkin, appears to provide evidence of coeval Lamar and Piedmont Hill Tribe occupations. Excavations here exposed two shaft and chamber burials containing prestige items associated with Mississippian elite exchange systems (i.e., two perforated spatulate celts, a marine shell gorget, and a Citico-style, Saltville variety marine shell gorget). It is Woodall's thesis that these burials evince the introgression of Lamar culture on an indigenous Late Woodland society, perhaps through marriage alliances. The fact that the recovered ceramics from the site are relatable to the Smyth series (i.e., net impressed, plain, brushed, fabric impressed, and cord marked), while only 1% of the collection exhibited complicated stamped treatments suggestive of the Burke series, strongly supports Woodall's thesis.

Between the Great Bend and Trading Ford on the Yadkin, surveys have generally recovered ceramics of the Dan River series, not Caraway. The majority of this pottery contains fine-to-medium sand temper that Woodall (1984, 1990) equates with P. Gardner's (1980:63) Dan River, variety Stokes, which he dated between AD 1550 and 1725. Recent investigations suggest that medium-to-fine sand tempering becomes dominant in Dan River pottery across the western Piedmont after AD 1400 and extends into the protohistoric period, where it transitions into the Oldtown series (see Davis et al. 1998a:25; Eastman 1999). Folded rims occur in variable proportions within late Dan River assemblages across Virginia and North Carolina (see Davis et al. 1997a, b, c, 1998a, b; P. Gardner 1980:45), further demonstrating the late persistence of the Dan River tradition. Folded rims are present at the Hardy site (31Sr50), located at the northern end of the Great Bend, and at the Donnaha site (31Yd9), about 15 km farther downstream (Woodall 1990:77). Given what is known about the temporal trajectory in Dan River/Old Town ceramic series, it is likely that folded rims are associated with late prehistoric and protohistoric occupations, which would suggest contemporaneity with Guatari.

A cemetery in the western section of the Hardy site appears to date to this time as well (Woodall 1990:46–52). Wood charcoal from one of these burials (59–5) returned a conventional radiocarbon age of 340 ± 70, which produces two 1-σ options, AD 1479–1530 (23.5%) and 1539–1635 (44.7%) (Oxcal 4.3.2). Another burial (60–2) in this cluster contained a Stokes Cob Marked vessel with a conoidal base and a flaring rim resembling pots interred with burials at the Fredericks site (1680–1710), believed to represent the refugee settlement of the Occaneechi as they fled southward from the Roanoke after the 1676 Bacon Rebellion (see Ward and Davis 1993:416). This may suggest an even later age for the cemetery, but the chronological distribution of vessel forms in Piedmont Siouan assemblages is not yet well defined. The radiocarbon date from burial 59–5, however, argues for an earlier age, somewhere between the middle sixteenth and very early seventeenth centuries.

Overall, there is reason to believe that the Yadkin and upper Great Pee Dee valleys supported a dense, culturally heterogeneous population in the sixteenth century, rivaling the densely packed contemporaneous occupations of the Wateree and Catawba valleys. Clearly, some of the Stokes occupation in the Great Bend was contemporaneous with Guatari and these populations may have been subjects of this possibly multicultural chiefdom.

Lower Pee Dee, Lower Santee, and Winyah Bay

Much less is known about the archaeological and ethnohistoric landscape of the middle and lower reaches of the Great Pee Dee River and associated tributaries. A review of the South Carolina state site files indicates that sites containing Mississippian complicated stamped pottery are continuously distributed from the state border to Waccamaw Neck, as well as in the tributary valleys of the Black, Lynches, Little Pee Dee, and lower Waccamaw rivers, at least below Horry County. The Eastern SAM culture area butts up against the Late Woodland cultures of the southern coastal zone of North Carolina in this region. Unfortunately, the occupation history of most of these sites is known only from unsystematic surface collections that have yielded small and poorly understood artifact samples.

Because the region was outside of the search path of the Soto and Pardo expeditions, the sixteenth-century ethnohistoric picture is just as blurry. What we are left with is the late seventeenth- and early eighteenth-century accounts of the English settlements and three relatively intensive archaeological investigations conducted at sites that might be expected to yield evidence of protohistoric and/or early contact occupation (i.e., Pedeetown, the factory at Yauhannah Bluff, and the Wachesaw landing site). None of this information successfully bridges the gap between the archaeologically documented SAM occupation and the historical record of tribes present in the region at the turn of the eighteenth century. Both lines of evidence are reviewed below.

Jack Wilson (1983:193) characterized the early eighteenth-century landscape of the Pee Dee valley as an “almost uninhabited wedge of land” buffering the Tuscarora and the slave-trading Westos and Occaneechis to the north, the Catawba to the west, and the South Carolina colony to the south. According to William Byrd’s (1966:398) account, into this vacuum respectively came the Sara and Keyauwee, from the Dan and Yadkin valleys around 1703. They settled adjacent to one another near present-day Cheraw, South Carolina, based on the Barnwell map of 1711. About the time that the Keyauwee were fading from history, after 1716, a new tribe identified as the Pedee appeared in the historic record and was credited with occupying a location farther down the Great Pee Dee River near Pine Bluff as depicted on both the 1711 Barnwell map, published in 1722, and the Mosely map of 1733. Wilson (1983:195) argues that this was not a coincidence and that Pedeetown represented the resettlement of the Keyauwee.

Alternatively, some (Hudson 1990:76; Rudes 2004a:403; Rudes et al. 2004: 302, 317) have suggested that Pedeetown was established much earlier, in the sixteenth century, principally because of the phonetic similarity between Pedee and "Vehidi." This name was transcribed on a list of *orata* in Bandera's long relation (Hudson 1990:260), after he came to Canos (Cofitachequi) to meet with Pardo in 1567. The actual location of Vehidi was not discussed in any of the Pardo accounts, so its placement at Pedeetown is highly speculative. Added to this mix were the remaining Saxapahaw, who resided near the Tuscarora town of Nahantes on the Neuse River until 1711 when they were attacked by the Tuscarora for refusing to join in their campaign against the South Carolina colony (Wilson 1983:193). They first sought refuge with the Waccamaw on the lower Pee Dee, but before the end of the Yamassee War in 1717 they had incorporated with or settled close by the Sara near Cheraw.

Historic documents and maps record the presence of three groups in the area of Waccamaw Neck between the mouths of the Great Pee Dee, Waccamaw, and Santee Rivers. These are the Waccamaw, Winyah, and Sewee. The Waccamaw enter the historic record between 1712 and 1715 (Goddard 2005:27–28; Mooney 1894:76; Rudes 2004a). Swanton (1946:203) states that the 1715 census credited the Waccamaw with 6 villages situated along Waccamaw Neck near the mouth of the Great Pee Dee River, with a total population of 610 individuals. The Woccon, who were settled on the Neuse River near its confluence with Contentnea Creek, disappear from the historical record after 1710 and this coincidence suggested to Taukchiray (in Rudes 2004b:8) that the Waccamaw constituted the southern resettlement of the Woccon who were fleeing the outbreak of the Tuscarora War (1711–1713). Both groups appear to have spoken Siouan (see Rudes 2004a:374) and Lawson indicated that at the turn of the eighteenth century the Woccon had two villages supporting 120 warriors, which would translate to a total estimated population of 500 to 600 individuals (see Mooney 1894:65). This figure is nearly identical to the 1715 census for the Waccamaw, which provides additional support for Taukchiary's reconstruction.

There is evidence that would suggest that the Winyah came to reside in the area of Waccamaw Neck by the 1670s and probably earlier. The Santee Indians told Lawson (in Mooney 1894:76) in 1701 that the Hooks and Backhooks, with whom they were warring at the time, occupied the mouth of Winyah Bay. These names appear corrupted from their original forms (see Mooney 1894:76), and Goddard (2005:23; see also Rudes 2004a:402 and Swanton 1922:38) speculates that they might correspond to "Huaq" and "Pahoc," two tribes listed in the Ayllón, Gordillo, and Quexos voyages along the Carolina and Virginia coasts between 1521 and 1526. The names "Hooks" and "Backhooks" seem phonetically out of place when compared to names given to surrounding groups, which have "-ee" endings. Blair Rudes (2004a:402; see also Swanton 1922: 38) suggested a connection between these names and Uca *orata* (chief), who traveled to Ylasi in 1568 to meet with Pardo. He further states that "Uca" may derive from the Catawban words for "black" (*ha-wu-ka*) or "something black" (*pa ha-wu-ka*), which coincidentally are synonymous with the name of a tributary (Black River) that flows into Winyah Bay. Certainly, the phonetic correspondence between these Catawban words and "Hooks" and "Backhooks" is clear and strengthens the contention that the Winyah may have been indigenous to the region in the sixteenth century.

Hudson (1990:Figure 13, 302) places two towns, Usi and Sanapa, in the general area of Waccamaw Neck based on indirect information from the long Bandera relation of 1567–1568. Bandera stated that these towns were 60 leagues from Santa Elena at a place where salt is made next to the sea. Translated into miles, this distance would be somewhere between 190 and 210 miles from Santa Elena, depending on which unit Bandera was using (i.e., Spanish nautical league or the *legua comuna*). Either measure would bring the towns very close to Winyah Bay, or along the Waccamaw River. Stepping forward in time, the Winyah appear to be the group identified as the Waniah who were settled along Winyaw River in the 1670s (Gallay 2002:59–62). At the end of the Westo War in 1680, they became targets for the planters' ambitions to acquire native slaves through a trade pact with the Savannahs. Consequently, the Waniahs' numbers were greatly reduced. In 1715 (Swanton 1946:207), the Winyah occupied a single village located at the mouth of the Pee Dee River on Winyah Bay, with a population of 106.

Although the traditional homeland of the Sewee has consistently been identified with the area between the Wando and Santee Rivers, the first definitive location for one of their villages comes from Maurice Mathews (in Waddell 1980:289–291), who places it on the north side of the "St. Pa" (Santee) River in 1671. Within the first decade of colonization, however, the Sewee were said to be living on a reservation with other "Settlement Indians" about three miles from the mouth of Wando River (Waddell 1980:289). The 1685 Thornton-Morden map identifies only the "Sewee Settlement," but in the 1695 version the Wando and Sampa are also shown to have resided in this general area. In this later version, the Sewee settlement is referred to as "Sewel [sic] Indian Fort." Waddell (1980:290) speculates that the fort may have been erected with assistance from the colony as a defense against the Westos. In the early leg of his journey in 1701, Lawson (Lefler 1967:16–17) landed at a location on Bulls Bay, where he spent the night near a temporarily vacated plantation house on the site of "a deserted Indian Residence, call'd *Avendaughbough*." This residence has traditionally been linked to the Sewee, but it is not known from Lawson's description whether it contained standing ruins that might suggest a recent occupation or merely the scattered debris associated with archaeological sites in the area.

The historic maps from the era do not identify a Sewee settlement on Bulls Bay. Lawson's first encounter with the Sewee came two days after his landfall in Bulls Bay, when the party had travelled between seven and fifteen miles up the Santee River from its mouth. Here, several Sewee were firing the cane swamps to drive game. Land warrant records show that a Sewee chief, King Jeremy, occupied a plantation between Collins and Washasha creeks near the mouth of the Santee in 1701 and that the Sewee were settled on Wadmacon Island in this general area in 1698 (Waddell 1980:319, 338, 361). The census of 1715 indicates that the Sewee had one village of 57 persons situated 60 miles northeast of Charleston. Waddell (1980:297) notes that this distance would place the Sewee near the Black River, about 20 miles north of the Santee. He discounts this distance as an overestimate because in 1716 the Sewee were situated near the French settlements on the Santee. All of these accounts situate the Sewee along the lower Santee River, with the exception of the time corresponding to the Westo and Waniah Wars when they were confined with other "Settlement Indians" along the upper Wando River.

Rudes (2004a:401–402) equated the Suya—whose *orata* visited Pardo at Gueca in 1567—and Joye or Xoje of the Ecija voyages in 1605 and 1609 with the Sewee and suggests that the main village at that time was situated somewhere on the southwest side of the mouth of the Santee River. Rudes and colleagues (2004) plot Suya very near where Lawson came up on *Avendaughbough*, but this placement is purely conjectural. Waddell (1980:287) indicates that Ecija's accounts place Joye/Xoje on the Santee River upstream from the mouth. An association with the mouth of the Santee River also appears to be confirmed by Rojas (Waddell 1980:287), who was sailing up the coast in search of survivors of the Ribault expedition in 1564. Rouffi informed him that a province called Suya was located about 30 leagues (100 miles) from Port Royal, a distance that Waddell considered consistent with the Santee River. Again, these very early historic references, as vague and imprecise as they are, suggest that the Sewee's homeland was located on the lower Santee River.

Based on this review, it would appear that the ethnohistoric groups occupying locations on the middle and lower Pee Dee River, Waccamaw Neck, and the mouth of the Santee in the late seventeenth and early eighteenth centuries fall into three classes. These are: (1) those that can be documented to represent early eighteenth-century immigrants (the Keyauwee, Sara, and Saxapahaw); (2) those that were probably early eighteenth-century immigrants (the Waccamaw); and (3) those that might represent continuity with indigenous late prehistoric and protohistoric groups (the Sewee, Winyah, and Pedee). Clearly, all of these groups might represent separate waves of immigrants moving southward in response to the pressures set into motion by the expansion of the American colonies and the disruptions that this imposed on the native populations. Yet, as poorly described as they are, Spanish accounts from voyages along the coast and from vague allusions to visitations of chieftains reported by members of the interior expeditions of Soto and Pardo strongly suggest that Native Americans were present somewhere near Winyah Bay and the mouth of the Santee in the sixteenth century. Precisely what this presence means is a question that can only be satisfactorily answered through extensive archaeological investigations in the region, but available evidence can provide some perspective.

Paul Hoffman's (1984, 1990, 1994) discovery of Alonso de Chaves's rutter (handbook of sailing directions and bearings) for the Ayllón voyages has produced a "rosetta stone" for unraveling the mystery surrounding the modern-day locations of the rivers and bays discussed in the accounts of all three voyages, when juxtaposed against the 1526 Vespucci and 1529 Ribeiro maps. Much of the controversy surrounding the search for Ayllón's ill-fated settlement of San Miguel de Gualdape in 1526 has concentrated on the true location of the closely proximate Rio Jordan and Cabo de San Nicolas (Roman). Prior to the discovery of the rutter data, two research positions had emerged, those favoring an association with the area around Cape Fear and those that would place this first landing at the mouth of the Santee River or Winyah Bay (Hoffman 1994:43–46). Once the solar latitudinal readings made at the time of the voyages were converted to modern values and the direction and distances between rivers and bays were translated from leagues to miles (see Hoffman 1990:Appendix A), Hoffman was able to argue in favor of the Santee River/Winyah Bay alternative. The corrected latitude that Quejo recorded in the first voyage undertaken in 1521 corresponds precisely to Winyah Bay and the closely clustered occurrence of a river mouth, a cape, and a bay is uniquely consistent with this geomorphic setting (see Hoffman 1994:44).

Hoffman's resolution of this problem is of special importance here, because the Rio Jordan/Cabo de San Nicolas anchor point is where Francisco de Chicora, the native slave whose testimony was presented to Peter Martyr, was captured in 1521. During the first expedition the Spanish encountered a large group of curious natives on the shoreline and, after a 22-day stay, as many as 60 of them were coaxed onto the ships and taken hostage as slaves (Hoffman 1984:420). In the expedition of 1525, Quejo, one of the two pilots involved in the first encounter, landed at this same location, which was probably Winyah Bay where better anchorage could be achieved, and sought out the same native group. Initially the Spaniards were met with hostility, but Quejo assured them that their relatives would soon be returned (Hoffman 1990:53). This relieved some of the tension and the party spent several days there, giving out gifts and crop seeds before continuing the voyage. If Hoffman's reconstruction is correct, the accounts of the first and second expeditions, taken together, most probably confirm the presence of a stable village in the general vicinity of Waccamaw Neck, although there is no indication from the documents that the Spanish actually visited it on any of their three landfalls.

Hoffman (1990:9, footnote 20) claims that numerous villages were present upstream from the porting location, of which Chicora's was but one. Clearly, this inference overreaches both the historic accounts, which never actually mention any villages, and the archaeological evidence he cites in support of it. Hoffman's authority on the latter was a personal communication from Robert Stephenson, the former director of the South Carolina Institute of Archaeology and Anthropology (SCIAA), who stated that complicated stamped pottery dating between AD 1300 and 1600 had been found on sites in the Waccamaw Neck area. Although accurate as a sweeping generality, none of these sites have yielded irrefutable evidence (i.e., folded rims) of post-AD 1500 occupation. Based on what is actually known from historic accounts, the area appears to have been only sparsely populated. The Spaniards thoroughly reconnoitered the adjoining coastline for a few days when they were anchored in the mouth of the Santee during the 1521 voyage. Hoffman (1990:8) suggests that they searched as far south as modern Cape Romain before steering their ships northward into Winyah Bay, where they spent at least two weeks without reporting that they had actually visited Chicora's village.

It is unclear whether the ships of the 1526 voyage anchored in the mouth of the South Santee or Winyah Bay, but during that visit Ayllón spent at least a month building a replacement for the main cargo ship, which sank on a sand bar (Hoffman 1990:67). Scouting parties were sent inland during this respite, perhaps to search for Chicora and the other native translators who had escaped after the ship sank, but when they returned it was reported that the area would not be suitable for establishing a colony. It is Hoffman's (2014) hunch that this meant the search party had not found native settlements, as the Spanish interpreted the presence of large indigenous population concentrations as positive proof of successful places to live. Whether this means that Chicora's village had been vacated or that the residents by this time were so wary of contacting the Spanish directly that they hid is not known.

This was a theme throughout the Ayllón voyages. The Spanish very seldom made contact with Native Americans along the coast. It is sometimes forgotten that most of the ethnological information supplied to Peter Martyr came via the second-hand testimony of Francisco de Chicora when Ayllón took him to Spain and not as a consequence of direct Spanish observations. Oviedo's (in Swanton 1922:48) description of conditions surrounding the Gualdape colony in 1526 can be more generally applied to the totality of the Spanish experience with indigenous populations: "The Spaniards who came with the licentiate Ayllón did not see the villages; they only met with a few isolated houses or cabins forming little hamlets, at great distances one from the other." Again, if Hoffman's reconstruction is accurate, Ayllón's settlement of San Miguel de Gualdape should be proximate to Sapelo Sound, probably on one of the mainland creeks near the Guale settlements (Hoffman 1994:42).

Once it was determined that Winyah Bay was an unsuitable place to establish a colony, Ayllón sent out scouting parties southward along the shore in ships and inland on foot to search for an appropriate place to build his settlement (Hoffman 1990:68–70). Three ships differentially combed the shoreline from Winyah Bay to a point approximately 100 leagues to the south that Hoffman identifies as the Rio Corrientos, about 10 leagues south of present-day St. Augustine. Each of the major bays and river mouths was reconnoitered, including Charleston Bay, Port Royal Sound, and the Savannah River. According to Hoffman, after discussions with the pilots of the ships, it was determined that the area around Sapelo Island presented the most promising aspect for a settlement. This suggests to Hoffman (1990:72) that there was no evidence of significant population concentrations between Winyah Bay and Sapelo Island, which is consistent with Thomas's (1993:42–43) observation that the South Atlantic coast above the Guale occupation of Sapelo Sound was virtually abandoned in the mid-fifteenth and sixteenth centuries. When they reported back, the inland scouting parties did not find any evidence of a better prospect either.

Ayllón then sent the ships southward and took a small contingent of healthy men and horses with him overland, planning to meet up with the larger colony within a matter of weeks. Hoffman suggests that the inland party may have taken an historic trail arcing from Winyah Bay and running across the major streams somewhere near the present-day US 17 Alternate route, where fording would have been more manageable. What Ayllón witnessed along the inland march was never reported, suggesting that he did not encounter significant Native American population concentrations along the way.

Thus, although there was probably early sixteenth-century occupation at Winyah Bay, the evidence from the Ayllón expeditions does not suggest that it represented a teeming community with multiple villages, hamlets, and numerous farmsteads similar to that of the thirteenth- through fifteenth-century Mississippian occupation that is everywhere evident across most of South Carolina. It should be clarified that the mere presence of sixteenth-century occupation does not controvert the existence of the GDO, however, as my argument for it is not based on the complete abandonment of regions but rather on significant population reduction and functional reorganization of land use patterns. For instance, Anderson (1994, 1996:167) was able to show a significant diminution of Middle and/or Late Lamar occupation in survey data from the lower and middle Savannah River valleys, but he was unable to demonstrate the complete absence of such components. Some of the latest components may very well represent historic intrusions by the Westos, Shawnee, and others (see M. Smith 2002), but even during the Soto entrada small bands of natives were encountered along the Broad River and points to the north (Hudson et al. 1984:72–74). In addition, several structures interpreted by the Spanish as huts for hunters/fishermen were noted near Soto's encampment at the confluence of the Broad and Congaree rivers. Anderson suggests that the scattered occupations potentially dating to the later fifteenth and sixteenth centuries in the lower and middle Savannah Valley are best explained as ephemeral, seasonal visitations of this sort. Certainly, the weight of the evidence suggests that these areas were depopulated and that permanent residential occupation by Mississippian groups had ceased or was significantly reduced, even though seasonal hunting, fishing, and gathering forays were probably commonplace.

Available archaeological evidence from the middle and lower Pee Dee drainage system, the mouth of the Santee, and Winyah Bay is scant, but locations where the most intensive investigations have taken place are especially relevant to the ethnohistoric record reviewed herein. These include Pine Bluff on the middle Great Pee Dee River, where numerous eighteenth-century maps plot Pedeetown; Yauhannah Bluff near the mouth of the Pee Dee, where the Black River trading post was established in 1716; and the site of Wachesaw Landing on Waccamaw Neck, where seventeenth- and early eighteenth-century occupations hypothesized to represent a Waccamaw settlement have been identified through archaeological testing. Added to this list is the extensive survey and testing conducted between the Santee and Wando Rivers in the Francis Marion National Forest.

If the larger Pee Dee drainage system was part of the GDO, it would be anticipated that sixteenth-century occupation there would be muted and that settlement continuity with the earlier Mississippian florescence would be disrupted. Based on the model of Eastern SAM decorative modes, sixteenth-century ceramic assemblages should contain significant proportions of folded rims and Lamar Incised ceramics. An objection could be raised that applying this model might result in missing a potential sixteenth-century occupation if the resident groups were not participating in the Eastern SAM interaction sphere. However, this is highly improbable since all of the known sixteenth-century occupations contain this horizon and, in addition, all of the neighboring contemporary Piedmont Siouan (see Davis 1987; Eastman 1999) and Coastal Plain Cashie (see Heath and Swindell 2011; Millis 2009) assemblages exhibit this decorative style horizon in the protohistoric period. Consequently, the presence and abundance of this style horizon should provide the most effective basis for evaluating the character of sixteenth-century occupation in any of the proposed river valleys within the GDO.

Folded rims and Lamar Incised body treatments are known to have wide temporal distributions in Eastern SAM ceramic sequences, so the presence of these traits would not necessarily establish sixteenth-century occupation. Incising begins in the Early Lamar period, sometime around AD 1425, and extends into both the Middle and Late Lamar periods, while folded rims appear to be restricted to Middle and Late Lamar in most locations, although there is some evidence that they emerged in the fifteenth century in the Dan Valley. So, the mere presence of this horizon would probably not be sufficient to distinguish sixteenth- from seventeenth-century occupations, but its presence would be necessary to entertain the possibility of a sixteenth-century occurrence.

Historic accounts indicate that the region containing the most complex occupation history encompasses the lower drainages of the Central and North coasts of South Carolina. Extensive survey coverage of the Francis Marion National Forest strongly establishes that sixteenth-century occupation between the Wando and Santee Rivers was extremely light or nonexistent, as no folded rims and only very sparse occurrences of Lamar Incised have been identified. Such a conclusion is consistent with the Ayllón voyages of the 1520s, the accounts of Ecija at the turn of the seventeenth century, the records of the South Carolina colony in the 1670s, and Lawson's observations in 1701. All of these accounts agree that sightings of Native Americans in this region took place from the mouth of the Santee northward into Winyah Bay and not southward into the Bulls Bay region. The sites with possible sixteenth-century occupation in the Pee Dee drainage system are discussed in detail below.

Wachesaw Landing

Investigations at the Wachesaw Landing site, located on the east bank of the Waccamaw River about 17 miles north of Winyah Bay, provide a window into the archaeology of this region. Survey, testing, and examination of collections in the possession of the landowner led Trinkley (Trinkley and Hogue 1979; Trinkley et al. 1983:25–32) to hypothesize that the Native American occupation at the site consisted of two components. The earliest one, which he estimated spanned the period from AD 1650 to 1700, was represented by Pee Dee pottery abundantly distributed across the site. The later component consisted of a small cluster of flexed inhumations associated with trade beads and other European artifacts that were unearthed during the construction of a cabin in the 1930s. He dated the burials to the first third of the eighteenth century and associated with this component a less abundant ceramic series that he named Wachesaw. He further speculated that this component represented an historic Waccamaw Indian occupation and might be lineally related to the earlier occupation. A third (more rare) ceramic series, Kimbel, was characterized by fine sand temper and predominantly plain and burnished exterior surfaces. Trinkley drew correlations between this series and the Caraway and Hillsboro series of North Carolina, suggesting that its presence might be attributed to trade with the Keyauwee or Sara (Cheraw) who were settled on the middle Pee Dee by this time.

This reconstruction does not present an impediment to the GDO model, as all of the occupation would have occurred after the sixteenth century. However, there is reason to believe that much, although not all, of the occupation Trinkley describes is prehistoric and that the site occupation history has been conflated into a narrow historic time frame. The Pee Dee series, which I contend represents the deposits from a prehistoric Eastern SAM occupation, was assigned by Trinkley to the last half of the seventeenth century based on the presence of traits that would suggest a later chronological position than the aggregate collection of Pee Dee series ceramics from Town Creek (Trinkley et al. 1983:25–30). The Pee Dee series ceramics from Wachesaw Landing were characterized by higher percentages of plain, lower percentages of complicated stamped, rare occurrences of Lamar Incised, a much larger representation of filleted rim strips, and an unusually high percentage of simple stamping. The latter two characteristics suggested to Trinkley a connection with the Contact period Siouan Hill Tribes of North Carolina. He further drew correlations with the sequence at the Mulberry site, where filleted rim strips and percentages of plain increased in popularity in the later deposits, which also contained Lamar Incised.

His very late temporal placement of the Mulberry assemblage was based on Caldwell's (1974:89–94) contrasting descriptions of the ceramics of the premound humus under Mound A at the Mulberry site and the later Upper Village deposits, along with a radiocarbon date on charred corn cobs from a pit in the premound deposit below Mound A. The cobs yielded an uncalibrated age of 430 ± 200 BP, or AD 1520 (see Ferguson 1973b:18 and Trinkley et al. 1983:29). Caldwell (1974:89) indicated that rim decoration in the premound assemblage consisted of the typical Town Creek-Irene axis decorative modes (i.e., hollow reed punctations, rosettes, pellets, riveted nodes) and occasional reed punctated strips, which today we know, as an aggregate, date to the period from about cal AD 1325 to 1425. A sixteenth-century date for this complex is clearly too late, which suggests that the date on the corncob was not adjusted for fractionation, a common oversight in the early days of radiocarbon dating. When fractionation adjustments for C4-pathway plants like corn are not made the resultant radiocarbon assays will reflect substantially younger ages than the events targeted for dating, by as much or more than 200 years. Bender (1968) indicates that an average fractionation (13C/12C ratio) for corn is -12.3 ‰. When this adjustment is applied to the material from the submound pit, a conventional age of 638 ± 206 BP is obtained, which produces a calibrated mean of 1324 ± 192 AD (OxCal 4.3.2). The most probable 1σ option (65.6%) for the sample is 1153–1484 AD, a result that is much more consistent with the types of rim elaborations Caldwell found in his premound sample.

Based on what we know today about the chronological ordering of Eastern SAM decorative modes, then, Trinkley's dating of the Pee Dee series collection from Wachesaw Landing is incorrect. The two principal decorative modes consist of filleted rim strips with segmented/notched (or "pinched" as Trinkley described them) and hollow reed punctations positioned close to the lip and below the lip, and nonstrip hollow reed punctations arranged in rows below the lip (see Trinkley et al. 1983:24, Table 1). These traits would place the assemblage within the range of variability associated with the fifteenth-century patterns seen at the Mulberry and Fogarty Creek sites (Table 5.1). Minority modes include rosettes and shoulder ticks. Although not a single-phase assemblage, the decorative mode associations suggest a relatively narrow range of occupation primarily confined to the period between AD 1375 and 1450 for the Pee Dee series. Moreover, an absence of lip notches argues strongly that pre-AD 1375 occupation is only marginally represented. Additionally, the low percentage of shoulder ticks would order the collection between the Fogarty Creek and Wateree-Town Creek phases on the one hand and McDowell I on the other. A small Lamar Incised representation of 0.6% (6 of 981 Pee Dee series sherds with recognizable surface finishes) is also consistent with this chronological placement.

Although the Pee Dee series ceramics from Wachesaw Landing conform to the generalized eastern SAM decorative mode seriation and fit well within the late fourteenth- to early fifteenth-century time frame, the presence of such high percentages of simple stamping contrasts starkly with the neighboring sequences of the Central South Carolina coast, the Wateree Valley, and Town Creek. All of these sequences contain very little simple stamping in deposits of comparable age, while simple stamping makes up 30.8% (303 of 983 sherds) of the

Table 5.1. Comparison of Decorative Mode Percentages between Wachesaw Landing and the Late Fourteenth- and Fifteenth-Century Phases from the Central Coast and the Wateree Valley

Site	Phase	Notched Lip	Punctated Lip	Hollow Reed Punctated	Solid Punctated	Rosettes/Pellets	Nodes/Lugs	Rim Strip	Shoulder Ticks	Folded	Lip Ticks	Frequency
Mulberry	McDowell II		0.9	1.8			2.7	62.5	28.6		3.6	93
Mulberry	McDowell I			8.9			20.0	37.8	31.1	2.2		42
Wachesaw Landing[a]				38.5		6.6		50.8	4.1			122
Fogarty Creek	Fogarty Creek			16.7		25.0		58.4				12
Adamson	Wateree-Town Creek	23.5	5.9	5.9	5.9	29.4		23.6				17

[a]Data from Trinkley et al. (1983:Table 1).

identifiable exterior surfaces in the Pee Dee series from the Wachesaw Landing collection. Although these specimens could represent terminal Late Woodland and transitional Mississippian components, when simple stamping is actually a predominant type in neighboring Eastern SAM sequences, two details mitigate against this explanation. First, if a significant terminal Late Woodland component were present, it should be accompanied by substantial representations of notched and simple stamped lips, neither of which was noted by Trinkley. Second, Trinkley and colleagues (1983:Figure 12, D, E) illustrate examples of simple stamped rim sherds with punctated and segmented appliqué strips and nonstrip hollow reed punctations, which again supports the thesis that most of the simple stamped sherds assigned to the Pee Dee series in the collection make up an integral part of the AD 1375 to 1450 assemblage.

Given these factors, a strong connection with the sequences of Piedmont and coastal North Carolina is indicated, although the time frame must be much earlier than Trinkley suggested. Simple stamping becomes the predominant surface treatment type in a number of eastern North Carolina sequences during the protohistoric and Contact periods, including the Algonquian-affiliated Colington II and III phases (Heath and Swindell 2013), the Iroquoian-affiliated Cashie phases (Heath and Swindell 2011), and the Piedmont Siouan-affiliated Hillsboro, Mitchum (Sissipahaw?), and Jenrette (Shakori?) phases (see Ward and Davis 1993:410–418). The origin of simple stamping as a major assemblage component in these regions is not yet well understood, so the direction of trait adoption is not certain.

Simple stamping comprised a major element of Cashie I assemblages as early as the thirteenth century (Heath and Swindell 2011:10–103, footnote 89). Stratigraphic and radiocarbon evidence from the Jordan's Landing site, moreover, suggests that simple stamping replaced fabric impressing as the dominant exterior surface treatment in Cashie I assemblages at some point before AD 1450 (Heath and Swindell 2011:10–46). Moving westward, Hillsboro Simple Stamped emerges in association with late Haw River assemblages in the Eno and Haw River drainages between AD 1350 and 1450 (Ward and Davis 1993:409–410). The earliest known, full-blown Hillsboro assemblage comes from the Wall site on the Eno River, dated to the mid-fifteenth century (Ward and Davis 1993: 412). It has been suggested that the Wall site Hillsboro component represents an intrusion into the region by a group other than the population associated with the earlier Haw River phase (Davis and Ward 1991:42), which was characterized by nearly 90% net-impressed pottery. The Hillsboro assemblage from the Wall site is dominated by simple stamped (73%), check stamped (14%), and plain (11%) exterior surfaces. Clearly, then, the popularity of simple stamping was spreading across eastern North Carolina as early as the fourteenth century and could have served as the source of elevated simple stamped percentages in the fifteenth-century Pee Dee series from the Wachesaw Landing site. Exactly how this may have come about is not clear, but present evidence suggests that the residents of Wachesaw Landing were engaged in overlapping interaction spheres. An older one had been established with neighboring Eastern SAM cultures to the south and west and a younger one appears to have been centered to the north, made up of a multiethnic network of Piedmont Siouan groups, outer Coastal Plain Iroquoians, and coastal Algonquians.

The association between the Wachesaw series and an early eighteenth-century Native American occupation is, in Trinkley's own words, "circumstantial" (Trinkley et al. 1983:30). The cluster of burials exposed when a cabin was built on the property in the 1930s manifested clear evidence of a late seventeenth- or early eighteenth-century Native American presence. The inhumations were removed quickly by the Charleston Museum, but a sketch of one of them depicts a flexed position (Trinkley et al. 1983:60), which is entirely consistent with the seventeenth-century burial patterns of Piedmont Siouan groups (Ward and

Davis 1993; Ward 1987). Recovered along with the burials, in which a minimum of eight individuals were represented, were trade beads broadly dated to the seventeenth through early nineteenth centuries, a brass spoon with silver plate most common in the last half of the seventeenth century, and a solid brass or copper C-bracelet (Trinkley et al. 1983:39–42). Again, these items are consistent with later seventeenth-century grave offerings in Piedmont Siouan burials (Carnes 1987).

What is not so clear is the association of the European ceramics recovered from the excavations. Trinkley makes the case that these ceramics can be grouped into two components, with respective mean ceramic dates of 1719 and 1819 (Trinkley et al. 1983:46–47). While he felt the later component to be associated with a Federalist period plantation complex, the earlier one was argued to date within the range of the Native American occupation. Later, Michie (1984:53–54) conducted a survey across the Richmond Hill/Wachesaw plantation grounds, which included the Wachesaw Landing site, and made a strong case that the earliest occupation associated with European ceramics had a mean date in the late 1730s and is probably associated with European settlement. This conclusion was based on the elimination of ceramic types with either uncertain or chronologically wide manufacturing dates.

The evidence that is presented to demonstrate an association between the Wachesaw series and the burial cluster consists of: (1) its higher stratigraphic position relative to the Pee Dee series in test square 160R30 and (2) its association with nine trade beads found in the fill of Feature 1, an oval pit (Trinkley et al. 1983:32). A direct association with the burials cannot be established because aboriginal ceramic vessels were not included in the grave offerings. Moreover, the fill of Feature 1 contained sherds from all three ceramic series, 76% of which were identified as Pee Dee. Therefore, strong independent evidence that the Wachesaw series is associated with either the trade beads in Feature 1 or the burials does not exist, so the mere presence of the series in the fill of the pit does nothing more than establish an undefined *terminus post quem* relationship. That is, the pit was filled after the initial manufacturing date of the Wachesaw series, but how long after is undetermined.

The Wachesaw series actually bears a strong resemblance to the Pee Dee 2.5 paste series defined for the Central Coast sequence, which emerges in the fourteenth century and increases proportionally in the early fifteenth century (see Figure 2.4). Wachesaw series ceramics are distinguished from the Pee Dee series at Wachesaw Landing by greater densities of coarse sand inclusions, along with the occasional occurrence of pebbles up to 4 mm in diameter (Trinkley et al. 1983:78). This description corresponds closely to the clastic distribution of the Pee Dee 2.5 paste series (see Figure 2.3c). In addition, complicated stamped motifs of the Wachesaw series are described as sloppily applied. Although this characteristic is commonly seen as a diagnostic criterion distinguishing protohistoric and Contact period assemblages from earlier prehistoric Mississippian ones, the collections from the midden at Fogarty Creek establish that this phenomenon emerged much earlier, at the beginning of the Early Lamar horizon. It is important to distinguish sloppy application from "exploded" or enlarged motifs, as the latter are more likely an indication of a late association while the former appears much earlier. In addition, the Wachesaw series exhibits rim elaborations that would not differentiate it from the Pee Dee series ceramics from the site (Trinkley et al. 1983:30). Only two decorated rims were identified, one with fingernail punctations below the lip and another with a row of hollow reed punctations, characteristics that emerge in the fourteenth and fifteenth centuries.

Trinkley (Trinkley et al. 1983:Figure 13) also notes that several rims of the Wachesaw series exhibit a type of exterior thickening that is morphologically distinguished from rim folds and strips. Thickened rims of identical construction are found in both the Pee Dee 2 and Pee Dee 2.5 paste series from fourteenth- and fifteenth-century Mississippian ceramics found in the Central Coast sequence (see Figure 2.7d). Trinkley further confirms a close lineal relationship between the Wachesaw and Pee Dee series at the site by observing that "the proportions of surface treatments are nearly the same" (Trinkley et al. 1983:30). Based on the formal similarities and the identical stratigraphic relationships with earlier manifestations of the Pee Dee series in both regions, it is virtually certain that the Wachesaw and Pee Dee 2.5 series are correlated and chronologically positioned in the late fourteenth and early-to-middle fifteenth centuries. Their terminal chronological range, however, is not known, as the Central Coast lacks a substantial post-AD 1425 occupation and the Wachesaw Landing prehistoric occupation appears to have ended before the turn of the sixteenth century.

This leaves the Kimbel series as the only possible representative of the late seventeenth/early eighteenth-century Native American occupation at Wachesaw Landing. That there was such an occupation is borne out by the burials, at least one of which is flexed, containing European trade items that are consistent with this time frame. It is unlikely that a cemetery would have been maintained in isolation, so there should also be evidence of residential habitation in the form of structures, other features common to residences, and Contact period Native American pottery. Apparent postholes were identified at the base of the plow zone in two of the 10-×-10 ft excavation squares. However, their association with the Contact period occupation is unclear as the prehistoric component was extensive and characterized by high artifact density, suggesting that most of the structural remains contained in the excavated area are associated with the Pee Dee/Wachesaw component. This inference is also supported by the recovery of two urn burials clustered in another part of the site that are consistent with fourteenth- and fifteenth-century Mississippian burial patterns (see Caldwell and McCann 1941:38). Michie (1984:50–52) exposed a gently arcing line of four postholes in a 2.5-×-2.5 ft test square at Wachesaw Landing during his survey. The arc looks suspiciously like a portion of an exterior wall of an aboriginal structure, but two of the features yielded tabby mortar, which suggests that they are associated with the European settlement complex.

One feature exposed during test excavations that can be considered a strong candidate for membership in a Contact period occupation is Feature 1, the large, steep-sided oval pit that produced nine trade beads (see Trinkley et al. 1983: 19–22). The pit measured 3.7-×-4.2 ft in plan and extended to a depth of 1.59 ft. Pits of similar morphology and dimensions are commonly found at Contact period Siouan settlements, where they are generally interpreted as roasting pits or food preparation facilities (Dickens et al. 1987; Ward and Davis 1993). Although European ceramics and other artifact types were abundantly present in the overlying plow zone of the test unit (160R190), none were found in the fill of the pit, other than the trade beads. In addition, the pit fill contained a sparse collection of 172 aboriginal sherds, less than 35 gm of animal bone, and charred corncob fragments that were scattered throughout the fill. The contents strongly argue that the feature was a roasting pit. Two Kimbel series sherds were also found in the fill of Feature 1. Although the contemporaneity of the pit and these sherds can only be tenuously demonstrated, the formal similarities between this series and Late Contact period Piedmont Siouan ceramics suggest that they could easily fit into this time horizon.

The salient characteristics of the Kimbel series are: (1) fine-textured paste; (2) proportional dominance of plain surface treatments (78.7%); (3) secondary contributions of bold, parallel simple stamped (6.4%) and complicated stamped (14.9%); (4) undecorated rims; (5) straight or excurvate orifices; and (6) cazuela and hemispherical bowl forms (Trinkley et al. 1983:78–79). Only 47 sherds were assigned to the series in the excavated collection, so there is a danger that this description may mischaracterize the representation of rare traits, such as minority surface treatments and rim decorations. Surface treatment types are similar to the Pee Dee and Wachesaw ceramics from the site, which might suggest that the Kimbel series merely represents another paste variant within the larger prehistoric assemblage at the site. However, the large percentage of plain sherds reveals a commonality with Contact period Siouan assemblages of the seventeenth century that is not evident in the Pee Dee and Wachesaw series. An apparent lack of, or more likely a decrease in the incidence of, rim decoration is also a trend in Contact period Siouan ceramics (Davis 1988; Eastman 1999:128–129; Ward and Davis 1993:414–416).

We know that three groups were present in Winyah Bay and the lower Waccamaw River at the turn of the eighteenth century, according to historic accounts: the Waccamaw, the Saxapawhaw (Sissipawhaw), and the Winyah. If the Waccamaw represent the resettled Woccon from the Neuse River and the Saxapahaw represent the southward migration of the residents of the Mitchum site from the Haw River, then it is reasonable to expect that their ceramic assemblages would reflect a heavy signature of the Siouan Hill Tribe tradition. Table 5.2 summarizes the exterior surface treatment patterns for Hill Tribe assemblages, most of which emerged out of the net-impressed dominated Haw and Dan River ceramic traditions of the late prehistoric period. The Mitchum site exhibits some similarities to the Kimbel series, as plain and simple stamped comprise the entire Contact period assemblage from this site. However, complicated stamping, a primary constituent of the Kimbel series, is absent. In fact, none of the Contact period Piedmont Siouan assemblages contains significant proportions of complicated stamped. The Woccon are not yet known archaeologically, but it is likely that their Contact period signature does not contain significant quantities of complicated stamped either, given their close geographic proximity to contemporary Piedmont Siouan tribes of the Haw and Eno rivers. Simple stamping dominates mid-seventeenth-century southern Tuscarora assemblages as well, while complicated stamping is essentially absent (Heath and Swindell 2011:10–40; Millis 2009:278).

When exterior surface treatment is considered in isolation, the Kimbel series is most similar to the fifteenth-century Wachesaw Landing assemblage, which strongly suggests cultural continuity with late prehistoric groups in the region. If the Kimbel series is actually associated with a Contact period Native American occupation, the Winyah, who were documented to be in the area in the 1670s and could represent the descendents of Francisco de Chicora's people of the 1520s, would stand alone as the best candidate for the identity of the Contact period occupation at Wachesaw Landing. Measured by the unusually large contribution of simple stamping relative to other SAM assemblages, the inhabitants of Winyah Bay appear to have maintained differential and significant interaction with the Cashie archaeological culture, much as the Piedmont Hill tribes of the Haw and Eno river valleys did during the protohistoric and Contact periods.

It is of some interest in this regard that Francisco de Chicora referred to a number of groups along the Carolina coast, probably inclusive of his own people (who were never directly named by the Spanish), as "under the dominion" of the province of Duharhe. This province was located on the other side of Winyah Bay to the north or northwest of Chicora's homeland (Hoffman 1990:9–10, 40). The ethnographic identity of Duharhe has long been debated, but Blair Rudes's (2004b:9–12) phonetic and linguistic analysis of the Ayllón accounts

TABLE 5.2. CONTACT PERIOD CERAMIC TYPE PERCENTAGES FROM NORTH CAROLINA

Site	Location	Period	Ceramic Series
William Klutz	Dan	Late Prehistoric	Dan River
Power Plant	Dan	Late Prehistoric	Dan River
Lower Saratown	Dan	Late Prehistoric	Dan River
Upper Saratown	Dan	Late Prehistoric	Dan River
Hairston	Dan	Late Prehistoric	Dan River
Hairston	Dan	Protohistoric	Oldtown
Power Plant	Dan	Protohistoric	Oldtown
Hairston	Dan	Early Contact	Oldtown
Lower Saratown	Dan	Middle Contact	Oldtown
Upper Saratown	Dan	Middle Contact	Oldtown
William Klutz	Dan	Late Contact	Oldtown
Upper Saratown	Dan	Late Contact	Oldtown
Hogue	Eno	Late Prehistoric	Uwharrie
Wall	Eno	Protohistoric	Hillsboro
Jenrette	Eno	Contact	Jenrette, Fredericks
Fredericks	Eno	Late Contact	Fredericks, Non-Occaneechi
Holt	Haw	Late Prehistoric	Haw River, New River, Hillsboro
Webster	Haw	Late Prehistoric	Haw River, Hillsboro, Caraway?
Guthrie	Haw	Late Prehistoric	Haw River, Hillsboro, Caraway?
George Rogers	Haw	Protohistoric	Haw River, Hillsboro, Caraway
Edgar Rogers	Haw	Protohistoric	Haw River, Hillsboro, Caraway
Mitchum	Haw	Contact	Jenrette
Poole Site	Caraway Creek	Contact	Caraway

[a]Smoothed-over net impressed
[b]Primarily smoothed-over simple stamped

Table 5.2. *continued*

Plain	Roughly Smoothed	Burnished	Simple Stamped	Check Stamped	Complicated Stamped	Net Impressed	Brushed	Cob Impressed	Cord Marked	Scraped	Sherd Frequency	Source
4.0						88.8	2.8	0.4	4.0		250	Ward and Davis (1993)
8.0						79.4	1.0	1.7	9.9		928	Ward and Davis (1993)
6.0						84.4		7.5	2.1		1387	Ward and Davis (1993)
8.5						87.2	0.6	0.6	0.8		632	Eastman (1999)
3.3		0.3	2.8		0.8	80.2	3.9	1.1	7.7		363	Eastman (1999)
8.2		10.0	12.0		18.6	40.4	0.7	0.2	9.5		441	Eastman (1999)
92.1		5.6	0.9	1.4							214	Ward and Davis (1993)
42.5		15.2	1.9	8.6	1.4	20.8	3.9	0.7	2.5		433	Eastman (1999)
88.5		0.4	0.8	2.9	0.1		7.4				1383	Ward and Davis (1993)
51.0		13.9	2.5	9.1	1.1	4.5	12.0	4.4	1.1	0.4	2739	Eastman (1999)
41.4		1.8	6.8	26.5	1.5	20.7	1.3				1712	Ward and Davis (1993)
44.4		9.4	4.7	20.6	1.2	14.5	1.8	1.5	1.1	0.7	3509	Eastman (1999)
	2.4[a]					45.3	33.2		19.0		289	Ward and Davis (1993)
11.2			72.8	13.9		1.3	0.1	0.5	0.1		3545	Davis (1997)
12.3	29.3[b]		32.9	6.2			9.7	3.6			982	Ward and Davis (1993)
34.1			6.1	52.9	0.4		3.7	0.4	2.4		3521	Davis (1988)
4.6			0.1			91.9	2.9		0.5		885	Ward and Davis (1993)
4.0			2.3		0.3	92.1	0.1		0.8		1195	Ward and Davis (1993)
19.3			6.7	2.9	1.3	56.9		1.3	3.8		239	Ward and Davis (1993)
22.4			36.5	7.4	0.3	31.5	1.1	0.3	0.5		1932	Ward and Davis (1993)
14.2			27.6	53.8	6.8	27.1	0.2	0.4	3.1		457	Ward and Davis (1993)
58.5			41.0						0.5		3155	Ward and Davis (1993)
57.0		5.0	< 0.5	7.0	7.0	9.0	8.0	5.0	2.0		1252	Wilson (1983)

strongly aligns it with the Tuscarora, which is consistent with a northward direction from Winyah Bay. His synonymy includes Tuscaroran words with Spanish filters (i.e., phonetic approximations) for beans (*xathi* for *Saugh-he*), king (*Datha* for *Teeth-ha*), and several other terms. In addition, Chicora described a form of pyrotechnic burial ceremony practiced in Duharhe that is identical and unique to the Tuscarora and some Northern Iroquois of later historic accounts. Finally, Rudes draws a correlation between the name of the province and an important Tuscaroran town on the southern Neuse called *Naur-hegh-ne*, which was visited by Lawson and identified as Narhuntes on the Barnwell map. In the southern Tuscaroran dialect, these names correspond to Teyurhehte. According to Rudes, then, Duharhe is the Spanish attempt in Peter Martyr's history to phonetically translate the name of this town.

The Ayllón accounts (Hoffman 1990:12–13) identify three provincial seats that held influence over neighboring groups scattered across the early sixteenth-century landscape of the Carolinas. These included Duharhe, which Hoffman associates with the Cherokee but following Rudes more likely represents the Tuscarora; Xapida, which was situated inland and bordered Duharhe to the west; and Orista inisiguanin, which clearly represents the Province of Orista in Port Royal Sound. It would not be much of a stretch to suggest that Xapida was associated with the dense populations on the Wateree, Catawba, and Yadkin rivers, based on the claim that this land had pearls and terrestrial gems. Soto found abundant stores of pearls in the temples of Cofitachequi along the Wateree River and "gems" that the natives, at least, found precious, including mica and copper (DePratter 1989:134–135).

It is difficult to piece together a consistent identification of the native groups that may have fallen under the influence of Duharhe from the various linguistic synonymies that have been proposed, but it is clear that the greater province included groups west and inland of the Tuscarora heartland and east of Xapida. In addition, there were a number of groups straggling along the coast to the south. Hoffman (1990:13) suggests that Duharhe's influence may have spread as far south as Charleston Harbor, which we know from Ecija's accounts of 1609 was home to groups that spoke a different language than that of the Province of Orista. One of the inland groups west of Duharahe was identified as *Anicatixe*, which sounds similar to the *Aconechy* visited by Lawson in 1701 (Lefler 1967:61) and to the *Ackenatzy* of the earlier Lederer accounts, which are believed to respectively represent different town locations of the Occaneechi on the Eno and Roanoke rivers (Ward and Davis 1993:416; Wilson 1983:82).

Based on these disparate lines of evidence, it is possible to suggest that the Winyahs represented a surviving Eastern SAM branch that was gradually being brought under the sphere of influence of Duharhe in the fifteenth century.

During the protohistoric and Contact periods the main population concentration of coastal Native American groups north of Port Royal Sound was situated in eastern North Carolina, where a loosely held confederacy of Iroquoian-, Algonquian-, and Siouan-speakers was engaged in an interaction sphere that extended westward to the Haw River and southward to at least Winyah Bay, if not farther. The linguistic affiliation of the Winyah appears to have been Siouan/Catawban (Mooney 1894:76), but this is by no means established with certainty.

Summarizing the occupation history of Wachesaw Landing, then, the following observations can be made. First, there is no evidence of a sixteenth-century occupation in the artifact collection. The primary occupation spans the period from the late fourteenth through the middle fifteenth centuries. A second occupation of unknown scale occurred sometime between about 1670 and the early 1700s. This Late Contact settlement was probably a small village or hamlet, as the archaeological correlates consist of a cluster of inhumations, an oval pit, and possibly Kimbel series ceramics. The Kimbel series exhibits strong continuity with the fifteenth-century Pee Dee/Wachesaw assemblage, which supports the inference that Winyah Bay may never have been completely abandoned by native groups in the sixteenth century. Based on regional correlations, it is most likely that the Late Contact period component at Wachesaw Landing is associated with the Winyah Indians, who may have been the group that was encountered in this general area during the 1520s by the Ayllón expeditions and subsequently identified by the Spanish as Chicora. Michie's (1984) survey of portions of the Richmond Hill and Wachesaw plantations, of which Wachesaw Landing is a part, also failed to locate evidence of a sixteenth-century occupation (i.e., no folded rims). These findings suggest a break in settlement continuity between the fifteenth and sixteenth centuries and support the proposition that the region experienced a severe reduction in population in the protohistoric period.

Yauhannah Bluff

Near Wachesaw Landing on the south bank of the Great Pee Dee River is the site of Yauhannah Bluff (38GE18), which contains the remains of a portion of an eighteenth/nineteenth-century plantation, the short-lived Yauhannah Trading Post established in 1716. It also contains a range of prehistoric and Native American occupations. Archaeological investigations at the site were first conducted by Coastal Carolina University between 1992 and 1994 under the direction of James Michie. This resulted in systematic shovel test sampling and the excavation of 99 dispersed 1-×-1-m test squares (see Adams et al. 2006:51) and eight 5-×-10-ft test units, most of which were contiguously placed to define a line of historic posts. New South Associates (Adams et al. 2006:54) later mechanically stripped much of the area that Michie tested and excavated 187 cultural features, including 18 postholes and 15 pits of prehistoric or Native American association. Although some European artifacts with manufacturing dates that included the early eighteenth century were recovered at this time, none of the historic features and structures exposed by excavation could be firmly associated with the trading post.

Since the trading post was intentionally located in the heart of the Waccamaw and Winyah settlements, it would seem reasonable to expect evidence of Contact period occupation to be present at the site either during or before the post's establishment. However, only small numbers of ceramics that could be considered potentially affiliated with late prehistoric, protohistoric, or Contact period occupation were identified. Table 5.3 displays the inventory of ceramics recognized as Mississippian from the Coastal Carolina investigations. The data represent only preliminary categories taken directly from the analysis forms housed at the university. There was a series of design and motif descriptions for stamped Mississippian types that I grouped into complicated stamped or simple stamped. The original descriptive categories for each are also supplied in the table. Plain sherds were not categorized by culture-historic type, so the collection, as structured, is incomplete.

At face value, the Mississippian ceramics here could be characterized as reflective of pre-fifteenth-century occupation. Folded rims and rim strips are absent and complicated stamping predominates. However, two characteristics of the collection contradict this simple interpretation. First, the primary decoration consists of nodes. This would be unusual for Savannah period assemblages, which are dominated by nonstrip hollow reed punctations, punctated lips, pellets, and rosettes, and only rarely nodes. Just two of the seven nodes were recovered from a single test unit, which indicates that the trait is widely distributed at the site and not the result of a chance recovery from just one or two pots. The second contradiction is the relatively high incidence of incising, a trait that begins to emerge only after AD 1450 in Eastern SAM assemblages and occurs in association with substantial quantities of rim strips and shoulder ticks. Adams and colleagues (2006:213) also report a strong representation of incising within their collection from feature excavation, where 5 of the 27 Pee Dee or Ashley series sherds exhibited incising.

Although nodes are not abundant in Contact period North Carolina assemblages, they generally occur as the only kind of appliqué decoration other than rim strips. Pellets and rosettes are characteristically absent (see Davis 1987; Eastman 1999; Ward and Davis 1988, 1993). In addition, typical nodes from Contact period sites were constructed differently than those from Eastern SAM assemblages. SAM-associated nodes are finely smoothed balls of clay that are often punctated and attached to the wall of a vessel by riveting, while the Contact period forms in North Carolina are lumps of clay that are smoothed into the vessel wall. Information on the formal characteristics of the nodes in the Coastal Carolina collection was not recorded on any of the analysis forms, but the occurrence of so many nodes in the absence of other appliqué decorations strongly suggests a connection to Contact period North Carolina.

The details of the incised designs may also point to a North Carolina Contact period association. Incised motifs on two of these sherds are described in the Coastal Carolina analysis forms (see Table 5.3). One plain sherd classified as Mississippian displayed a "zigzag multi-line strand with parallel filler," while an unclassified rim sherd exhibited "zigzag incising" over fabric impressing. Although similar design elements do occur in Late Lamar contexts (Hally 1994a: 153), the most common motifs of Lamar Incised consist of curvilinear scrolls, ovals with brackets, loops, and pendant festoons. Incised zigzag motifs are much more abundant in North Carolina Hill Tribe sequences, where curvilinear designs are rare. Two types consistently occur, true zigzags pendant to the rim with closed peaks and "inverted v's," the latter of which consist of multiple parallel straight lines that converge toward a peak but remain unconnected (Wilson 1983:235–236). The true zigzag can be filled with parallel lines in a manner similar to that described for the plain sherd from the Coastal Carolina analysis, or it can occur without further elaboration. This family of designs has a wide temporal distribution, occurring on late prehistoric Dan River through Contact period assemblages (Wilson 1983:251, 327, 388–390, 427, 457, 469–470). In the Contact period, incised decoration on the body of vessels is restricted to plain and burnished plain types (Davis 1988:41; Ward and Davis 1993:130, 353), while in the earlier Uwharrie and Dan River series incising is commonly applied over net impressed exteriors (Eastman 1999:51–52). Fabric impressing is not common in any of these phases.

Incising over a fabric-impressed exterior may reflect interaction with the Cashie and/or Collington culture areas farther to the east where this combination commonly occurs (Heath and Swindell 2011:10–45, 2013:footnote 2). Similar variants of incised fabric impressed are commonly found in the Chickahominy and Townsend series of the Virginia Coastal Plain (Egloff and Porter 1982:107–109). Temper in the fabric-impressed sherd from Yauhannah Bluff was described as medium sand, which would contrast with Cashie (coarse sand and pebbles) and the various shell-tempered series mentioned. Clearly, then, the sherd could be of local manufacture. Its chronological placement, however, is unclear as the temporal distribution of this combination of attributes is not well defined.

The Native American occupation at Yauhannah Bluff, then, appears to consist of a pre-fifteenth-century Mississippian component and a probable Contact period component that may have preceded the establishment of the trad-

TABLE 5.3. FREQUENCIES OF LATE PREHISTORIC CERAMIC TYPES FROM YAUHANNAH BLUFF

Season/Stage	Complicated Stamped, Curving Lines	Complicated Stamped, Linear	Complicated Stamped, Ladder	Complicated Stamped, Concentric Circles	Complicated Stamped, Curvilinear	Complicated Stamped, Rectilinear	Incised[a]
Weeks Survey					1	3	
1992	23	3		5			1
1993		6		1			
1994		4		2			
Undated Blocks	2	3	1	6	3	2	
Totals	25	16	1	14	4	5	1

[a]Zigzag multiple strand w/parallel filler.
[b]Incised notches below lip or notching on lip.

ing post. A more detailed analysis of the paste composition of the pottery might allow us to differentiate successfully which elements belong to each component. The fact that finer paste examples of simple stamping inferred to represent Late Woodland types were distinguished from Mississippian specimens in the Coastal Carolina analysis suggests that the Mississippian sherds contained a coarser paste consistent with the Pee Dee series. Neither Coastal Carolina nor New South Associates (Adams et al. 2006:222) specifically identified Kimbel series ceramics in their collections. Decorations identified in the collections include two hollow reed punctated sherds, a "Camden" rim (which I take to be either a notched or simple stamped lip), and an incised rim (which I would interpret as either ticking below the lip or incised lip notching). These modes are common to both prehistoric Eastern SAM and Contact period assemblages from North Carolina and as such are not diagnostic of either. The Contact period occupation may very well be associated with the Waccamaw Indians, but this is a supposition yet to be demonstrated. What can be concluded from this review is that sixteenth-century Eastern SAM occupation as evidenced by the occurrence of folded rims has not been identified at Yauhannah Bluff.

Table 5.3. *continued*

Simple Stamped, V-Shaped Lines	Indeterminate	Very Thin	Nodes	Incised Rim[b]	Camden Lip[c]	Late Woodland Simple Stamped	Fabric Impressed/ Incised[d]	Unidentified Incised	Pipe Stem	Pipe Bowl
					1	1		1		
12	6	2	3	1		11	1		1	
4			3			3			1	1
1			1			1				1
3	2									
20	8	2	7	1	1	16	1	1	2	2

[c]Notched or simple stamped lip.
[d]Zigzag incising over fabric impressed.

Pedee Town

A final project in the region of sufficient scope to allow an evaluation of occupation history was conducted by Steen and colleagues (1998) on the extensive Pee Dee Town tract. This tract encompasses a bluff along the middle course of the Great Pee Dee River that measures about 3 km in length and 0.3 to 0.5 km in width, corresponding to the presumed location of the early eighteenth-century Pedee settlement as mapped by Barnwell and others. The tract was characterized by extensive surface exposures in the form of plowed fields and timbering trails, and a field strategy incorporating shovel testing and surface collection at 30 m intervals was deployed to define 28 artifact concentrations of varying sizes and component associations. Remains of historic Pedeetown should be expected to occur at this location because the topographic details supplied by Barnwell and, later, DeBrahm seem to perfectly match the distinctive features of the bluff. However, no clear evidence of a ceramic assemblage that could be reasonably associated with the Contact period was recognized. The ceramics classified as "Mississippian-Late Wares" by the authors include not only complicated stamped but also fabric impressed, cord marked, and check stamped. None of these types, including the complicated stamped sherds, exhibited any kind of appliqué decoration. Given what is now known about the decorative patterns of the Eastern SAM seriation, the absence of appliqué elements (i.e., rosettes, pellets, riveted nodes, rim strips, and folded rims) strongly suggests that the overwhelming source of the late prehistoric occupations on the tract corresponds to the Early Mississippian period, prior to AD 1375.

Whether the noncomplicated stamped surface treatments, other than cob marked, are associated with the Mississippian/Contact period occupation of the bluff or, alternatively, with Middle Woodland components is still very much an open question. As Steen and colleagues (1998:74) note, coarse sand- or grit-tempered ceramics with cord marked, fabric impressed, and check stamped patterns are commonly associated with Middle Woodland occupations in central South Carolina (see Anderson et al. 1982; Cable 2002a). The Late Contact period Oldtown assemblages from William Klutz and Upper Saratown, which are the likely sources of the migrants who settled Cheraw Town (not to be cofused with Charraw Town at Nations Ford on the Catawba Reservation) near present-day Cheraw, South Carolina, are dominated by plain and burnished pottery, do not contain fabric impressed ceramics, and exhibit very little if any cord marked (see Table 5.2). The same is true of the Caraway collection from the Poole site, which is generally assumed to represent the town of Keyauwee on Caraway Creek. The inhabitants of this town first "incorporated" with the Sara at Cheraw Town in the very early eighteenth century and then resettled at Pedeetown after about 1716 (Wilson 1983:195). Check stamped and net impressed exteriors are substantially represented in all three of these assemblages and these surface finishes might provide a basis for identifying the location of Pedeetown on the tract. In addition, a more detailed paste analysis of the collection would allow differentiation of Mississippian and Contact period plain and plain burnished wares, as the latter should be characterized by much finer pastes.

The terminal Late Contact period (1750–1759) assemblages from Nassaw-Weyapee and Charraw Town at Nation Ford on the Catawba Reservation have greater representations of cord marked exterior surfaces (Fitts 2006:31, Table 4), but they are also characterized by significant occurrences of folded rims (Davis et al. 2015:35–39), an association that is not demonstrated in the Pee Dee Town tract where it is believed, but not yet confirmed, that the eighteenth-century settlement of Peedeetown is located. There has been a tendency to historically link the residents of Cheraw and Charraw towns because of the obvious phonetic similarity of the names (cf. Beck 2013:246). However, it would be difficult to draw a direct link between the Charraw Town assemblage from Nation Ford and the Late Contact period Sara assemblages on Dan River unless Sara ceramics underwent extremely abrupt changes in the early eighteenth century once the Sara settled at Peedeetown (see Riggs 2010:35).

It should be appreciated that folded rims drop completely out of the Oldtown series assemblages in the Dan River sequence by the Middle Contact period and are only rarely represented in the Early Contact period. Neither the high incidence of cord marked sherds nor the general abundance of folded rims on a variety of exterior surface types in the Charraw Town assemblage are consistent with the expected descendants of the Late Contact period William Klutz and Upper Saratown sites. It is likely that the close interaction between the Sara and Keyauwee once they were incorporated on the Pee Dee River would have resulted in a convergence of Oldtown and Caraway ceramics, but it is unlikely that such a blending process would have resulted in the kind of assemblage described for Charraw Town, which is dominated by cord marked pottery.

Reprise

It can be concluded from this discussion that no firm evidence of sixteenth-century Middle Lamar occupation has been demonstrated at Wachesaw Landing, Yauhannah Bluff, or the Pedee Town tract. All three locations contain Mississippian occupation dating to the fourteenth and fifteenth centuries, and a Late Contact period component that could be associated with the Winyah Indians is present at Wachesaw Landing. The configuration of ceramic traits at Yauhannah Bluff strongly suggests that there is a Late Contact period presence at this site as well but, based on the low density of potentially associated ceramics, it is very limited in extent. Historic accounts indicate that the Late Contact period Pedeetown is located somewhere on the Pedee Town tract, but the ceramic description from the survey is not sufficient to determine if it is represented in the artifact collections.

The absence of folded rims also suggests that the town of Vehidi mentioned in the Pardo accounts was either not located on this tract or was of very limited extent. Based on this finding, it is further concluded that Vehidi and Pedeetown were not historically linked. At present, the more likely scenario would stipulate that Pedeetown represents the eighteenth-century relocation of the Keyuawee from a location farther upstream near Cheraw Town. These lines of evidence by no means establish that the region was abandoned in the sixteenth century, but they do supply strong support for the contention that the lower Pee Dee valley experienced significant population decline and settlement reorganization during this period.

Cape Fear

The Cape Fear River valley forms the northeastern boundary of the GDO. This region contains a dense and widely distributed population of archaeological sites associated with the Middle and Late Woodland periods. In contrast to the Eastern SAM culture area, grog-tempered Hanover, crushed rock-tempered Yadkin, and sand-tempered Cape Fear ceramics are characteristic of this region, and these series are dominated by fabric impressed and, secondarily, cord marked exterior surface treatments (Herbert 2009:164–176; see also Cable 2011a: 580–593). Overwhelmingly, these series yield rims and body sherds without any decorative elaboration. Joe Herbert (personal communication, July 22, 2014) observes that of the nearly 37,000 potsherds collected from the projects on Fort Bragg, only 47 exhibit any decoration, and, of these, 17 are punctated and 30 are incised. Folded rims and rim strips are absent. Less than half (n = 18) of the decorations are associated with the Middle and Late Woodland series in the region; the remaining examples occur on unclassified or earlier types. This pattern has convinced Herbert that the area encompassed by Fort Bragg was virtually abandoned in the latter part of the fifteenth century (see Gunn 2002:25).

A low incidence of decoration, and in particular the absence of rim strips or folded rims, is a pattern that is manifest in Late Woodland assemblages along the entire course of the Cape Fear River down to the Atlantic Ocean. Herbert's (2003:75) reanalysis of the grog- and clay-tempered ceramics from the Haag (1958) and Loftfield (1976) survey collections yielded only five sherds with decoration (i.e., punctations) out of a sample of 3,643. Haag (1958:66–68) found decoration to be rare on all of his paste series, including sand- and shell-tempered types. He identified only 15 incised and 3 punctated sherds in a sample universe of 7,381 sherds, and nearly all of these examples were applied over fabric impressed or cord marked exteriors. This is a trait most commonly found in Late Woodland assemblages dating no later than the fifteenth century across North Carolina. Excavations at a number of sites (i.e., Papanow, Pond Trail, and Riegelwood) in the middle and lower reaches of the Cape Fear have produced a similar dearth of decoration (Herbert 2009:67–96). Finally, the latest calibrated radiocarbon age means on charcoal from Hanover contexts fall in the early to middle fifteenth-century range (Herbert 2011:Table 4.1; see also Gray 2012:293–303 and Millis 2011:6-7).

A distinctive feature of the Cape Fear drainage is the sand burial mound complex (see MacCord 1966). Irwin and colleagues (1999:Figure 1) illustrate the locations of 19 of these features, which are primarily concentrated in the Sand Hills, but their distribution extends into the southern North Carolina Coastal Plain in smaller numbers. As a result of an archival analysis of the McLean Mound ceramic assemblage (Herbert 1999:45, 2003:134–145), along with modern calibrations of the radiocarbon dates that MacCord (1966) obtained during his excavations in the mid-1960s, Irwin and colleagues (1999:63–71) concluded that this mound was built and used sometime after AD 800 to 1000. My separate evaluation of the radiocarbon data indicates that the most likely use of the mound occurred in the twelfth and thirteenth centuries (Cable

2011a:589). The ceramic assemblage is clearly dominated by late Hanover ceramics (Hanover II), but a sand-tempered series (Cape Fear) that also appears to be associated with the life span of the mound makes up about 22% of the assemblage. Predominant exterior surface treatment types consisted of fabric impressed (54.6%) and fabric paddle edge stamped (31.7%), while minor types included smoothed (9.3%), smoothed over stamped (1.6%), net impressed (1.6%), and burnished (1.1%). Significantly, decoration was a common feature of the assemblage, including rows of reed punctations incidental to the lip on exteriors of three Hanover Smoothed bowls and multiple rows of parallel, horizontal incised lines below the rim on one Hanover Paddle Edge Stamped bowl. All of these decorative techniques are consistent with the contemporaneous and neighboring Dan River, Haw River, and Uwharrie series of the late prehistoric period in North Carolina.

Based on their review of available evidence, Irwin and colleagues (1999:79–80) place the sand burial mound complex within the Late Woodland period and speculate that the tradition may have extended into the latest indigenous occupation of the South Coastal Zone of North Carolina. Exactly when the use of sand burial mounds ended is not well established, but it would seem likely that most of it did not extend past the fifteenth century, based on the decorative mode patterns discussed above. The Buie Mound assemblage (Wetmore 1978) constitutes an exception to this general model. Atypical for the region, this ceramic assemblage is dominated by sand-tempered burnished plain and incised types that suggest a sixteenth- or seventeenth-century age. The mound is situated in a poorly drained area of Robeson County along a tributary of Lumber River on the outer reaches of the Great Pee Dee drainage basin rather than along the Cape Fear River. This location was part of a broad and porous cultural boundary zone between the Eastern SAM culture area and the Woodland tradition of the South Coastal Zone of North Carolina, which may very well explain the apparent disconnect between the Buie Mound and the larger sample of sand burial mounds that Irwin and colleagues (1999) discuss. Before the implications of the Buie Mound can be effectively addressed, however, it is necessary to understand more fully the cultural development of this boundary zone. What is known of this area is discussed below.

North Coastal Zone of South Carolina

During the Middle Woodland period, the ceramic sequence in this region borrows elements from both North Carolina and the Central South Carolina coast (Cable 1998). In the earliest phases, sand tempering dominates and, in respective order of popularity, cord marked, fabric impressed, and check stamped comprise the major exterior surface treatments. As a group, this pottery is identified as the

Cape Fear series. Sometime around AD 1, and probably much earlier (see Cable 2002b, Cable 2011a:584–585), the grog-tempered Hanover series emerges and sustains a prolonged period of temporal overlap with the Cape Fear/Deptford types. The same range of exterior surface types is extant in the Hanover series, but fabric impressed ascends to numerical dominance, while check stamping declines. Hanover gradually replaces Cape Fear as the dominant series after AD 500, a change trajectory similar to that of the South Carolina central coast (Anderson 1982; Cable 2002a) and the South Coastal Zone of North Carolina (Herbert 2009). Check stamping is virtually nonexistent in the Haag (1958) and Loftfield (1976) collections (Herbert 2003:63–75), but it is comparatively much more abundant in Middle Woodland assemblages at Fort Bragg. Check stamping, as an important feature of Fort Bragg ceramic assemblages, does not occur north of Little River, a tributary of Cape Fear; it is, however, comparatively abundant in collections south of the river (Cable 2011b). This suggests that during the Middle Woodland period the cultural boundary between these two coastal zones ran across the Sandhills in the area of Fort Bragg and then maintained a southeastwardly course along the tributaries of the Lumber and Little Pee Dee drainage catchments to the ocean.

Sometime around AD 900, at the beginning of the Late Woodland period, a major shift in interaction occurred between the North Coastal Zone of South Carolina and the emerging Eastern SAM polities to the south. On the Central South Carolina coast and the upper Santee and the lower Pee Dee rivers, grog-tempered and sand-tempered Middle Woodland assemblages give way to medium-to-fine sand-tempered assemblages dominated by simple stamping. A similar Late Woodland simple stamped stratum has not yet been identified in the North Coastal Zone of South Carolina. Instead, a number of sites in the region have yielded Hanover and Cape Fear assemblages with associated radiocarbon dates that extend into at least the twelfth and perhaps thirteenth or fourteenth centuries. Southerlin and colleagues (2005:53–55) report three dates on charcoal from features at 38HR243 and 38HR254 on Little River Neck that are associated with Hanover ceramics. The means range between AD 1045 and 1340. These results would be out of line with Eastern SAM sequences to the south and west but are consistent with the prolonged survival of Hanover in the South Coastal Zone of North Carolina. Added credence to the persistence of Hanover in the region is provided by the apparent lack of a Late Woodland simple stamped horizon.

Either due to acculturation or territorial expansion through migration, Eastern SAM occupation is present in the region during the fourteenth century. Southerlin and colleagues (1997:15–17) summarize 14 projects in Horry County conducted through the mid-1990s that recorded Mississippian components, as identified by the presence of complicated stamped pottery. Because most of these components are described only by cursory survey level data or chance recovery of disparate evidence not targeted by particular investigations, it is difficult to characterize the intensity of Mississippian settlement in the region or the larger record of its occupation history. Those projects producing a more definitive picture of Eastern SAM occupation in the region are discussed below.

38HR364

An exception to this general sparcity of detail is supplied by investigations at 38HR364 located east of Conway, South Carolina, on the Waccamaw River (Steen and Errante 1996). Large numbers of Mississippian ceramics were recovered from the site, which, as a group, suggest a mid-to-late fourteenth-century habitation similar in structure to, but somewhat earlier than, the Wachesaw Landing site discussed earlier. Complicated stamped (n = 153) comprised the main element of the ceramics identified as Mississippian, but plain sherds affiliated with the occupation went undifferentiated. It is also likely that contemporaneous simple stamped sherds were associated with the occupation, but the magnitude of this association in Georgetown and Horry counties was not fully appreciated at that time. Decorations consisted of punctations (n = 7), noded rims (n = 6), and appliqué rim strips (n = 5). This distribution would fit somewhere between the Wateree-Town Creek and Fogarty Creek assemblages in the Eastern SAM seriation model (see Table 2.10), which is consistent with occupation in the last half of the fourteenth century.

Tidewater

Excavation at the Tidewater site (Southerlin et al. 1997), situated on Little River Neck near the North Carolina state line, provides a glimpse into Eastern SAM-derived occupation that postdates the late fourteenth- to middle fifteenth-century component at the Wachesaw Landing site. Field operations consisted of a staged deployment of (1) 52, 0.5-×-0.5 m test units spaced at 10-m intervals across a maximum area measuring 70-×-70 m; (2) judgmental placement of 15, 2-×-2 m test units within this grid; and (3) mechanical stripping at seven locations, each measuring 2-×-6 m. Five cultural features were identified, including two shell-filled pits (Features 601 and 602), two pits with diffuse shell (Features 612 and 613), and a cob-filled smudge pit (Feature 603).

In total, 5,464 sherds and ceramic artifacts were recovered during excavations, of which 4,477 were classified as residual and another 178 were described as either eroded or displaying unidentifiable decoration. Exterior surface treatment and temper were recorded for the remainder of the collection, but only 229 sherds representing portions of 93 vessels were assigned culture historic

Table 5.4. Ceramic Type Percentages and Decorative Mode Frequencies from Excavated Contexts at the Tidewater Site

Contexts	Plain	Scraped	Roughened	Burnished Plain	Simple Stamped	Textile Wrapped	Complicated Stamped
Vessels	15.9	3.2	4.8	1.6	34.9	3.2	3.2
Shovel Tests	62.2			0.8	28.4		0.8
Locus 1 Test Units	64.0				30.0		
Locus 2 Test Units	50.5			1.6	32.5		11.7
Locus 3 Test Units	62.2[2]			0.2	26.3		0.6
Locus 4 Test Units	75.5			2.1	16.0		
F-601 Locus 1	100.0						
F-602 Locus 2							100.0
F-603 Locus 3							
F-612 Locus 3	42.9[a]				53.6		
F-613 Locus 3					100.0		

Note: Decorative mode data are reported in frequencies.
[a]Includes scraped and roughened plain specimens.
F = feature.

identifications. All vessels had at least one rim sherd in the aggregated sample. Seventy of these vessels were assigned to either the Pee Dee series or indeterminate Mississippian (Southerlin et al. 1997:49–51, 55–70). Table 5.4 displays the relative proportions of exterior surface treatments from the Tidewater vessel sample. If it were not for the fifteenth-century assemblage from Wachesaw Landing, it would be very difficult to fit the components of this sample into a framework of Eastern SAM ceramic development, as only 3.2% of the vessels have complicated stamped exteriors. The strongest correlation with Wachesaw Landing is the large contingent of simple stamping, which is manifest in no other region of the Eastern SAM culture area. Perhaps the most remarkable feature of the sample is the unprecedented representation of incising, which is present on fully a third of the vessels. However, since incising occurs on the rim,

Table 5.4. *continued*

Incised	Incised/Punctated	Check Stamped	Total Sherd Count	Punctations	Segmented Rim Strip	Cockle Impressed	Nodes
28.6	4.8		63	5	2	1	
7.9			127	3			
6.0			150	5			1
2.7	1.1		188	1	2		
10.8		0.2	502	2			1
6.4			94				1
			3				
			14				
3.6			28				
			3				

neck, and shoulder of otherwise plain or burnished pots, this percentage clearly overestimates the incidence of incising that would normally occur in a generalized sherd assemblage.

To transform the ceramic data into a more comparable data set, it is necessary to generate type proportions from the larger collection. Since most of this collection was not classified by culture-historic type, however, only an approximation of such patterns can be cobbled together. Sherd percentages in the remainder of Table 5.4 were generated by segregating from the artifact appendix all of the sand-tempered sherds with exterior surface treatments conforming to the Pee Dee/Mississippian vessel sample. This procedure has undoubtedly resulted in the incorporation of some earlier sand-tempered types, but the overall impact is negligible, resulting in only a slight overestimate of plain sherd representation

Table 5.5. Radiocarbon Dates from the Tidewater Site

Sample No.[a]	Feature	Ceramic Association	Material	Conventional ^{14}C Age B.P. (±1σ)
85971	601	Hanover Fabric Impressed and Cord Marked	Shell	1020 ± 80
85972	601	Hanover Fabric Impressed and Cord Marked	Charred Material	960 ± 130
85973	602	Pee Dee Complicated Stamped	Charred Material	660 ± 60
85974	602	Pee Dee Complicated Stamped	Shell	810 ± 60
85975	603	None	Charred Maize	310 ± 60

Note: Calibrations performed using OxCal 4.2.3 Program (Bonk Ramsey 1995, 2001), IntCal13 Atmospheric and Marine 13 Curves (Reimer et al. 2013).

brought about by including small numbers of Thoms Creek and Mount Pleasant Plain. The data are presented as a series of different samples, including the dispersed shovel tests, the combined test unit excavations from the four different loci defined by the researchers (Southerlin et al. 1997:32–36), and feature excavations.

The shovel test sample constitutes the most unbiased estimate of the representation of Pee Dee/Mississippian types. It reflects an assemblage dominated by plain and, secondarily, simple stamped types. An occurrence of 7.9% incised sherds is comparable to percentages characteristic of Middle and Late Lamar assemblages across the Eastern SAM culture area. Combined test unit data from the four loci indicate that this particular assemblage occurs consistently across the site, which suggests that the artifact universe of the site was deposited in a relatively limited span of time. Decoration is rare and consists of punctations (n = 11), nodes (n = 3), and segmented rim strips (n = 2). A single check stamped sherd was included in the Pee Dee/Mississippian assemblage because it exhibited punctated decoration. A cockleshell-impressed sherd from the vessel sample is more likely associated with the Thoms Creek series.

Based on the high percentages of incising and evidence from radiocarbon dating, Southerlin and colleagues (1997:86, 91) argued that the Pee Dee/Mississippian occupation at the Tidewater site most likely occurred in the sixteenth century, "although possibly a little earlier and/or a little later." A problem

Table 5.5. *continued*

Calibrated Age AD (±1σ)	1σ Range (cal AD)	Area under Probability Curve (1σ)	2σ Range (cal AD)	Area under Probability Curve (2σ)
1351 ± 61[b]	1298–1416	.682	1226–1466	.954
1060 ± 129	970–1220	.682	777–794 800–1277	.012 .942
1333 ± 44	1279–1320 1350–1391	.341 .341	1261–1411	.954
1518 ± 36[b]	1446–1560	.682	1425–1644	.954
1577 ± 82	1495–1601 1616–1647	.527 .155	1446–1669 1781–1798	.929 .025

[a]Beta-Analytic lab numbers.
[b] Reservoir effect unknown and proxy not applied.

with this interpretation, however, is that a sixteenth-century occupation would be expected to contain substantial contributions of folded rims, which this collection does not have. Folded rims are only rarely found on incised and plain vessels, but they should be abundantly represented in the decorative constellation of simple stamped and other stamped and impressed exterior surface types from this period. Folded rims make up about 17% of the Burke series rim assemblage from the Berry site (Moore 2002:76), for instance. This holds true even for the protohistoric Piedmont Siouan sites of the Haw and Eno River valleys. At the George Rogers site, folded rims occur respectively on 14.6 and 21.4% of Hillsboro Simple Stamped and Check Stamped (Ward and Davis 1993:97–98). Similarly, the rim assemblages of Hillsboro Net Impressed, Simple Stamped, and Check Stamped from the Wall site exhibit incidences of folded rims ranging between 17.0 and 26.6% (Davis 1987:200). Twenty-two Pee Dee Simple Stamped vessels were identified in the Tidewater collection, each represented by at least one rim sherd. At least three or four of these should have displayed folded rims if the collection was part of a sixteenth-century occupation, yet none did.

Table 5.5 presents the radiocarbon dates from the Tidewater site along with modern calibrations. Based on the apparent uniformity of the ceramic assemblage, Southerlin and colleagues (1997:77) concluded that all five of these dates, which exhibit widely overlapping probability distributions, fall within the range of the Pee Dee/Mississippian occupation. Further, they argued that the span of the occupation should probably be narrowed to the mid-to-late sixteenth century.

We know today, however, that radiocarbon dates on marine shell cannot be used interchangeably with charcoal dates as was done here, unless an additional correction is made to adjust for what is known as the "reservoir effect" (see Thomas 2008). Because carbon resides much longer in the ocean than in the atmosphere, it is necessary to adjust for isotopic fractionation to compare the ages of terrestrial and marine radiocarbon samples. Fractionation can be directly measured by calculating the $^{13}C/^{12}C$ ratio for an individual sample. Generally, fractionation on marine samples results in pushing the radiocarbon age back in time about 400 years, which approaches the average age of carbon in ocean water at the surface. This disparity is registered as the difference between the measured and conventional radiocarbon age. The radiocarbon dates on the marine samples from the Tidewater site were corrected for fractionation, as is shown in Table 5.5. However, the reservoir of oceanic carbon is variable from one location to the next due to localized patterns of upwelling. Differences in circulation, fluctuation in atmospheric carbon exchange at the ocean surface due to climatic shifts through time, and localized differences in coastal geomorphology and estuarine composition bring about this variation. Most often, this localized variation is so great that it requires an additional adjustment that can push the age of a calibrated sample as much or more than 200 years older or younger beyond fractionation-corrected conventional ages. This adjustment is what is referred to as the "reservoir effect." Unfortunately, this effect cannot be directly measured through isotopic analysis. Instead, its calculation needs to be empirically derived from either paired terrestrial and marine samples from the same context or by comparisons of radiocarbon ages from modern era shell samples whose pre-bomb (the era before nuclear testing commenced in the 1940s) year of death is known.

Southerlin and colleagues (1997:77) comment that when charcoal and marine shell dates were paired from the same feature—as was the circumstance with four of the five dates from the Tidewater site—the marine date was consistently younger than the charcoal date. This same phenomenon was observed on paired samples from features at two other sites on the Little River Neck (Southerlin et al. 2005:55). In general, the calibrated marine shell dates range between 150 and 300 years younger than the paired charcoal dates, a disparity that highlights the negative magnitude of the reservoir effect in the area. Until such a correction can be rigorously generated, and at least a start to this endeavor is provided by the paired samples from the Little River Neck mitigations, marine shell assays from the region cannot be used to accurately date archaeological deposits.

Most importantly, using them interchangeably with terrestrial samples will produce an incoherent chronology. When only the terrestrial dates from the Tidewater sample are considered, statistical comparisons strongly suggest that three different occupations are represented. Feature 601, which is associated with Hanover ceramics, is statistically different ($t = 4.390244, \chi^2_{.05} = 3.84, df = 1$) from Feature 602, which contained 14 sherds from a Pee Dee Complicated Stamped jar. In turn, Feature 602 is statistically different ($t = 17.01389, \chi^2_{.05} = 3.84, df = 1$) from Feature 603, a cob-filled smudge pit that did not contain artifacts.

Feature 601 yielded a calibrated date in the mid-eleventh century, which is too old to be associated with the Eastern SAM time span in coastal South Carolina but should occur in association with Late Woodland, sand-tempered simple stamped assemblages if the change trajectory paralleled cultural development on the Central Coast. The dominance of Hanover Fabric Impressed and Cord Marked sherds in the feature, however, constitutes strong evidence for the persistence of the Hanover series into the Late Woodland interval on the North Coast and for the existence of a cultural boundary. Sherds with identifiable surface treatments from the pit consisted of four grog-tempered fabric impressed and cord marked (i.e., Hanover) and three unidentified, coarse sand-tempered plain. Given the age of the radiocarbon sample, it is likely that the sand-tempered sherds are affiliated with a pre-Mississippian series that was included in the fill through mixing.

The association of the partial Pee Dee Complicated Stamped vessel with a radiocarbon mean in the mid-fourteenth century strongly suggests that there are two late components at the Tidewater site, not just one. Nearly all of the complicated stamped ceramics on the site were recovered from the test units in Locus 2 where the complicated stamped vessel was recovered (see Table 5.4). Adjacent TUs 402, 407, and 409, as well as Feature 602 in TU 407 from this locus produced 90% (36 of 40) of the complicated stamped sherds recovered from the entire project. The two segmented rim strips came from this same excavation locus (TUs 407 and 415). Although exterior surface treatments on these sherds could not be recognized, it is likely that they too are associated with this mid-fourteenth-century occupation, given their close spatial association with the concentration of complicated stamped ceramics. Rim strips begin to appear in the last half of the fourteenth century in the Central Coast sequence, and the later one-sigma option for the radiocarbon date from Feature 602 of AD 1350 to 1391 is consistent with this correlation (see Table 5.5). Based on the restricted distribution of complicated stamped at the site, it can be estimated that this component covered an area of only 40 to 50 m in diameter, and the low density of associated ceramics would further suggest that it was of short duration, consistent with the expected correlates of a farmstead or other specialized occupation type.

When the contribution of complicated stamped is subtracted from the other Pee Dee/Mississippian types (i.e., plain, incised, and simple stamped), the remaining assemblage fits only marginally within the general trajectory of Eastern SAM ceramic development. As was the case for the Kimbel series from Wachesaw Landing, the closest known correlate for what will be called henceforth the Tidewater phase is found at the Mitchum site on the Haw River (see Table 5.2). Decorations that are shared include stick and reed punctations and nodes. The Mitchum and Tidewater assemblages are also similar in the near absence of folded rims. Only one folded rim was identified in the Contact period assemblage from the Mitchum site, out of a sample of 242 rims.

Without more information, it is difficult to seriate the Tidewater assemblage and arrive at a precise date range for the phase. The temper of most of the assemblage is recorded as coarse-to-very coarse sand, but small numbers of medium-to-fine sand specimens are also noted. Although other aspects of the paste are not revealed, the dominance of coarse sand would suggest a closer affinity to the Pee Dee series than to the finer-pasted Kimbel series. Jenrette ceramics from the Mitchum site are tempered with a variety of clastics, including, most predominantly, coarse-to-fine crushed feldspar, but also sand, mixed crushed quartz and feldspar, medium-to-fine crushed quartz, and grit (Ward and Davis 1993:130–132, 349–354). Overall, then, the Jenrette series is characterized by coarser paste than that described for the Kimbel series, which may constitute an attribute state distinguishing some seventeenth- from some eighteenth-century Contact period assemblages.

Based on the paucity of European trade items, Ward and Davis (1993:413–414) argue that the Mitchum site was occupied after AD 1650. The recovered artifacts included seven gunflints, a single frizzen spring fragment, brass bells, rolled brass or copper beads, and white and blue small glass beads. Failure to recover lead shot and iron tools suggested to the authors that the occupants of the site, who are believed to have been the Sissipahaw, were only marginally situated within the European trade system. No historic trade items were recovered from the excavations at the Tidewater site, but this is not sufficient to argue that the Tidewater occupation is pre-Contact in age. The Tidewater site would have been even more isolated from European contact than the Mitchum site, which was located near the Trading Path. Significant trading opportunities with Europeans may not have occurred in the North Coastal Zone until after the establishment of Charleston in the 1670s. The radiocarbon date from Feature 603 is likely associated with the Tidewater phase and yields a date mean positioned in the last quarter of the sixteenth century, while the two-sigma range extends to 1669 on the late end (see Table 5.5). Given the absence of folded rims and the very low occurrence of rim strips, which are more likely associated with the earlier Mississippian occupation at the site, it is probable that the Tidewater occupation occurred after AD 1575 and ended before 1670.

Another indication that the Tidewater phase is chronologically positioned in the first half of the seventeenth century is the relatively high incidence of clay pipes. Thirteen clay pipe fragments of apparent aboriginal manufacture were recovered from excavated contexts at the Tidewater site (Sutherlin et al. 1997:46). These included both bowl and stem fragments of unspecified form or style. Ward and Davis (1993:367) note that the incidence of clay pipes increases dramatically in Contact period Piedmont Siouan components when compared to prehistoric and protohistoric deposits and suggest that this change reflects a shift from restricted ceremonial use of tobacco to mundane, quotidian use in the seventeenth century. In keeping with this general trajectory, clay pipe fragments are only rarely recovered from prehistoric Mississippian sites along the South Carolina coast. Trinkley and colleagues (1983:36) identified only a single pipe stem of aboriginal manufacture at Wachesaw Landing, for instance. As a means of comparison, the ratio of pipe fragments to nonresidual sherds of Mississippian or Contact period affiliation at Wachesaw Landing is 0.0007, while at Tidewater the ratio (.0183) is 38 times greater.

Although the Tidewater assemblage manifests similarities with the Mitchum site, the two are far from identical. The similarities have been reviewed above, but the characteristic that most substantially sets Tidewater apart from Mitchum is its high incidence of incising. Incising on the neck and body of pots is rarely present in Jenrette series ceramics from the Mitchum site, where it occurs on only 0.3% of plain sherds (Ward and Davis 1993:130–132). By contrast, incising is found on 13.6% of plain and burnished plain sherds in the Tidewater assemblage, which is an underestimation because an unknown quantity of earlier types of plain and possibly simple stamped are included in the calculation. Moreover, the dominant decorative techniques characterizing the Mitchum phase are lip and rim edge notching on plain rims and stamped notching on simple stamped vessels. Such techniques were not noted for the Tidewater assemblage.

These important differences indicate that the North Coastal Zone of South Carolina supported a regional subtradition that was somewhat separate from the cultural developments of the protohistoric and Contact period Siouan occupations on the Haw and Eno River valleys. Although there are clear connections between the Tidewater phase and the late fourteenth- and mid-fifteenth-century Fogarty Creek/McDowell-like occupation at Wachesaw Landing, there is a gap between them. A possible hiatus in this same general time frame is also evident in the occupation sequence at the Tidewater site. Precisely how this seeming break in continuity figured into the settlement dynamics of the region is unclear. Since the region appears to have supported its own distinctive and culturally continuous subtradition in Mississippian and Contact period times, it is likely that resident populations were present in the sixteenth century. The occupational disjunction at the site level, however, suggests that there were land-use shifts and settlement reorganization between these phases and that the sixteenth-century population density was comparatively low.

Buie Mound

It is only with the rearrangement of the sequences from Wachesaw Landing and the Tidewater site that a better understanding of the assemblage recovered from the Buie Mound can be achieved. The Buie Mound is located near the community of Red Spring in Robeson County, North Carolina. Archaeological investigations there spanned a period of four years beginning in 1971, with test square excavations conducted by the North Carolina Archaeological Society in collaboration with David McLean of St. Andrews Presbyterian College and J. Ned Woodall of Wake Forest University. Students at Pembroke and St. Andrews colleges and members of the North Carolina Archaeological Society continued excavations at the site between 1971 and 1974, which were variously directed by Bennie Keel, Jeffrey Gordon, and David McLean. Fieldwork proceeded in a coordinated manner, resulting in the exposure of a contiguous block comprising 55 squares of 5 feet on a side covering an area of 1,375 ft^2. Ruth Wetmore (1978) synthesized all of the investigations several years later and it is her efforts that serve as the basis for the following discussion.

According to local informants, the roughly circular mound had once stood 2.5 to 3 ft above the general ground surface, but it had been the subject of repeated pot hunting disturbances over the years, particularly in its central core. At the time of archaeological investigations, the feature was only vaguely recognizable as a mound, but, as Wetmore noted, the recovery of portions of seven human bone concentrations in the excavations establishes that it once functioned as a sand burial mound. The surviving outline of the mound consisted of an oval measuring 30×45 ft.

While about 12% (n = 65) of the sherds with recognizable exterior surfaces represented Middle and Early Woodland Hanover, Cape Fear, and Deptford types, the remaining collection (n = 486) was made up of a sand-tempered series of burnished plain and incised types (Wetmore 1978:44–48). Incising occurs on 5.6% of the burnished plain sherds from the excavation squares. Henceforth, this assemblage will be referred to as the Buie phase, which is unique but reminiscent of the Tidewater phase. The numerical dominance of burnished plain and incised ceramics, coupled with the large sizes of many of these sherds, indicates that the burial mound was used during the interval in which the Buie assemblage was being manufactured and deposited. A major contrast between it and the Tidewater assemblage is the absence of simple stamping. Two opposing explanations for this difference can be suggested. Either the late ceramics from the Buie Mound represent a specialized funerary assemblage that is contemporaneous with Tidewater, or it is emblematic of a separate phase. Although these alternatives are not necessarily exclusive, several characteristics of the Buie assemblage argue in favor of a later placement in time.

First, the temper appears to be finer than the coarse sand typical of most of the Tidewater assemblage. Wetmore (1978:44) indicated that the paste of the burnished plain and incised sherds from the Buie site was "generally clayey, sometimes gritty." Although Keel (1970:20) recovered only 60 sherds from his excavations at the site, he states that the predominant temper consisted of "small amounts of sand or crushed quartzite." Had the pottery resembled that of the Tidewater site, it is likely that Keel would have described the material he recovered as "sugary," since this was the standard term for the paste of Pee Dee ceramics from Town Creek. A shift toward finer pastes appears to be a change that characterizes the Piedmont Siouan sequences of North Carolina through the seventeenth century, and its presence in the Buie assemblage would be consistent with this trajectory. The Buie assemblage, then, would seriate later in time than the Tidewater phase, which would suggest similarities with the Kimbel series, the latter of which is a candidate for a turn-of-the-eighteenth-century ceramic phase on the lower Waccamaw.

Second, Wetmore (1978:Figure 7, A) illustrated the bowl and upper stem fragment of a burnished plain pipe that she described as an elbow pipe. However, the stem appears to intersect the bowl on the same axis of orientation, which would suggest that it is in fact a classic example of an onion pipe (see Coe 1952:Figure 166j; Ward and Davis 1993:Figure 9.49a–c). Although plain onion pipes have been found in protohistoric contexts at the Wall Site (Coe 1952:311), Ward and Davis (1993:367) indicate that most examples in Piedmont Siouan sites date to the later half of the seventeenth century. Many onion pipes found in Piedmont Siouan Contact period contexts were made with pastes described as terra cotta or white clay (Ward and Davis 1993:203–205), which are very fine in texture and may have been manufactured by both Native Americans and Europeans in early colonial times. Wetmore indicated that this particular pipe fragment was nontempered, which could suggest that it represents a terra cotta pipe. Since she did not note mold seams on the pipe, however, it is assumed that it was of aboriginal manufacture. A resident of Red Spring reports finding colonial pipe stems at the site in the 1920s or 1930s (Wetmore 1978:33). These could very well have been terra cotta, white clay and/or kaolin, as all of these occur in association with aboriginal pipes in Late Contact period contexts at the Fredericks site (Occaneechi) and Upper Saratown (Ward and Davis 1993:367).

Third, Wetmore (1978:Figure 7, C) illustrates a base sherd with a sharply flattened contour that appears to resemble the bottoms of bowls or cups from the Federalist period Catawban villages of Ayers Town (Davis et al. 2015:161) and New Town (Riggs et al. 2006:71). The base, perfectly circular in outline, is sharply perpendicular to the body orientation and the basal contour is slightly concave. A similarly shaped flat base is a common feature of the Courtland series of southeast Virginia, which Binford (1964:305) identified as "Colono-Indian." The Catawba assemblages are no doubt later than the Buie Mound, but, as Cobb and DePratter (2012) argue, various elements that can be considered "Colonoware" in its broader definition found their way into Native American ceramic traditions during the Contact period throughout eastern North America. It is their contention that Native American groups living in frontier zones where contact was restricted to remote trading outposts and forts would not have participated so intensively in acculturative European contact and their ceramic technologies would have incorporated few "Colonoware" elements into their traditional inventory of vessel forms. The presence of a "Colonoware" form in the Buie assemblage is another indication that the mound was used during a period of increasing European contact in a marginal location.

Absolute dates for the mound do not exist, but the information on temper, clay pipe associations, and the presence of what appears to be a "Colonoware" vessel all suggest a date no earlier than about AD 1650. The fact that Wetmore characterized the pottery as burnished rather than plain also supports a late seventeenth- or early eighteenth-century age. Most probably, the North Coastal region supported a multicultural mix of aboriginal groups that represented Winyah, Waccamaw, and the poorly known Cape Fear Indians. Certainly, if the Buie assemblage and the Kimbel series are contemporaneous, then there is enough of a difference in the ceramic composition to suggest separate phases.

The relationship between Tidewater and Buie could very well represent lineal cultural continuity. Although the latter does not contain simple stamped pottery, the assemblages exhibit broad similarities in incised motifs. Figure 5.2 displays line tracings of incised sherds with recognizable motifs from the Buie Mound and the Tidewater site. A motif common to both assemblages (Figure 5.2a, c, d) is a running festoon of concentric semicircles pendant to a series of parallel horizontal lines. It differs from the more common festoon treatments found in Lamar assemblages in which individual curvilinear and straight-line elements are fused and then nested in a running pattern (see Hally 1994a:Figure 14.5). Several designs associated with the Tugalo and Barnett phases incorporate isolated concentric circles in the center of festoons, but they are generally bracketed by fused elements along the bottom. Riggs (2010:Figure 2e) illustrates a festoon motif from the Hardins site on the South Fork of the Catawba River that com-

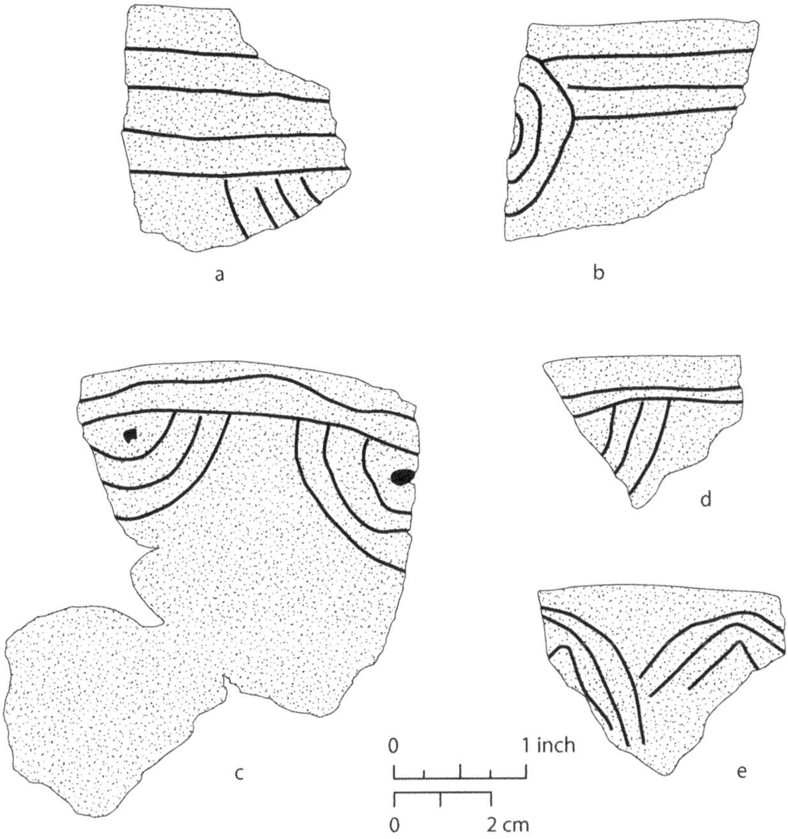

Figure 5.2. Lamar Incised sherds from the Buie mound and the Tidewater site: (a, c, d) running festoons with concentric semicircles; (b) running spiral/Figure 9 with disconnected parallel lines; (e) inverted "V" or running arc.

bines disconnected and fused elements within a single motif. The other motif from the Buie Mound is only partially revealed (see Figure 5.2b). The observable portion resembles either a running spiral or "figure nine" that is continuous to the left but not the right, or a spiral/concentric circle element separated by disconnected parallel horizontal lines. The former occurs in Georgia Lamar assemblages and in the Burke series of the Upper Catawba (see Moore 2002:Plate 15), but there it is most commonly accompanied by one or two fused elements along the bottom of the motif. The final motif (see Figure 5.2e), from the Tidewater assemblage, resembles the inverted V or running arc motifs that constitute a primary family of designs in the late prehistoric through Contact period

Piedmont Siouan phases (Eastman 1999; Wilson 1983; Ward and Davis 1993; Davis 1987:Figure 8.11, B). This particular application incorporates both arcs and inverted Vs.

These regional correlations suggest that both the Tidewater and Buie phases were simultaneously interacting with Piedmont Siouan cultures in North Carolina and Lamar horizon Eastern SAM cultures to the west. Neither phase exhibits folded rims, suggesting that they represent, in the main, seventeenth-century phenomena. The presence of simple stamping would suggest an Early Contact period association for the Tidewater phase (AD 1600–1650) and the various colonial associations of some of the ceramic elements in the Buie assemblage, along with an absence of simple stamping and increased burnishing, suggest that the Buie phase is of Late Contact period affiliation (AD 1650 to 1700). How they are related is an open question.

Concluding Thoughts

This review cannot demonstrate that the lower Pee Dee, Winyah Bay, the coastal margins of the Grand Strand, the swampy Coastal Plain of the Little Pee Dee and Lumber Rivers, or the Cape Fear Valley were completely abandoned in the sixteenth century. In fact, it is probable that pockets in these regions supported aboriginal populations. What seems clear, however, is that the sequence of all of these regions contains an apparent hiatus or decline during this period, sandwiched between extensive fourteenth- and fifteenth-century Mississippian polities and Late Woodland Hanover settlement systems, on the one hand, and relatively widespread seventeenth-century occupation on the other.

An argument could be raised that cultures in this region may have been so isolated from neighboring traditions that they simply never incorporated folded rims into their decorative repertoire. Hence, sixteenth-century occupation may be widespread but undetected due to the false assumption that this time horizon is always identifiable by the substantial presence of folded rims. A counter to this line of argument is that such barriers to interaction were not manifest in the incised motifs of the Tidewater and Buie assemblages, regardless of whether they date to the sixteenth or seventeenth centuries. Nor did seeming isolation bar the participation of the Wachesaw Landing site in the larger decorative mode patterns of the mid-fifteenth-century Eastern SAM culture area. Therefore, it would be unreasonable to expect that folded rims would somehow not have entered this particular area in substantial quantities if dense populations had been there.

Much of the identified or proposed seventeenth-century occupation in the region may represent migrations of groups moving southward and eastward in response to pressures brought about by the progress of European colonization and the selective supply of firearms to certain Native American groups to advance the slave trade (i.e., the Westo, Savannah, and Occaneechi). The details of the larger adaptive framework of the GDO are yet to be worked out, but this review clearly identifies the sixteenth century as a time of broad occupational hiatus and reorganization in these affected areas. Chapter 6 considers the story of what was happening in the regions surrounding the GDO.

6

Migration to the Ring of Drought Resilience

The study of the present surroundings is insufficient: the history of the people, the influence of the regions through which it passed on its migrations, and the people with whom it came into contact, must be considered.
—Franz Boas (1887, in Stocking [1974:64])

Thus far, I have been concerned with what ceramics can tell us about the occupation histories of regions that I suspect experienced major abandonment events in the fifteenth and sixteenth centuries in the Eastern SAM culture area. Focus on the actual processes of abandonment was necessarily light because we do not yet have sufficient detail or adequate theoretical resolution to make sense of the archaeological patterns that would represent the harbingers of this large-scale event. Unless you know *what* you are looking for, it can be difficult to make sense of what you are looking *at*. For now, I have set some broad parameters for how to go about studying the regions of the GDO in the future.

In this chapter, I describe the archaeological patterns that emerge in the regions surrounding the GDO, where the climatic effects of megadrought were less severe but in which demographic readjustment ushered in sweeping changes to the organizational structure of the surviving Eastern SAM polities. Unlike the subdued archaeological record of the GDO, these surrounding regions, which as a group I will call the Ring of Drought Resilience (RDR) because it rings the depopulated lands, display a rich history of occupation that more immediately yields insights into the societal reorganization that must have come about in response to the external forcing factors of extreme climate change and immigration.

The maps produced by Milner and colleagues (2001) of late prehistoric, protohistoric, and Contact period population aggregates in eastern North America provide a bird's-eye view of the consequences of depopulation in the context of two major cycles of MCA megadrought. These droughts effectively burned a hole through the center of the Eastern Woodlands, driving a wedge between the densely packed thirteenth- and fourteenth-century cultural systems that had earlier been connected in a vibrant network of interaction and exchange. The Northeast appears to have weathered the MCA without significant disruption or declining population, but the distribution of aggregates across the lower Southeast is significantly fragmented and reduced. That the Northeast was relatively untouched in the sixteenth century is understandable when we consider the geographic implications of the PDO/AMO scenarios described by McCabe and colleagues (2004, 2008). The Northeast was essentially immune to the drought-forcing consequences of positive AMO, whereas the populations in the RDR situated below the Vacant Quarter and the GDO survived even though they were likely stressed by extended drought conditions (see Figure 4.2a, c). Contrary to what might be expected, however, it appears that population actually increased in the regions of the RDR during the fifteenth century and I will show below that this demographic pattern most likely chronicles the arrival of waves of immigrants from the GDO. Such patterns would certainly be expected to occur if the regions that comprised the GDO were experiencing out-migration. Since the regions of the RDR were already occupied, though, these migrants would have needed to negotiate the problems of culture contact through conflict, extinction/replacement, or assimilation into pre-existing societies (Rouse 1986; Willey et al. 1955).

If assimilation were the dominant response, and this would appear to have been the case, then the emerging social formations would fit the model of "coalescent societies" described by Kowalewski (2006). Such societies tend to emerge as a response to external pressures or threats, such as regional warfare, population decline, migrations, or large-scale abandonments. According to Kowalewski, processes of integration under such conditions tend to be associated with the formation of multiethnic and multilingual groups, population aggregation into larger towns, the construction of fortified settlements, intensification of subsistence production, expansion of trade relationships, and the repurposing of social institutions that promote collectivism and regional confederacies at the expense of centralized hierarchical authority. In the Southeast, the concept of coalescence has been customarily applied to seventeenth- and eighteenth-century ethnohistoric groups that formed through the consolidation of refugees of diverse linguistic and ethnic heritages in response to the external pressures of European colonization (see Ethridge and Hudson 2002). It is generally believed that it was within this coalescent context that the Contact period confederacies emerged and replaced the centralized authority of chiefdoms in the Southeast. Given what can be inferred from the consequences of megadrought, however, it is now possible to suggest that the processes of coalescence were well underway in the fifteenth century, prior to European contact, and that they emerged initially in response to a different set of external pressures.

Figure 6.1 provides an imprecise but informative picture of the arrangement of the mid-fifteenth- through sixteenth-century settlement ringing the GDO. The boundaries and identities of the various polities that are illustrated were derived from a number of sources, including Davis and Ward (1991), Hally (1994a, b), Milner and colleagues (2001), Moore (2002), M. Smith (2000), Woodall (1999), and Worth (1994, 2003). When aggregates could be confidently linked to ones visited by Soto or Pardo, the mid-sixteenth-century ethnohistoric name was used in accordance with Hudson and colleagues (1984) and DePratter and colleagues (1983). The others are identified by their associated archaeological phase names, when available. As it stands, the map is not complete and the boundaries of known polities represent only rough approximations that deserve better definition. For instance, the boundaries of Ocute probably encompass the territories of multiple local polities, which may have been subjects of a paramount authority. In this regard, Hally (1994a:167–172) argues that as many as four separate polities occupied the Oconee Province during the fifteenth and sixteenth centuries. An accounting of the occupation histories of each of these polities is beyond the scope of my knowledge base, but several examples with which I am more familiar will suffice to establish a baseline discussion of the conditions that human populations in each of these regions must surely have confronted, as well as the tenor of their responses. These are the Yadkin and Wateree-Catawba valleys to the north, the still hazy picture of the northeast perimeter encompassing the upper stretches of the Neuse, lower Haw, and Northeast Cape Fear drainages and Oconee Valley on the southern side of the RDR.

Yadkin, Catawba, and Wateree Valleys

In both the Yadkin and Catawba valleys, there is evidence for northward migration of Eastern SAM populations. Robin Beck (2013:52–57) argues that a "massive demographic realignment of the Carolina Piedmont" occurred in the first half of the fifteenth century. In support of this observation he notes that only

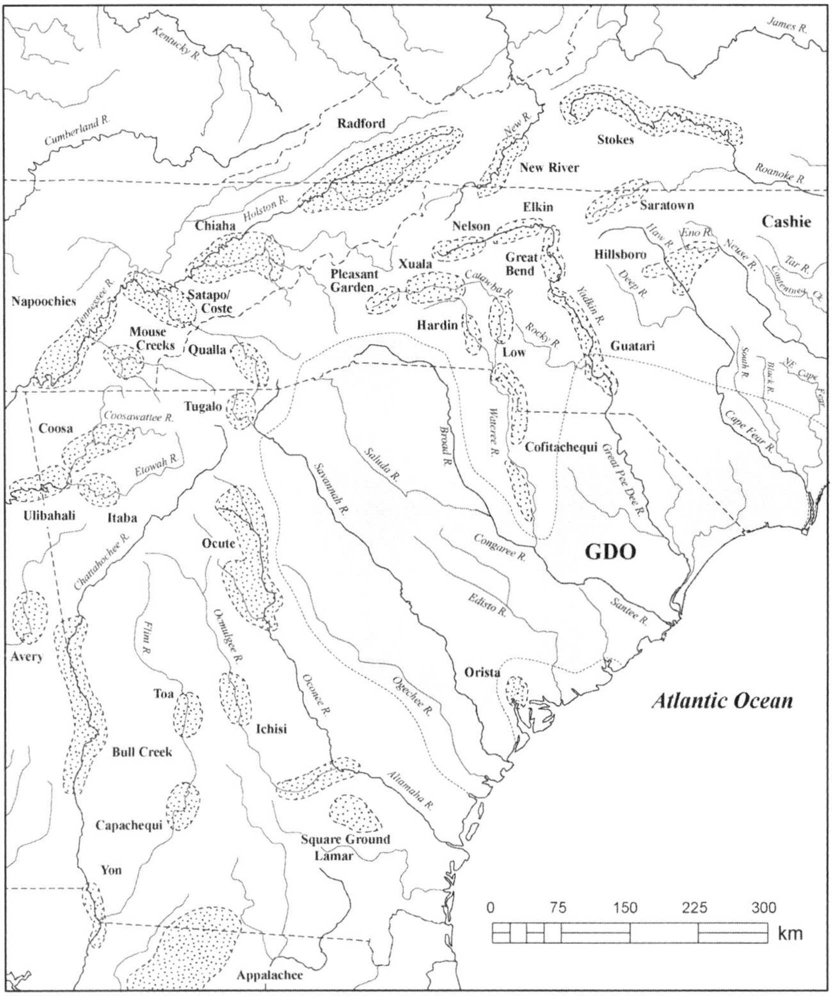

Figure 6.1. Fifteenth- and sixteenth-century polities making up the settlement halo.

very scant evidence of occupation has been identified in the upper and middle Catawba valleys prior to AD 1400 and that it likely did not give rise on its own to the dense and intensive protohistoric occupation of the Burke phase chiefdom known to Soto as Xuala. Most of the Burke phase sites are situated along the Johns River and Upper Creek tributaries of the Catawba next to broad floodplain formations. A range of four site types based on size was recognized from survey data. The Berry site (presumptive Xuala) is the largest and uniquely contains a platform mound. The other classes occur in clusters at regularly spaced distances and appear to represent the remains of secondary towns, hamlets, and farmsteads (Beck and Moore 2002). Beck suggests that much of the rapid population growth in the protohistoric period here was a consequence of refugees migrating into the middle Catawba from locations being abandoned along the middle Savannah, Saluda, and Broad rivers in the fifteenth century. In fact, the primary source may have been much closer.

Domingo de Leon's recounting, in 1584, of experiences he had witnessed as a member of the Pardo expeditions 16 years earlier indicates that the entire span of the Wateree Valley to the north and south of Cofitachequi was tightly packed with settlements of various kinds (Worth 1994:6). Leon described a stretch of 12 leagues between the first town (i.e., Hymahi) above the confluence of the Congaree and Wateree rivers and Cofitachequi, in which settlements were spaced at distances of one-half to one league apart on the west bank of the river. Another 20-league stretch north of the capitol situated on the opposite bank contained even more densely packed settlements. When leagues are recalibrated using the *legua comuna* standard of 5.56 km and the Mulberry site is used to anchor Leon's observations, the northern wing of the settlement system actually abuts against the southern borders of the Low and Hardin polities in the Catawba drainage system (see Figure 6.1). I shortened the extent of the northern boundary of the Cofitachequi settlement cluster as described by Leon to reflect a buffer zone between these various polities. Thus, there is evidence to suggest that pressure from population crowding was being applied from within the larger drainage system and that a major source of the rapid population expansion during the Burke phase was likely the consequence of northward migration of populations from positions downriver in search of unoccupied arable land. It is of interest that the two settlement concentrations around Cofitachequi were situated on different banks of the river and were buffered by the capitol town. This suggests that they were separate social divisions. Although this is purely speculation at this stage, the more southerly division may represent migrations from the Scott's Lake mound complex on the upper Santee.

The appearance of Burke (AD 1400–1600) and Happy Valley (AD 1600–1700) phase occupation on the upper Yadkin River most likely represents a continuation of this northward migration (Moore 2002:180). To distinguish it from Xuala, I have named the upper Yadkin polity "Nelson," after the T. F. Nelson Mound and Triangle. Unlike the case of the middle Catawba, migration into the upper watershed of the Yadkin appears to have represented a site-unit intrusion into an already occupied territory. Prior to the mid-fifteenth century, the upper Yadkin above the Big Bend was populated by a regional variant of the Smyth culture, which is characterized by steatite-tempered pottery that is typically net impressed or plain. Smyth ceramics and the sand-tempered variant called Dan River were manufactured by the late prehistoric groups occupying vast portions of Piedmont North Carolina and southern Virginia, which are generally considered to represent the ancestors of the protohistoric and Contact period Siouan tribes. In the Dan River valley, this understratum gave rise to the protohistoric Saratown phase and the Oldtown ceramic series (Wilson 1983).

In the upper Yadkin valley, the late Smyth assemblage of the protohistoric period is referred to as the Elkin phase (Moore 2002:181). It is not known if a segment of the Elkin phase occupation was displaced by the Burke phase intrusion, but there is evidence for culture contact between the two polities at the Porter site (Woodall 1999). The Porter site is situated approximately 30 km east of the Nelson polity; excavations there recovered a ceramic assemblage referable to the Elkin phase. Two shaft and chamber burials containing prestige items associated with Mississippian elite exchange systems (i.e., two perforated spatulate celts, a marine shell gorget, and a Citico style, Saltville variety marine shell gorget) were also recovered. It is Woodall's thesis that these burials evince the introgression of Lamar culture on an indigenous Late Woodland society, perhaps through marriage alliances.

A late prehistoric site-unit intrusion into the already occupied territory of a second Piedmont Hill Tribe polity appears to have occurred on the lower Yadkin. As discussed previously, the seat of power of the Eastern SAM chiefdom once located at Town Creek appears to have shifted northward to the Pardo town of Guatari in the last half of the fifteenth century. This dating is based on the character of the ceramic assemblage that Coe reportedly recovered from his excavations at the Trading Ford site, which, by his description, was well within the range of variability associated with Early and Middle Lamar pottery. Hudson (1990:93–94) has suggested that Guatari was a young chiefdom at the time of the Pardo expeditions in the 1560s, because there is no documentary or archaeological evidence that confirms the presence of a substructure mound. Certainly, had there been a mound of any size, Coe (1995:160) would have discovered it when High Rock Lake was drawn down and excavations at the Trading Ford site were undertaken in 1938. The absence of a mound might, of course, support the idea that the chiefdom was young; still, it more directly establishes that the town itself was young and not necessarily that the lineage of political leadership was new. The use of Town Creek appears to have shifted from primarily a domestic residential hub to a specialized cemetery complex in the fifteenth century, a shift that was accompanied by a reduction and cessation of efforts to appreciably expand the volume of the mound (Boudreaux 2007:109–111). These patterns may very well constitute logistical abandonment behaviors associated with the establishment of a new political center at Guatari. In this way, the founding of Guatari may represent the vanguard of a migration northward of the former Town Creek polity. Ricciardelli's (2014) survey data indicate that the hinterland around Town Creek was still occupied and this too may evidence an additional migratory wave of Eastern SAM populations moving out of the GDO from locations farther downstream.

Town Creek is about 95 km south of the presumptive site of Guatari, which suggests that there may have been two or three spatial aggregations of population along this transect. Guatari was situated in a culture contact position relative to the late Dan River phase (Stokes, var.) occupation that extended north to the westward turn of the Great Bend of the Yadkin (see Woodall 1984, 1990). The Pardo accounts indicate that 39 chiefs were subject to the female chief of Guatari (Hudson 1990:93), which suggests that this list most probably included the Piedmont Hill Tribe-affiliated towns of the Great Bend region on the Yadkin. Guatari's location suggests a similar position in the regional settlement pattern to the frontier settlements of the Ucayali basin of eastern Peru that DeBoer (1981) describes. Tribal groups within the basin were distributed along linear stretches of the floodplains and each maintained buffer zones of uninhabited land to ensure separation between competing groups. Frontier settlements were situated at the termination of territories to defend the population against hostile attack. These defensive settlements were large and housed a disproportionate number of warriors, taking on the appearance of garrisoned embattlements. Guatari may have initially served as a fortified frontier settlement during the initial stages of intrusion, but later developed into the seat of a chiefdom as hostilities subsided and new alliances were solidified. Pardo's chroniclers do not say whether the town had a defensive palisade wall in the 1560s, but such fortifications were a common feature of late prehistoric and protohistoric towns in both the Piedmont Siouan and Eastern SAM culture areas. The presence or absence of palisaded villages at particular points in time most probably evidences shifts in regional relations and relative levels of violence.

The archaeological record of both the Catawba-Wateree and Yadkin valleys evidences the presence of dense and culturally heterogeneous populations during the fifteenth and sixteenth centuries. This situation is in stark contrast to the GDO, where it is difficult to identify any components dating to this period. Moreover, the apparent movement of Eastern SAM populations northward in both valleys is consistent with the notion that early fifteenth-century groups were migrating away from the GDO. Such a dynamic must have resulted in demographic crowding and impingement on former buffer zones, creating tensions between neighboring groups and conditions that could have accelerated the formation of coalescent societies.

Haw and Neuse Valleys

A similar process of intrusion and assimilation appears to have been underway in the upper watersheds of the Haw and Neuse river valleys. Davis and Ward (1991:43) note that the ceramic assemblage from the Wall site is so different from the earlier Haw River phase (AD 1000–1400) settlements in the region that the former most likely represented a site-unit intrusion. Late Haw River phase ceramics are dominated by net impressed exterior surfaces, which achieve roughly 90% popularity in sherd assemblages. By the mid-fifteenth century, however, a new, distinctively different ceramic series known as Hillsboro appears in the region, a development that is not easily derived from the Haw River phase using a gradualist model of ceramic change. At the Wall site, which dates to the mid-fifteenth century, the associated assemblage is composed almost exclusively of Hillsboro Simple Stamped, Check Stamped, and Plain (see Table 5.2). Net impressed is nearly absent and could represent mixing from an earlier Haw River occupation. Other features of the ceramic assemblage that have no antecedents in the region are cazuela bowls and carinated jars, both characteristics shared with Eastern SAM assemblages to the south and west. Although not dispositive on its own, the fact that the Wall site was palisaded provides additional credence for the site-unit intrusion hypothesis.

The Wall site rests on the Eno River, a tributary of the Neuse, but two later, sixteenth-century Hillsboro phase sites (Edgar and George Rogers) located on tributaries of the upper Haw River contain what appear to be hybrid ceramic assemblages that recall characteristics of the earlier Haw River series (Ward and Davis 1993:412–413). The most salient feature of these assemblages is the significant contribution of net impressed pottery (see Table 5.2). Although these examples are typed as Haw River Net Impressed, interior finishing characteristics and paste attributes align them more closely with Hillsboro. Ward and Davis (1993:413) suggest that the ceramic contrasts between the Wall site and these two later sixteenth-century sites can be "viewed as a manifestation of a broader cultural process through which Hillsboro peoples adapted and possibly incorporated the indigenous population into a new cultural complex" from which "emerged the Sissipahaw, Shakori, and possibly Eno tribes." The George and Edgar Rogers sites are closely proximate to the Wall site, but in a different valley, suggesting that they were the products of slightly different historical processes. Consequently, the described contrasts in the ceramic assemblages may very well reflect cultural as well as chronological differences.

The Wall site assemblage is clearly foreign to the region and the GDO model would predict a source to the south. Prior to the analysis presented in this book concerning the importance of simple stamping in the Wachesaw Landing and

Tidewater site ceramic assemblages of North Coastal South Carolina, archaeologists would have pegged the simple stamped-dominated ceramic traditions to the north (Gaston) or east (Cashie) as the source for the Wall site assemblage. Now, however, a source to the south becomes more plausible. Unfortunately, the potential source areas, which include the Neuse between the confluences of the Eno River and Contentnea Creek, the lower Haw valley, and the northern tributaries of the Cape Fear (South, Black, and Northeast Cape Fear rivers), have not received sufficient archaeological coverage to directly address this question (see Figure 6.1). Some information bearing on this subject, however, is supplied by ethnohistory. It is generally held that the late prehistoric Haw River and Hillsboro cultures were ancestral to historic Siouan groups. Moreover, there is evidence that at least one tribe regarded as Siouan, the Woccon, was settled in two villages near the confluence of the Neuse and Contentnea Creek when Lawson visited the region at the turn of the eighteenth century (Lefler 1967; Mooney 1894:65). Lawson commented that villages of the Algonkian, Tuscarora, and Woccon were not more than 10 leagues distant in this location and that the Woccon were only 2 leagues away from the nearest Tuscarora village further up Contentnea Creek. He also observed that owing to their great numerical dominance in the region, the Tuscarora language was a sort of lingua franca, being "understood by some in every Town" (Lefler 1967: 233). During the late Contact period, then, the Neuse was a culture contact zone and the sharing of simple stamped-dominated ceramic assemblages across these three linguistic groups, even in the late prehistoric period, provides testimony to just how interactively linked the populations of eastern North Carolina and the North Coastal Zone of South Carolina were throughout this long history. The intrusion of the Hillsboro phase into the upper Neuse and Haw valleys appears to chronicle an expansion of this interaction sphere into more traditional Piedmont Siouan regions.

Perhaps the most telling difference between Hillsboro assemblages and those of Gaston and Cashie is the substantial representation of check stamping. The Hillsboro sites discussed here had check stamped representations ranging between 7.4 and 53.8%. Check stamping was present at the Gaston site, but it made up only 1.1% and 0.8% respectively of the Gaston associated ceramic types from the control tests and feature excavations (South 2005:59, 155, 158). Much of the Gaston occupation may be later than the Hillsboro site cluster, so this cannot be thought of as a synchronous comparison. More to the point, however, check stamping is probably absent from Cashie assemblages throughout their long development (Herbert 2011:4–22) and it is so rare in Virginia that Evans (1955:128, 133, 143) explained its sporadic occurrence as an outcome of trade. Check stamping is also not found in substantial quantities in the mid-fifteenth-century Wachesaw Landing assemblage or in the Tidewater assemblage from the Little River Neck, which appears to date to the late sixteenth or early seventeenth century.

Thus, the Hillsboro tradition appears to be a unique manifestation that developed in place along the middle course of the Neuse and possibly the lower reaches of the Haw River and the northern tributaries of the Cape Fear. Moreover, its apparent continuity with groups thought to be historic Siouan suggests that Hillsboro was the antecedent of one division of the Piedmont Hill Tribes as defined and elaborated by Coe (1952) and Hudson (1970); however, in this case the geographic origins may very well have been located on the Coastal Plain. In this light, the sudden appearance of Hillsboro in the upper Haw and Eno valleys in the mid-fifteenth century is consistent with a northward migration away from the GDO as groups from more southerly homelands began to crowd into the RDR.

Oconee Valley

Shifting focus to the southern part of the RDR, there is evidence of population movement and reorganization in the Oconee Valley of Georgia, where the chiefdom of Ocute was located (Hudson et al. 1984). The direction of movement, again, was upstream and to the north (see Figure 6.1). Survey data provide strong evidence of dramatic population growth across the upper Oconee Valley beginning in the late fourteenth or early fifteenth centuries along the river terraces and throughout the uplands, at least from the Wallace Reservoir northward (Kowalewski and Hatch 1991; Rudolph 1994; M. Williams 1994). Based on the Wallace Reservoir sample (see DePratter 1976), Rudolph (1994:111) argues that there was a 20- to 90-fold increase in site frequency between the earlier Savannah period and the Duvall phase of the late fourteenth and early fifteenth centuries. In addition, Kowalewski and Hatch (1991) synthesized survey and excavated data from Lamar sites outside of the reservoir situated in upland tracts on both the east and west sides of the Oconee River. For a long time it was assumed that small sites in the hinterlands would represent deposits from ephemeral special purpose camps and resource extraction loci, but excavations at a number of these upland scatters have demonstrated that many of them represent the remains of permanently occupied single family homesteads and small hamlets. Much of this occupation is seen as occurring late in the Lamar sequence, after AD 1520 (i.e., Dyar and Bell phases).

Kowalewski and Hatch (1991) constructed eight different simulations of population growth and decline in the uplands based on assumptions concerning the relative mix of Dyar and Bell phase components and the magnitude of the base population at the beginning of the Dyar phase at AD 1520. The simulations with the greatest assumed growth rates, which begin late or have a low base population, returned rates so high that they would have been impossible without substantial in-migration. Because the ceramic styles were homogeneous within the site sample, however, these outcomes were considered improbable. The simulation they considered most realistic, which required much lower growth rates (2–3%) to account for the projected site universe, assumed a median base population at AD 1540. In this outcome, the projected population increase could be explained simply by the natural processes of reproductive growth rather than substantial in-migration.

Kowalewski and Hatch's evaluation of alternatives might be taken to mean that the Oconee Valley was not the recipient of significant immigrant populations during the Lamar period. However, the expansion of settlement into the uplands that they have identified would not directly address the question of whether the valley had received high levels of immigration from the GDO, because this influx should have occurred much earlier. It is difficult to gauge what happened in the fifteenth century, however, since the implications of the chronology, which was still being worked out at the time, were not fully accommodated in the component assignments of the survey data. Rudolph's (1994:111) analysis of the Wallace Reservoir data suggested to him that there was a definite interval of accelerated growth between the fourteenth and early fifteenth centuries, which he associated with the Duvall phase, and possibly a later interval in the late Dyar and early Bell phases, which would correspond to the patterns Kowalewski and Hatch (1991) discussed.

The chronology that Rudolph used to identify Lamar components consisted of only three phases: (1) Duvall (AD 1350–1450), (2) Dyar (AD 1450–1600) and (3) Bell (AD 1600–1650). Refinements to the chronology made by Marvin Smith (1983) and Mark Williams (1983) entail two modifications that significantly impact the evaluation of growth rates. First, the Duvall phase has been shortened and its beginning date is shifted forward to a seventy-five-year period between AD 1375 and 1450. Secondly, the former Dyar phase has been partitioned into early and late divisions. The early division is now identified as the Iron Horse phase (AD 1450–1520), while the later one retains the original Dyar phase (AD 1520–1580) designation.

Rudolph (1994:99) was aware of the Dyar phase divisions but was unable to distinguish them in his component identifications. If an attempt is nevertheless made to achieve a division, and the current date ranges for the phases are used, a population curve in line with the conclusions of both studies can be offered. Figure 6.2 displays my reworking of the Wallace Reservoir data, in which the original Dyar phase components are allocated equally to the earlier Iron Horse phase and the new Dyar phase. As it stands, there is no basis for distinguishing these two phases and a decision to weight them one way or the other would run

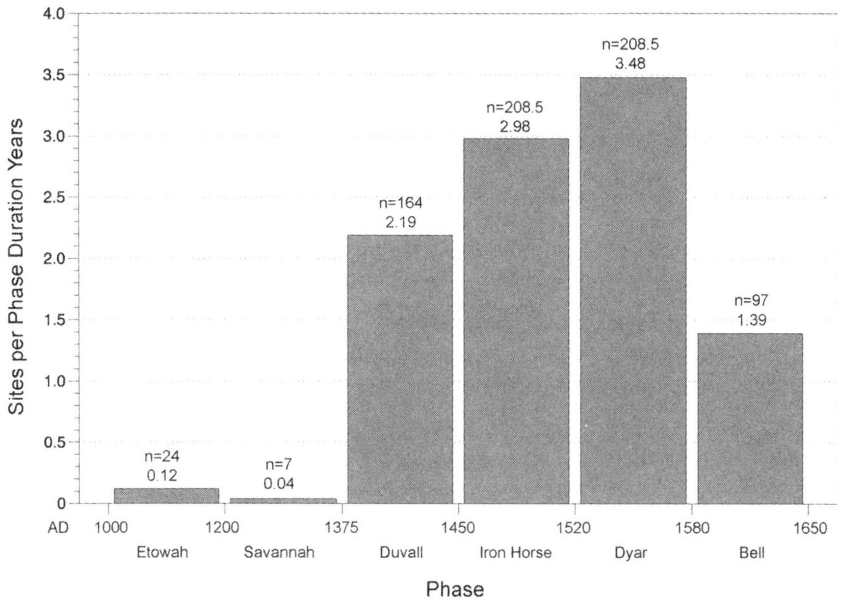

Figure 6.2. Bar chart of Mississippian site frequencies by phase for the Oconee Valley, Georgia.

the risk of biasing the conclusions without proper cause. The bar graph illustrates an average of sites added or subtracted per year, calculated by dividing the total number of components for each phase by the associated phase duration. There is a steep increase in component population associated with the shift from the Savannah period to the Duvall phase, but thereafter the rate of growth, although positive, begins to flatten out. The accumulated rate of increase from the Duvall to the Iron Horse phase is 36%, which if spread across the entire phase would yield a mean growth rate of 0.5% per year. From Iron Horse to Dyar, the accumulated rate of increase is 16.8%, or a mean growth rate of 0.28 percent per year. From Dyar to Bell, population begins to decline, which may represent a contrast between the river terraces and the uplands because Kowalewski and Hatch (1991) see an increase in the former.

These rates of growth are in no way comparable to that seen between the Savannah period and the Duvall phase. The Duvall phase represents an accumulated 5,375 percent increase over the Savannah period. To achieve such a rate, the population would have had to increase by an average of 71.7 percent per year. It is difficult to see how this increase could be achieved by natural rates of reproduction alone. Thus, during the early fifteenth century there is ample evidence from survey data to identify a period of substantial in-migration. Mark Williams (1994:190) has suggested that much of this influx came from the abandonment of the Savannah River chiefdoms. Owing to their proximity, I would guess that the most likely sources from the GDO would have been the Coastal Plain chiefdoms on the lower Savannah (i.e., the Lawton, Red Lake, and Irene polities) and the Ogeechee Valley. I believe Williams favors immigration from the Fall-Line chiefdoms on the Savannah and this source could have contributed to immigration as well. Although the stylistic and technological composition of the fifteenth-century ceramic assemblages from these regions are similar, detailed trait and attribute studies could provide a basis for recognizing the identity of "foreign" occupations and enclaves in the larger Oconee Province.

Here, again, is evidence for a dramatic influx of population into a river valley immediately adjacent to the GDO in the fifteenth century. In addition, there are clear signs of crowding, as settlement expanded into the suboptimal reaches of the upper watershed. Since most of this expansion occurred during the sixteenth century under conditions of natural population growth rates, it can be suggested that equilibrium between immigrants and the original resident population had been achieved by this time and that integration of possibly disparate ethnic groups was well underway by the early 1500s.

At the same time that this rapid infilling of the upper Oconee was underway, there was a contraction of the mound system (M. Williams 1994). By AD 1325, three uniformly spaced mound centers (see Figure 6.1) supporting separate population aggregates covered the entire expanse of what was to emerge as a larger paramountcy (i.e., the Oconee Province). However, by around AD 1400 the Scull Shoals and Shinholser sites were abandoned and the mound system had contracted to include only a very limited area in the center of the province marked by the reoccupation of the Dyar site and the continued use of the Shoulderbone site. In the mid-fifteenth century, the occupation at the Shoulderbone site was significantly reduced and the primary mound system shrank to include only the closely situated Scull Shoals and Dyar sites. M. Williams (1994:190) posits that this contraction implies a northward movement of population within the valley. Likewise, this northward contraction could very well indicate that immigrant populations were entering the Oconee Valley from the east or southeast, pushing the resident population northward.

Precisely how the fifteenth-century population influx affected the balance of chiefly power in the valley is a treatise yet to be written. What is known is that a contraction of the mound system is associated with the Duvall phase population influx. Subsequently, during the fifteenth and sixteenth centuries, there was a long period of stable population growth and expansion into almost every available upper watershed locale by dispersed homestead and hamlet settlements accompanied by an increase in the number of operating mound sites. Finally, this florescent interval was followed by a period of population decline in the seventeenth century, in which mound sites were abandoned and population shifted southward to the confluence of the Oconee and Ocmulgee rivers (M. Smith 1994:180; M. Williams 1994:191–193). In this light, the fifteenth century can be viewed as a period of population repositioning and political reorganization in the valley.

The contention that the Oconee Province was held together by a paramount chiefdom when Soto visited in the spring of 1540 is generally not disputed, but the re-emergence of multiple mound sites in the sixteenth century could just as easily imply decentralized fragmentation of authority structures into multiple power centers. Marvin Smith (1994:180) argues that after Soto's visit there was a rapid shift to egalitarian organization in the valley, a decline in settlement hierarchies, and a reduction in the magnitude of mound construction, which changed from the regular accretion of "thick mantles of earth to simple rebuildings of the summit structures." He notes in particular that radiocarbon dating of the latest summit structure of the mound at the Dyar site suggests construction was terminated in the mid-sixteenth century. Most researchers equate these changes with direct and indirect European contact and the introduction of European disease vectors. Certainly, the slave raids of the Chichimecos in the early 1660s were a factor (Worth 1995), but much of the organizational decline occurred prior to this event, in the early to mid-sixteenth century.

The Consequences of Coalescence

It is reasonable to expect that the combined effects of prolonged drought and population displacement in the fifteenth century would have acted to weaken the structural stability of Mississippian chiefdoms. Although drought years were not as frequent in the RDR, xeric conditions were nevertheless widespread, resulting in stressed wild resource yields, decreased predictability of domesticated food production, and reduced levels of annual production predictability across the Southeast. Over time, these conditions would have strained the managerial capacities of chiefdoms to increase surpluses through tribute collection as they attempted to maintain social hierarchies in the wake of decreased agricultural production. Simultaneously, the influx of immigrants would have compounded these administrative problems and, according to Kowalewski (2006), could have facilitated a breakdown in the authority structure of the nobility in favor of re-emergent, weakly ranked clans.

In the Southeast, the concept of coalescence has been almost exclusively applied by researchers to the multiethnic aggregations of refugees that were forced into reserves or hidden backwater locales by the disruptions of European contact in the seventeenth and eighteenth centuries (Beck 2013; Ethridge and Hudson

2002; Galloway 1995). By this time, however, chiefdoms were a distant memory to Native Americans in most areas of the Southeast. Vestigial Mississippian chiefdoms (i.e., the Apalachee, Caddo, Chitimacha, Natchez, and Timucua) survived into the late seventeenth and early eighteenth centuries in refugia at the margins of European contact (see Knight 1990), but within the established sphere of European influence Southeastern tribes such as the Creek, Choctaw, and Chickasaw had re-formed into devolved, weakly ranked or nonranked clan systems. In order to understand the collapse of chiefdoms in this "shatter zone," it is necessary to turn the dial back to the mid-sixteenth century, when "time's arrow" still preserved a trajectory of continuity with the powerful paramountcies of the thirteenth and fourteenth centuries. It is in this context that Kowalewski's model of political change becomes most relevant, because this was a time when the effects of the first wave of coalescent reorganization in the Southeast were well underway and the mechanisms of integration would have had sufficient time to disrupt the structural basis of chiefly authority.

An example of how assimilation might have occurred throughout the RDR is provided by Amanda Regnier (2006:244–256), who relied heavily on Kowalewski's model to explain the complex cultural landscape of the Alabama River. During the fifteenth century, this region received an overwhelming influx of people from three different cultural traditions (i.e., Pensacola, Moundville, and Lamar). Each occupied adjacent stretches of the valley, but crowding appears to have shrunk customary buffer zones, bringing the groups into intimate contact at territorial boundaries. Based on presumed synchronous ceramic style co-associations, Regnier has argued that individual towns contained unique mixes of people from these different traditions. Many of the larger towns were situated in defensible positions and some were fortified, suggesting that internecine hostility was a common feature of social interaction. In spite of this, the cultural heterogeneity of the towns led Regnier to posit that some overriding social institutions allowed individuals to move relatively freely across ethnic boundaries and perhaps this mobility was made possible by marriage alliances arranged through cross-cultural clan linkage.

Recapitulation

The archaeological record of the middle fifteenth and sixteenth centuries exhibits clear evidence of population movements away from the GDO and toward the regions of the RDR, as would be predicted by the geographic patterns of megadrought. Although the influx of population implies that different ethnic groups were coalescing in the regions of the RDR, the processes of integration that this would have entailed and the attendant impact on devolutionary trends are not yet well understood. Kowalewski's proposition that coalescence weakened the institutional foundation of Southeastern chiefdoms and gave rise to the development of confederacies by reinvigorating the power of descent groups should become a major focus of cultural reconstruction in the RDR. It is important to appreciate, however, that this reorganization was undertaken in an age of drought, which introduced additional stresses onto systems undergoing transformational change. The implications of these interacting forces are further explored in chapters 7 and 8 of this book.

7

Drought-Related Indigenous Disease Epidemics

A likely impossibility is always preferable to an unconvincing possibility.
—Aristotle, *Poetics*, c. 335 BC

The link between infectious disease outbreaks and extreme climate events is a major focus of modern epidemiological analysis, as droughts and flooding events have increased in frequency and magnitude (see Epstein 2001; Kovats et al. 1999; Kuhn et al. 2005; McMichael 2012; Morse 1995). Of particular significance for these studies is the drought-famine-starvation nexus, in which malnutrition, weakened immune systems, crowding, and social disorder expose human populations to a wide range of disease vectors. Similar conditions have prevailed in the past and we have historic records of some of these events, including the "black death" in mid-fourteenth-century Europe, the small pox epidemic associated with the Ming Dynasty collapse in the seventeenth century, and the Plague of Justinian in Constantinople in 542 AD. There has been a tendency to attribute all epidemics in the New World to European sources, but evidence is being pieced together that indicates that outbreaks of disease occurred at various times and places throughout the New World before the arrival of Europeans (cf. Herring 1992; Peros et al. 2009). One such instance may be represented by the purported pestilence of the 1530s that native informants of the chiefdom of Cofitachequi described to the chroniclers of the Soto expedition. I review the evidence for this seemingly impossible event below and consider the disorganizing impacts that such epidemics may have had on the larger Mississippian world system across an already destabilized environment.

During the Soto entrada, the Spaniards were directed by the casica of Cofitachequi to visit the grand village of Talimeco, the nearby town of her ancestors and former capital of the paramount chiefdom occupying the middle Wateree Valley in South Carolina (Elvas in Bourne 1904a:66). According to the casica, Talimeco had been abandoned one or two years before Soto's arrival in 1540 because of a pestilence that had killed many residents. Alonzo de Carmona, who visited the town with Soto, later informed Garcilaso that there were 500 vacant houses of quality construction, of which four were filled with the bodies of people who had died of the disease (Varner and Varner 1951:314, 325). If this can be believed, it would certainly suggest the likelihood that some kind of infectious disease had caused the abrupt abandonment of a densely populated town center.

Some have accepted the accounts at face value, tracing the possible source of the epidemic to indirect contacts with the Ayllón expeditions of the 1520s or some other indirect European source (Dobyns 1983:262–268; Hudson 1984:31–33, 1997:179–180, 421; Milner 1980:43–44). Others have attributed the observations made by the Soto chroniclers to different causes and have cited details in the accounts that appear to contradict the scenario of an epidemic (DePratter 1989:146–148, 1994:215–217; Widmer 1994:137–138). As I see it, two major hurdles must be cleared to accept the likelihood that the Soto entrada witnessed the recent aftermath of a disease epidemic in the chiefdom of Cofitachequi. The first and most challenging is the identification of an archaeological site that would satisfy the expectations for Talimeco. It would need to be a large mound site with an occupation history that terminates just prior to 1540. There are a number of known mound sites in the Wateree Valley, but only one could potentially yield an occupation history expressly consistent with the Talimeco portrayed by the chroniclers, and that is the Mulberry site. A fly in this ointment is that many regard Mulberry as the only known mound site that could possibly represent the mid-sixteenth-century town of Cofitachequi into which the chiefly retinue fled from the jaws of the epidemic. The second hurdle would be to identify and describe feasible alternative pathogen pathways for an epidemic of limited geographic distribution and noncatastrophic survivability rates relative to the larger polity. As of yet, a satisfying epidemiological explanation for the disease has not been formulated.

I address both of these topics below, beginning with a review of the issues surrounding contradictory observations in the historic accounts. This will be followed by a reconsideration of the occupation history of the Mulberry site. I argue that there is good reason to believe that the residential function of the site ended just prior to 1540, which strongly supports the notion that its true ethnohistoric identity is most consistent with Talimeco. Finally, I argue that the purported epidemic was most likely derived from a New World pathogen of the family of hemorrhagic fevers, which are spread by rodent populations. A characteristic of the epidemiology of these pathogens is that their transmission is aided and aggravated by cycles of prolonged drought followed by limited wet periods (Acuña-Soto et al. 2002, 2004, and 2005). It will be shown that just such a local cycle converged with the proposed abandonment date of Talimeco in the late 1530s.

The Ethnohistoric Record

Chester DePratter (1989; 1994) contends that the Mulberry site satisfies three criteria supporting its identification as Cofitachequi, and which no other known site in the valley meets. It is large, contains one or more platform mounds signaling a dominant position in the chiefdom and, finally, it exhibits a sizeable sixteenth-century occupation component. He (DePratter 1994:212) argues further that the Adamson site is the best candidate for Talimeco, for two reasons. First, it is situated 6.4 km (a little over a common Spanish league) from the Mulberry site, which is in keeping with Gracilaso's (Varner and Varner 1951:314) account of the distance between the two towns. Second, it was occupied up until about AD 1400 or 1450, which would bring its time of abandonment into chronological proximity to 1540. There are problems with this reconstruction, however, which DePratter himself discusses at length in a later paper (1998).

One problem (DePratter 1989:144; 1994:212) concerns the fact that sixteenth- or seventeenth-century European artifacts have not yet been recovered at the Mulberry site. Such items should be found at the town of Cofitachequi because not only were members of Soto's army lodged there, but an outpost (Fort Santo Thomas) was also built there in 1568 by Juan Pardo (Hudson (1990:151–152). Pardo stayed at the encampment with a contingent of about 63 men for 18 days, but neither men nor munitions were left when he departed. Soto remained at Cofitachequi for a period of 12 or 13 days before heading upstream along the Wateree River (Hudson et al. 1984:72–73). Because both the stays of Soto and Pardo were brief, it is likely that the accompanying discard of material culture by the Spanish was comparatively light. Most importantly, DePratter (1994:212) observes that the site has not yet been sampled sufficiently to establish the presence or absence of appropriately contemporaneous Spanish artifacts. Thus, the absence of Spanish artifacts does not yet present a disqualifying line of evidence for accepting the ethnohistoric identity of the Mulberry site as Cofitachequi.

There are two other problems, however, that pose serious challenges because they bring the association of Cofitachequi and the Mulberry site into direct conflict with the literal narrative of the Soto accounts. These are (1) the physical aspect of Talimeco and (2) the purported epidemic that led to the abandonment of Talimeco only two years before Soto arrived. Each of these is discussed in turn below.

Garcilaso (Varner and Varner 1951:314) writes that Talimeco was "situated on an eminence overlooking a gorge of the river." This description does not square well with the physical setting of the Adamson site, which sits on a low-lying ridge finger about 0.45 km distant from the Wateree River. In his latest re-evaluation of the equivalence of the Adamson site and Talimeco, DePratter (1998) notes that this description more closely aligns with the physical aspect of the Mulberry site, which "sits atop a 25-foot high bluff directly overlooking the river." If so, then he further predicts that the main town of Cofitachequi will be found about a league downstream. Hudson (1990:73–74) was also of the mind that the Mulberry site was Talimeco based on its physical aspect, but he posited that Cofitachequi could have been located upstream at the nonmound site of Ferry Landing, situated just below the Adamson site. The accounts do not say in which direction Soto traveled from Cofitachequi to get to Talimeco, so the prudent approach would include both directions in a search radius once an ethnohistoric identity can be securely established for one of these two sites.

The most intriguing and consequential detail for generating expectations about the occupation histories of the two towns is Elvas's (Bourne 1904a:66) claim that Talimeco was abandoned because of an epidemic that ravaged the chiefdom only two years before Soto's arrival. This information was supplied by the casica of Cofitachequi. I have looked at other translations of the Elvas narrative besides the Buckingham Smith version in Bourne (1904a, b), including those of Richard Hackluyt (1609) and James Robertson (1933), and all agree that Elvas specifically referenced two years as the time elapsed since the abandonment of Talimeco. On this point, then, there is no ambiguity. Whether Elvas comprehended exactly what the casica said or whether her statements were truthful are questions that cannot be answered within the corpus of the documentary accounts. Confirmative support for the timing of this event is provided by Garcilaso (Varner and Varner 1951:298), who wrote, "a great pestilence with many consequent deaths had ravaged their province during the past year, a pestilence from which their town [Cofitachequi] alone had been free."

A widespread epidemic occurring within one or two years of Soto's arrival that caused the abandonment of the former capital of the chiefdom (i.e., Talimeco), and perhaps other lesser settlements, would hold important implications for the occupation of the valley and in particular the histories of Talimeco and Cofitachequi. First, if Talimeco was abandoned only one or two years before

Soto arrived, then it could not be represented by the Adamson site, where occupation ended by about AD 1450 based on the results of the 1998 WARP testing project (Cable et al. 1999). Second, occupation at Talimeco would have terminated abruptly in the fourth decade of the sixteenth century. Third, there is a good possibility that the main occupation of Cofitachequi began just as abruptly as Talimeco's abandonment, between AD 1538 and 1539. Although Garcilaso's (Varner and Varner 1951:298) quote above suggests that the town of Cofitachequi was extant prior to the epidemic and somehow escaped its ravages, he may have conflated several different historical processes that he did not directly observe. Ranjel (Bourne 1904b:101) reports that Talimeco was the casica's ancestral village, an observation repeated by Garcilaso (Varner and Varner 1951:311), who quoted her as saying that Talimeco was the "residence and family seat of my own ancestors and the principal village of our province." It would seem a reasonable hypothesis, then, to suggest that Cofitachequi may have supported only a small population prior to the arrival of the former residents of Talimeco, especially the elite families, their retinues and associated lineages.

Of course, these scenarios hinge on the epidemic actually occurring. I have discussed the casica's testimony, but there are two other details of the accounts that appear to substantiate the event. Elvas (Bourne 1904a:66) stated that for a half league to a league off from Cofitachequi there were "large vacant towns; grown up in grass, that appeared as if no people had lived in them for a long time." Explaining why there was a scarcity of food in the chiefdom, Garcilaso (Varner and Varner 1951 298) offers further that the "inhabitants of the other villages of that province had fled to the forests without sowing their fields" because of the epidemic which had caused many deaths. DePratter (1989:146–147; 1994:215–217) finds other, well-argued interpretations that would not necessarily involve an epidemic. The food shortage, for instance, might be explained by the fact that Soto visited the chiefdom in May, at a time when the supply of stored crops would have been depleted simply as a normal function of the agricultural cycle. He also underscores the point that a number of mound sites (i.e., Adamson, Belmont Neck, and Boykin) that had been abandoned prior to AD 1450 dot the landscape in the proximity of Talimeco and Cofitachequi and that these sites more than likely represented the large vacant towns that Elvas described.

The other line of evidence is not as easily dismissed by alternative explanations. Garcilaso (Varner and Varner 1951:325), transcribing the testimony of expedition member Alonso de Carmona, stated "four large houses filled with the bodies of people who had died of the pestilence" were found while the party was at Talimeco. DePratter (1994:216) acknowledges that the piling of bodies in such a manner could very well provide evidence of an epidemic. However, he suspects that this description is mistaken because it was not repeated in any of the other accounts and it is unlikely that such a sight would have gone unnoticed by the other chroniclers or the raiding parties that were sent out to search the area for food and riches. He concludes that it is more likely that Carmona was actually referring to the bodies of past rulers laid to rest in the temple at Talimeco. If so, there would be little left in the accounts to support the occurrence of a pestilence, other than the testimony of the casica. The mere absence of an observation of major significance in more than one account does not negate its existence. Often times subject matter that is elaborately reported in one account goes by without mention in another. For instance, neither Biedma nor Elvas mentions the temple at Talimeco, but I do not think that scholars deny one existed there, or that it possessed large stores of weapons and sumptuary goods, or that Soto entered it.

The question of whether Carmona was confused about where and in what context he saw the bodies is not resolved. Talimeco was clearly a grand village in its heyday. Early in his description, Garcilaso (Varner and Varner 1951:314) states that the town had some 500 vacant houses and that the structures were large and "more distinguished than ordinary buildings." He further commented that the "physical appearance of this town made it very evident that as the seat and court of a powerful prince, it had been built with more neatness and embellishment than the common villages." The temple, which Ranjel (Bourne 1904b: 101) says was built on top of a mound, was situated in the center of the town, "in front of the chieftain's dwellings" (Garcilaso in Varner and Varner 1951:314). Garcilaso (Varner and Varner 1951:315) adds somewhat farther along in his narrative: "our men paused but a short time in these [abandoned] houses before proceeding to the temple." It is at the end of the description of the temple interior that Garcilaso (Varner and Varner 1951:325) inserts Carmona's statement that "in the town of Talomeco, where the rich temple and burial place was located, they found four large houses filled with the bodies of people who had died of the pestilence." Had Carmona stated that the bodies were seen in a single large building, the argument that he misremembered the incident and that the bodies were actually those of the elite that had been laid to rest in the temple would pose a strong counterpoint to those advocating for the occurrence of an epidemic. However, Carmona differentiated the temple from the four houses and specifically stated that they were situated in the town. The temple appears to have stood in isolation on the mound, as there is no mention of a temple complex of buildings. One can never be sure about the accuracy of historical documents, but the weight of the evidence suggests that Carmona was referring to houses in the residential sector of the town and not to the temple.

Another line of evidence that attests to a relatively abrupt and recent abandonment of the town is the fact that the houses were still standing, if Carmona is to be believed, and had retained sufficient structural integrity to be entered by the Spaniards. Had the houses been abandoned for any significant length of time, it is likely they would have fallen into ruin by the time the Spanish visited. Substantial rigid post structures with gabled or hipped roofs, which are commonly found in late Mississippian villages (Lacquement 2004) and would appear to correspond to the descriptions for the houses at Talimeco, have life spans of around seven to ten years in ethnographic contexts (Cameron 1991:160). However, this estimate would assume maintenance. Without repairs, the processes of decay and weathering would accelerate decomposition to even shorter life spans. I know of no studies describing and measuring the duration of deterioration for substantial Mississippian structures, but given the harsh weather conditions of the temperate humid subtropics it is not likely that unattended houses would be recognizable as anything more than a pile of ruins after a period of not more than 10 years.

The Archaeology of the Mulberry Site

It is feasible, then, that the historical documents of the Soto expedition accurately depict the abrupt abandonment of the main town of Talimeco only a short time prior to 1540 as a response to an epidemic of significant proportions. The veracity of this scenario, however, is only demonstrated by recourse to the archaeological record. We can predict that the archaeological site of Talimeco, if it was abandoned shortly before the arrival of Soto, should exhibit a long continuous history of prior occupation leading up to its abandonment and that its latest component should be consistent with an early sixteenth-century component assignable to the Mulberry I phase. Based on the AMS dates from TU 1, the McDowell II component at the Mulberry site should terminate sometime between AD 1515 and 1525. This would leave only a very limited span of years for a Mulberry I component to have been deposited. If the Mulberry site can be tied to the ethnohistoric identity of Talimeco, then, it would be reasonable to expect the attendant expression of a Mulberry I component to be subdued, since it would have been terminated long before the full duration of the phase had played out. This expectation is precisely the picture that emerges when the known archaeological patterns associated with the site are examined. Ceramics from excavated contexts have been unevenly analyzed and reported, but several contexts supply sufficient detail to inform on the occupation history of the site. Their locations are indicated on a map of the Mulberry site in Figure 7.1.

The joint University of Georgia/Charleston Museum expedition of 1952 was the first modern archaeological excavation conducted at the site. During this project the exposed face of Mound A at the riverbank was cleaned and straight-

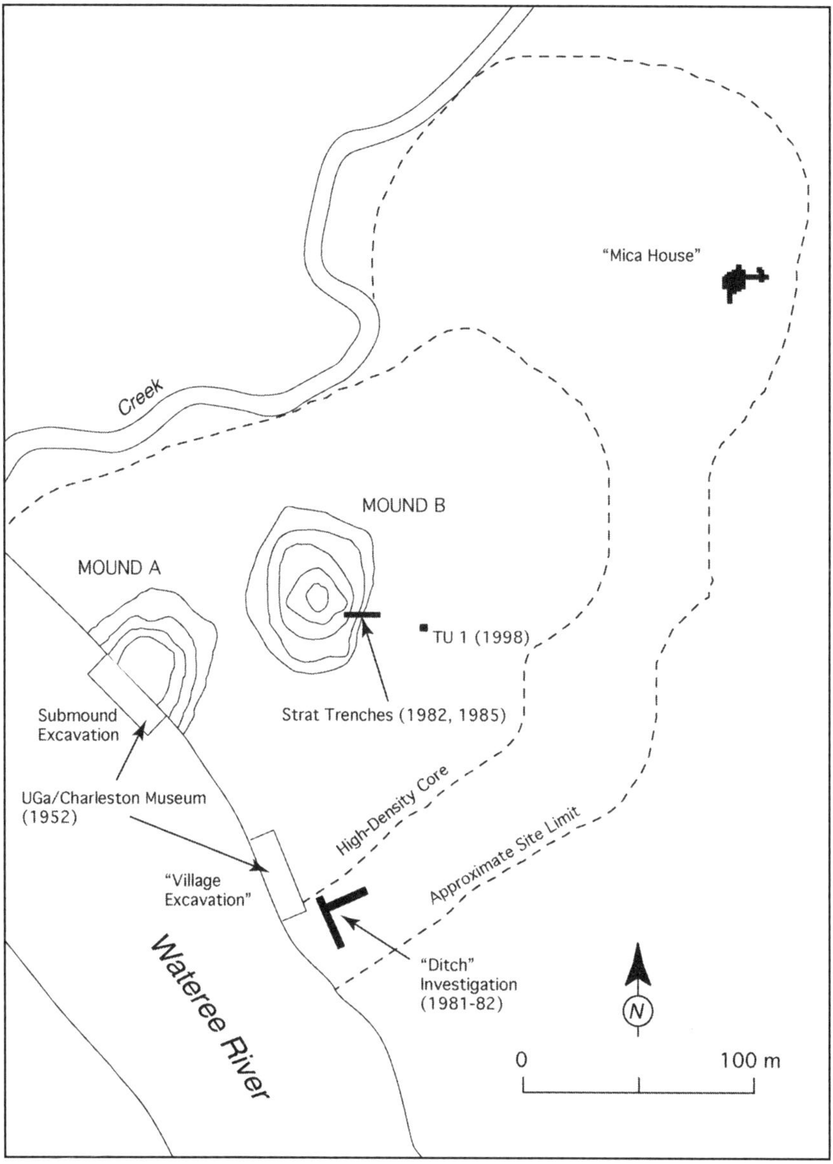

Figure 7.1. Plan map of the Mulberry site.

ened, the submound deposit along the bluff edge was leveled and stripped, and another block referred to as the "village" located about 200 feet to the southeast of the mound and measuring 20-×-100 feet was excavated in 5-ft squares (Kelly 1974:72). Joseph Caldwell (1974) performed the ceramic analysis. His efforts were abbreviated and type counts were provided for the contents of only one 5-ft square located in the village block. However, he personally inspected two-thirds of the entire collection of sherds, which included several thousand from the village block and all of the sherds recovered from the mound investigations.

The submound assemblage is consistent with the Adamson and Town Creek phases of the modern chronology (see Table 2.10). What he described from the village excavation, which postdated the deposits below Mound A, fits comfortably within the McDowell I and II model provided by the TU 1 midden excavations (Cable et al. 1999; see also chapter 2 of this book). Principal types consisted of Lamar Complicated Stamped and Plain, with minor contributions of Lamar Bold Incised. Caldwell noted that most of the plain pottery was burnished and cautioned that it would probably be necessary in the future to segregate these specimens into their own typological category. Respectively, 58 and 76 percent of the plain sherds from the McDowell II (n = 854) and McDowell I (n = 301) levels in TU 1 were burnished (Cable et al. 1999). Segmented rim strips comprised the bulk of the incidental rim decoration in the later village deposit and in the McDowell I and II assemblages from TU1.

Caldwell does not mention the presence of folded rims, although George Stuart (1974:109) observed that they were present at the site based on his surface collected data. Stuart (1974:108) also argued that Lamar Bold Incised was more frequent than Caldwell's sample reflected. Since Stuart gained his differing perspective by examining surface collections from the site, it is likely that he had a better view of the latest occupation. When excavations are placed into other localities of the site with relatively late deposits, however, both folded rims and incised sherds are found in very low proportions, as Judge's (1987:49–50) description of the material he recovered from the eastern flank of Mound B demonstrates. Judge reports incised representation of only 2 to 3 percent of the sherds with recognizable surface treatments. Furthermore, the folded rim to rim strip ratio is only 0.08:1, which would suggest that most of the material is associated with the McDowell phases and that only a very small percentage of the sherds could be assigned to a division of the Mulberry phase. Shoulder ticks were also present in the Mound B collection, representing one percent of the body sherds. Percentages from the McDowell I and II deposits in TU 1 are slightly higher, ranging respectively between 1.9 and 2.0 percent of sherds with recognizable surface treatments. Other than a small number of folded rims, then, the Mound B excavations that Judge reports are indistinguishable from the McDowell I and II phases. Moreover, as discussed earlier, Judge's criteria for identifying folded rims correspond more consistently to earlier thickened rim types, suggesting that actual folded rims were absent or only sparsely represented in his study.

Merry and Pekrul (1981) excavated a slot trench measuring 21.32-×-1 m on the southwestern end of the site, for the purpose of examining a feature that William Blanding (in Squier and Davis 1848:105–108) mapped and described as an embankment that purportedly encircled the Mulberry site and was still visible in the early nineteenth century. Results of the excavation suggested that the feature was probably not an embankment, but rather the elevated edge of a low-lying slough cutting across the eastern perimeter of the site. Stratigraphy consisted of three zones (see Merry and Pekrul 1983) from highest to lowest: (1) overburden, (2) ditch fill, and (3) the muck zone. The overburden zone consisted of a mixture of flood plain alluvium mixed with redeposited mound fill and slope wash containing both historic and prehistoric artifacts. The ditch fill zone was composed of an upward-fining sequence of minor flood events occurring when the slough was active. Finally, the muck zone was composed of a rich, organically stained midden characterized by high artifact density. A small lens of the wider site midden underlies the muck zone, but it was not exposed during the excavation.

The age of the "ditch" fill is undetermined. Historic artifacts were not recovered, but prehistoric sherds were not numerous, consisting only of 10 plain or eroded sherds. The muck zone is clearly prehistoric in affiliation, yielding a large collection of sherds (n = 488). Decorative modes were recognized on 31 sherds, 70 percent of which were rim strips while the remainder were punctations incidental to the rim. Of the identifiable styles of rim strip, 60 percent were punctated and 40 percent would correspond to the segmented type. Based on the Eastern SAM decorative mode seriation, this assemblage appears to contain both Town Creek and McDowell phase components. Since incised sherds made up only 1.1 percent of the identifiable surface treatments (5 of 427), it is most likely that the majority of the later component is made up of McDowell I ceramics. A Mulberry I phase is absent or indiscernible in the muck zone sample.

All of these contexts are located on the southern half of the site, as is TU 1, so it could be argued that wider coverage would succeed in identifying the substantial and continuous Mulberry phase deposit spanning the sixteenth century that would be necessary to demonstrate that the Mulberry site is indeed Cofitachequi. As part of the 1998 testing project (Cable et al. 1999), however, the central core of the site was shovel tested at systematic intervals of 20 m, providing a small but unbiased sample of the ceramic inventory of the main part of the site.

Care was taken to punch through the thick historic deposit covering the southern end of the mound sector to ensure even coverage of the occupation midden. Shovel tests were extended to depths of 70 cm, which was adequate to recover artifacts from the complete midden deposit in the northern half and the upper 20 to 30 cm of the midden in the southern half of the mound sector. Consequently, the sample is slightly biased toward the later portion of the occupation sequence, which means that the deposits of interest are fully represented.

Table 7.1 displays comparative percentage data of decorative mode categories for the mound sector shovel test sample and the combined McDowell I and II assemblages from TU 1 (levels 1 through 4). In general, the mound sector sample contains modes that would position it slightly earlier than those from TU 1. Positive percentage differences in the third row of Table 7.1 identify comparisons in which proportions of a particular mode are higher in the mound sector sample, while negative differences represent modes that have higher representation in the combined McDowell I and II assemblage. Both early and late Town Creek-Irene axis modes are more prevalent in the mound sector sample, while those modes associated with the later Lamar horizon style occur in higher proportions in the TU 1 assemblage. Although the Brainerd-Robinson coefficient for the two contexts is not high (137.4), suggesting that they are dissimilar, the p-value is only 0.217, indicating that the two contexts cannot be statistically differentiated given the sample sizes.

A hint that there is a later component representative of a Mulberry I phase comes from the presence of a single folded rim in the mound sector sample. Much of the thickened portion of the rim has been damaged by plowing, but examination of its profile strongly suggests that it represents a classic Clarksville type fold. A higher percentage of incising in the mound sector sample also suggests the presence of a Mulberry I component. While Lamar Incised in the combined McDowell phase assemblage from TU 1 makes up 2.8 percent (66 of 2388) of the sherds with identifiable surface treatments, it contributes 3.8 percent (12 of 314) to the identifiable sherd inventory from the mound sector sample. Again, however, this difference cannot be demonstrated to be statistically significant. A two-sample difference of proportions test (see Drennan 2010:139–144) yields a z-score of only 1.052 and a p-value for a one-tailed test of 0.1464. Standard statistical decision criteria would require a p-value of less than 0.05 for the demonstration of a significant difference.

What we know of the ceramic patterns at the Mulberry site, then, suggests that a Mulberry I phase, although likely to be present, is only subtly registered in the cultural deposits making up the mound sector. Such a conclusion clearly provides strong support for the expectation that the Mulberry site represents Talimeco rather than Cofitachequi. If there were a substantial Middle Lamar

Table 7.1. Calculation of Brainerd-Robinson Similarity Coefficient Comparison of Ceramic Assemblages from the Mound Sector Shovel Test Sample and Test Unit 1 at the Mulberry Site

Mulberry Site Samples	Punctated Lips[a]	Notches[a]	Hollow Reed Punctations[a]	Nodes[b]	Rosettes[b]	Pellets[b]	Rim Strips[c]	Shoulder Ticks[c]	Folded Rims[c]	Frequency	Brainerd-Robinson Similarity Coefficient	p-Value
Mound Sector (%)	0.0	0.0	18.8	6.3	6.3	6.3	31.3	25.0	6.3	16		
TU 1 (%)	2.0	0.7	4.0	0.7	3.3	4.6	54.3	30.5	0.0	151		
Difference (%)	-2.0	-0.7	14.8	5.6	3.0	1.7	-23.0	-5.5	6.3		137.4	0.217

[a]Early Town Creek-Irene Axis decorative modes.
[b]Late Town Creek-Irene Axis decorative modes.
[c]Lamar decorative modes.

presence at the site covering most of the sixteenth century, the associated assemblage should be similar to the Burke series collection from the Berry site (Moore 2002:75), which has been confirmed to be the town of Joara visited by both Soto and Pardo (see Beck et al. 2016). Folded rims constitute the dominant decoration and outnumber rim strips by a ratio of 6.83:1. In addition, Lamar Incised percentages should be much higher than any of the analyzed contexts at the Mulberry site reflect. It is difficult to extract hard quantitative data on incising at the Berry site from Moore's discussion (2002:86–88), but my rough estimate from the information provided is that Burke Incised (a Lamar Incised correlate) makes up over 5 percent of the Middle Lamar Burke series sherds with identifiable exterior surface treatments. If there had been continuity of occupation at the Mulberry site extending throughout the sixteenth and into the seventeenth century, just such an assemblage should have been deposited over the top of the McDowell II deposit in TU 1. None of the other contexts cited contain a substantial Middle Lamar assemblage either. This might be explained as a phenomenon of under-representation of the latest deposits at the site, but it would seem counterintuitive to expect that the mound sector should not possess a ubiquitous presence of a Middle Lamar assemblage if the Mulberry site was the town of Cofitachequi.

One last detail of the Mulberry site occupation history must be noted. There is evidence of a Late Lamar component, singularly manifested by what is commonly referred to as the "Mica House" (see Cable et al. 1999), so named because of the unusually high concentration of mica sheets and fragments found in its fill. A single radiocarbon date from a pit within the house yielded a 2-σ calibrated (OxCal 4.3.2) age range of cal AD 1650 to 1955 (Beta-131286). Since we cannot confirm aboriginal occupation at the Mulberry site after the turn of the eighteenth century, it is likely that the age of this sample corresponds to the earliest 1-σ option, which is AD 1665 to 1700. Moreover, since we know that the chiefdom of Cofitachequi disappeared from the historic record only a short time after Henry Woodward's visit in 1670 (DePratter 1994:221), it is likely that this house was occupied at the very end of the cultural sequence.

Although a detailed analysis of the ceramic assemblage from the house is incomplete, preliminary counts of approximately 40 percent (n = 1886) of it (J. Rice 2000) suggest the existence of a ceramic phase that is unanticipated by the current Wateree sequence. It differs in significant ways from the Middle Lamar, Daniels phase description, suggesting that the assemblage represents an even later phase and/or that the Daniels phase definition will require modification in the future. Plain sherds are reported to comprise 90 percent of the sample. Minority types include complicated stamped (2.7 percent), simple stamped (3.0 percent), incised (2.8 percent), and burnished (1.2 percent). These percentages correlate quite well with seventeenth-century assemblages from Georgia such as the Bell phase in the middle Oconee River valley. The Bell phase assemblage is dominated by plain sherds and contains very little complicated stamped (M. Williams 1983). Incised vessels constitute the chief decorated type. The previously discussed Caraway assemblage from the area around the site of Town Creek in the upper Pee Dee and lower Yadkin valleys may also be correlated with this Late Lamar style horizon. I have only cursorily inspected some of the ceramic material from the house, but wide folded rims are present, in keeping with Late Lamar assemblages of the seventeenth century (see Hally 1994a).

The "Mica House" is not situated in the mound sector, but at a distance of 200 meters from it on the eastern perimeter of the Mulberry site (see Figure 7.1). Its ceramic assemblage is so different from that of the McDowell and Mulberry phase divisions that it can be argued that the site was probably reoccupied in the seventeenth century after a significant hiatus spanning the middle and late sixteenth century and possibly the early seventeenth century. The fact that the house is not situated near the mound sector suggests that prohibitions to resettlement in the former ceremonial and administrative capital of the chiefdom may have been in force throughout the sixteenth and seventeenth centuries. Was this a consequence of the memory of the epidemic and the desire of the population to escape its potential ill effects by avoidance? This is a question still to be answered, and the multiyear project undertaken by the University of South Carolina in collaboration with the South Carolina Institute of Archaeology and Anthropology may provide answers to a number of the issues I have raised here.

If the Mulberry site is Talimeco, then what would the expected archaeological correlates of Cofitachequi look like? In terms of the physical characteristics, the Soto accounts leave little information. Garcilaso (Varner and Varner 1951: 311) informs us that a temple containing burials, large quantities of pearls, and items from the Ayllón colony was situated in the upper part of the town. Such a positioning might indicate that the temple rested on a mound, but this is unclear. He further offers that the temple was "about the size of the usual ones to be in Florida." The only town in the chiefdom for which literal mention of a temple mound was made is found in Ranjel's description of Talimeco (Bourne 1904b:101): "This Talimeco was a village holding extensive sway; and this house of worship was on a high mound and much revered." If there were a mound at Cofitachequi, Ranjel's description would suggest that it was not nearly of comparable size to the one described for Talimeco. In addition, the contents of the temple at Cofitachequi were only a pale reflection of what was found at Talimeco. According to Carmona, as paraphrased by Garcilaso (Varner and Varner 1951:324), "this temple [at Talimeco] and burial place" was "the richest and most superb of all those that our Spaniards saw in Florida." These contrasts would be unsurprising if Cofitachequi was a newly established town or one that had only recently been transformed into the main administrative center of the chiefdom. Therefore, the town may not have had a particularly high aboveground profile, particularly if the chiefdom was in decline. Such an absence of a readily identifiable aboveground presence would make it all the more understandable that a candidate for Cofitachequi has not yet been found. The one characteristic that must necessarily be present is a dominant and extensively distributed Middle Lamar component (i.e., Mulberry I and II), a feature not yet identified at the Mulberry site.

A Possible Source of an Indigenous Disease Vector

Based on my reading of the accounts and the information available on the occupation history of the Mulberry site, a strong case can be made for its identification as Talimeco rather than Cofitachequi. The absence of ubiquitous Mulberry I and II components supports the contention that the site was abandoned just prior to 1540 and Soto's visit. That the temple and the houses were still structurally sound also suggests that the former residents were avoiding the town, as would be expected if a catastrophic epidemic had caused its abandonment. Still, this hypothesized event would have occurred before substantial contact with Europeans, which reduces the likelihood that such an epidemic would have been caused by an introduced European pathogen as has been advocated by some (Dobyns 1983: 262–268; Hudson 1984:31; Milner 1980:43–44; M. Smith 1987:57–59; Wright 1981:44).

It is unlikely but still possible that a contagion of European origin could have made its way to the chiefdom of Cofitachequi before Soto's visit via secondhand contact with infected objects from the Ayllón expeditions of the 1520s, as some (Dobyns 1983:262–268; Hudson 1984:31–33, 1997:179–180, 421; Milner 1980: 43–44) have suggested. In this regard, Elvas (Bourne 1904a:67) said that in "the town [Cofitachequi] were found a dirk and beads that had belonged to Christians, who, the Indians said, had many years before been in the port, distant two days' journey." The case, however, for these objects to have carried Old World pathogens capable of inflicting an epidemic on the chiefdom in the late 1530s is weak. Modern studies of disease transmissions from inanimate objects (fomites) indicate that respiratory viruses (e.g., influenzas) survive in infectious concentrations for less than a day to not more than a week, while enteric viruses (adenovirus, hepatitis A, rotavirus, etc.) can survive in this state for a week to several months (Boone and Gerba 2007). A laboratory study (Rao 1972) of the role of fomites in transmitting smallpox further concludes that it is nearly impossible for nonporous objects to sustain viable concentrations of the virus for any length of time under uncontrolled conditions and that the virus completely disappears within a period of 12 days or less, even in the most favorable of situations. Porous materials, such as blankets or clothing, provide a medium more conducive to survival, but even in these cases virus concentrations decrease rapidly over periods of one to only several months. Rao (1972:5) concludes that being infected with smallpox by contact with a nonporous fomite "would seem possible only if a person put his fingers, contaminated with the virus, into his nose." Thus, it is unlikely that the posited Ayllón fomites would have been infectious even if they had been transported immediately to Cofitachequi from the coast in the 1520s. With smallpox, as well as many other kinds of viral infections, the primary mode of transmission is person-to-person through exchanges of saliva or inhalation of aerosolized fluid droplets from coughing or sneezing (Dixon 1948; Milton 2012). The possibility remains that infected visitors from the coastal zone could have introduced into the chiefdom an Old World disease that they contracted from the Ayllón expedition, but if so it was likely not smallpox. Kelton (2002) provides strong reasons to believe that smallpox was not a factor in the Native American epidemiology of the Southeast until the pandemic of 1696–1700. Types of insect-borne Old World diseases capable of producing lesser epidemics such as typhus and malaria are a possibility, but there is not good evidence that they were factors in the precontact Southeast either and they certainly would not have been carried by fomites. Consequently, a New World origin for the pestilence that struck Talimeco is at least as feasible as the hypothesis that it came about through a remote and circuitous route of infection from an Old World pathogen before a sustained European presence was established.

The best-documented instances of possible New World pathogens capable of inducing epidemics come from sixteenth-century Mexico, where three catastrophic events occurred during the intervals of 1519–1520, 1545–1548, and 1576–1580 (Acuña-Soto et al. 2000, 2002, 2004). The earliest event was clearly associated with small pox, but the later two have been attributed to a disease identified as *cocoliztli*, which is the Aztec word for pestilence. In 1576 Francisco Hernandez, the Protomedico of New Spain and one of the most highly regarded physicians of his day, described the symptoms of the disease, which were unlike any known Old World diseases present in Mexico at the time. Symptoms consisted of high and sustained fever, severe headache, vertigo, black tongue, dark urine, dysentery, severe abdominal and thoracic pain, large nodules behind the ears, neck and face, acute neurologic disorders, and profuse bleeding from the nose, eyes, and mouth. Death would generally occur within 3 to 4 days and approximately 80 percent of the native population of Mexico died in the 1545 epidemic; about 50 percent of the remaining population died in the 1576 event. Acuña-Soto argues that these symptoms represent a type of hemorrhagic fever and speculates that it may have been caused by a rodent-borne arenavirus like those in modern day South America that have produced similar symptoms.

Working with tree-ring reconstructions of winter-spring precipitation, Acuña-Soto and colleagues observed that each of the *cocoliztli* epidemics occurred during a wet period immediately following a prolonged drought. This suggested to them a complex intertwining of events in which the resident rodent population aggregated around oases of available food and water during the drought years, fueling intraspecies competition and aggression, thereby accelerating the spread of the virus within the population. As climatic conditions improved with increased precipitation, the severely infected and exploding rodent population would have spread to the agricultural fields and farm houses, whose inhabitants would have become infected through aspiration of excreta droplets. Acuña-Soto and colleagues note a similar dry-then-wet period spread of the hantavirus in 1993 in the southwestern United States, also a New World pathogen. In this case, rainy conditions after an extended drought saw a tenfold increase in the local deer mouse population, which has been credited with carrying the virus. In another study, Acuña-Soto et alia (2005) speculate that other massive depopulation events in the prehistoric era may also have been propelled by hemorrhagic fevers, in particular that of the Terminal Classic period in Mexico (AD 750 to 950). It too was associated with a series of prolonged droughts.

A specific arenavirus or hantavirus that could have been the source of the sixteenth-century epidemics in Mexico has not yet been isolated and there appears to be a great deal of epidemiological variability in need of more robust explanation. For instance, the symptoms and degree of severity of the *cocoliztli* outbreaks varied by geographic location and records indicate that there were ten lesser outbreaks between 1559 and 1642, nine of which began under drought conditions rather than during wetter periods following a prolonged drought (Acuña-Soto et al. 2002). These differences suggest that there may have been more than one pathogen responsible for the hemorrhagic fever outbreaks.

However, the correlation of a dry-then-wet couplet with the two catastrophic events of 1545 and 1576 argues strongly that the various expressions of *cocoliztli* were tied to the population cycles of rodent reservoirs. J. Mills and Childs (1998) observe that arenaviruses and hantaviruses are associated with specific species of rodents from the *Muridae* family and describe the process of transmittal to humans in the following manner: The virus fosters a prolonged infection in an individual rodent host that only rarely results in death or even symptomatic expression. It is then spread across the environment by urine, feces, and saliva. Other individuals in the rodent reservoir become infected through a number of behaviors, including fighting and biting among adult males. Finally, it is believed that humans become infected by inhalation of aerosolized virus.

Rodent reservoirs capable of hosting such viruses are widespread across the Americas (J. Mills and Childs 1998; Peters and Kahn 2002), which suggests that outbreaks of hemorrhagic fevers may have occurred numerous times in the precontact past, especially in agriculturally dependent societies like the late prehistoric systems of the Southeast. Two conditions fostering transmission would have been met in such societies: high human population density and a ready and abundant agricultural feeding source. Because most SAM political economies depended on corn agriculture, there existed abundant potential food resource patches to foster rodent reservoirs in and around towns and villages. Under conditions of prolonged and alternating cycles of wet and dry periods, then, it is likely that rodent populations were subject to cyclical boom and bust periods based on the supply of food.

Stahle and Cleaveland (1992) have generated rainfall reconstructions for the states of the Southeast Atlantic Slope, providing researchers with an opportunity to see where the hypothesized abandonment of Talimeco in 1538 or 1539 is positioned on the dry-then-wet cycle that Acuña-Soto and colleagues have identified as a likely stimulus for rodent-borne infection in human populations. Figure 7.2 illustrates a portion of Stahle and Cleaveland's curve for South Caro-

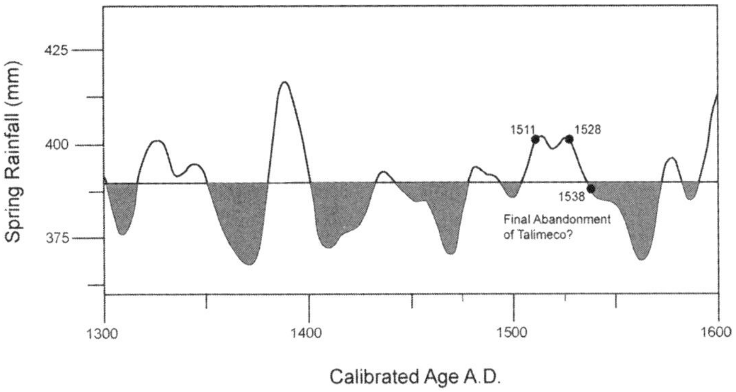

Figure 7.2. Spring rainfall curve for central South Carolina in relation to the hypothesized pestilence at Talimeco (adapted from Stahle and Cleaveland 1992: Figure 3C).

lina, covering the period from AD 1300 to 1600. This particular curve was constructed through the application of a cubic smoothing spline over 30-year intervals, which reduces noise from individual tree-ring measurements and produces curves expressing more generalized trends (E. Cook and Peters 1981). Shaded areas of the curve represent drought periods determined by their position below the horizontal line denoting long-term precipitation averages for the region.

Inspection of the precipitation curve indicates that the timing of the hypothesized abandonment in 1538 corresponds to the end of a significant wet period that began around 1504, peaked between 1511 and 1528, and rapidly declined to the theoretical average precipitation level by about 1537. In keeping with the model, this wet period was preceded by a prolonged, multidecadal drought spanning most of the fifteenth century. Slightly wetter conditions begin after 1480, but it is doubtful that climatic conditions would have been sufficiently favorable for a rodent population explosion until the wet period peaked between 1511 and 1528. There is a separation of as much as 10 years between the peak of wet conditions and the hypothesized abandonment of Talimeco, which could suggest a poor fit with the model. However, the rates of infection within rodent reservoirs are not yet well documented, nor are details of how long it takes human populations to become infected. This differential then, may describe a common progression of hemorrhagic fevers, the presence of which only become obvious in the final stages of development. The three-decade-long period of higher than normal precipitation at the beginning of the sixteenth century would seem to offer ideal conditions for the development of a hemorrhagic fever epidemic in an agriculturally based society like the Cofitachequi chiefdom.

Mississippianization was a relatively young phenomenon in the larger span of history in North America. Prior to this period, agricultural dependence and population aggregation was not common, as most Woodland period systems were seasonally mobile and relied on a diverse subsistence base of wild resource extraction and low-level cultivation of indigenous annuals. Mississippianism brought a whole new adaptation to the Southeast, in which stable villages and larger regional population densities were fueled at different times by chiefly manipulation of agricultural surpluses and tribute. Crowded conditions, agricultural dependence, and prolonged droughts during the MCA punctuated by shorter pluvial intervals might have created the conditions in which outbreaks of rodent-transmitted hemorrhagic fevers or similar diseases occurred regularly in Mississippian societies. Did such outbreaks play a larger role in the settlement shift from large nucleated villages in the thirteenth and fourteenth centuries to the dispersed farmstead pattern seen to have been so prevalent after the middle of the fifteenth century across much of the SAM culture area?

The Spaniards may have observed this process in motion when they passed through Cofitachequi. Dismissing the reports of abandoned towns, which were most probably mound sites abandoned over a century earlier, both Elvas (Bourne 1904a:66) and Garcilaso (Varner and Varner 1951:325) say that, after the pestilence struck Talimeco, many of the inhabitants of the chiefdom scattered to other towns or fled into the forest and had not yet returned. Whether this relationship was understood or not, such a response might very well have lessened the spread of an epidemic disease. Carmona's testimony that 500 abandoned houses could be seen at Talimeco, however, suggests that, before the epidemic occurred, Talimeco was indeed a residential hub with a large domestic population. Could similar epidemics have struck other fourteenth- and early fifteenth-century population centers like Moundville or Etowah? Were indigenous epidemics major factors in the settlement shift from nucleated mound centers to widely dispersed homesteads?

If outbreaks of rodent-transmitted pathogens were so abundant, it could be asked why possible evidence of such an epidemic was encountered only at Cofitachequi and not during Spanish visits to other polities as well. According to the Center for Disease Control fact sheet on viral hemorrhagic fevers, the geographic distribution of pathogens of this type is limited by the size of the rodent reservoir. Importantly, humans are not natural reservoirs for these viruses and they are rarely if ever transmitted from person to person. Consequently, outbreaks of rodent-transmitted hemorrhagic fevers occur only sporadically and irregularly. The typical buffer zones between SAM polities would have served as effective barriers to the geographic spread of infected rodent reservoirs, which would have been concentrated around agricultural fields. Moreover, humans travelling from one town to the next would not have been effective carriers of the disease. Finally, the much lower population densities of SAM polities relative to central Mexico would have worked to reduce the impacts of mortality on the resident human populations, thus explaining the seemingly high survivability rates witnessed by the Spaniards at Cofitachequi. Moving from a location of high agricultural field density around an established town to old fields or secondary forests at a distance might provide just enough isolation from the pathogen to avoid further infections. The purported league separation between Talimeco and Cofitachequi (Varner and Varner 1951:314) could very well have been effective in mitigating the extent of the outbreak. At the end of the day, then, the impacts of possible epidemics, especially those triggered by the dynamics of prolonged droughts, should be considered as an important factor in understanding the abandonments of town centers in SAM societies.

8

The Broader Implications of Late Prehistoric Societal Collapse and Transformation in the Southern Latitudes of the United States during an Age of Global Warming

What a "square chicken" accomplishes is, first, explain a large portion of the observed variability, and, second, organise the consideration of other causal factors by forcing the question: "What have we had to do to square this chicken?"
—Glenn Davis Stone (2001:164)

If one takes the footprint of North American megadrought during the MCA as a starting point, it is not difficult to begin constructing narratives about its human consequences (e.g., abandonment, migration, coalescence, and disease). This book has focused on a small subset of the larger story, endeavoring to predict the timing and distribution of such consequences as these were played out in the SAM culture area by examining the geographic patterns of occupation sequencing. What has yet to be done, however, is to ask, as Stone urges in the epigraph that starts this chapter, just what parts of the narrative push out of this box and what factors might require further consideration if we are to arrive at more satisfying explanations. I have identified four topics that push the chicken out of square and need further development.

The first concerns the question of why some regions in the southern latitudes of North America seem not to conform to the expectations for collapse and abandonment in the fifteenth century as would be predicted by the +PDO/+AMO model of chrono-geographic drought frequency. Perhaps the most glaring nonconformity is represented by the Hohokam of southern Arizona, wherein the Core Area of that irrigation society was subjected to processes of organizational decline and depopulation during wet conditions in the late fourteenth and early fifteenth centuries. Paleoclimatic data show that pluvial conditions and increased flooding during this time frame are actually consistent with the expectations of the model and I will further demonstrate how this finding provides support for a particular reconstruction of societal change linked to the ravages of flooding in the region (Graybill and Nials 1989; Graybill et al. 2006). Although playing out under different adaptive and climatic conditions, this sequence of collapse followed by decline and eventual abandonment should provide a basis for better understanding social responses to environmental perturbations during the MCA in the Mississippian world.

A second issue concerns the question of just how pervasive indigenous drought-related epidemics might have been in the Mississippian Southeast. If an indigenous hemorrhagic fever outbreak caused the abandonment of the former capital of the Cofitachequi chiefdom (Talimeco), it would seem likely that such an event process would have been repeated at some other densely populated Mississippian towns as well, without causing sudden regional abandonments. Accordingly, I ask and address two related questions. First, are there any other candidates where seemingly abrupt town abandonments occurred under similar dry-wet climate cycles? Second, if such candidates exist, how might we go about demonstrating that these events were actually caused by indigenous disease pathogens?

Third, if fifteenth-century Mississippian chiefdoms were subjected to the forces of coalescent societal reformation, then according to Kowalewski (2006) they should exhibit evidence of a breakdown in centralized authority structures. Circumstantially, I showed that the conditions favoring coalescence were operable in the RDR in the fifteenth century, but I did not discuss in much detail the social consequences that should accompany this new landscape of relations. This far-ranging topic deserves much greater attention than I can give it here. However, I isolate one avenue of the larger question of decentralization by comparing differences in the effort expended in the construction of monumental architecture between prehistoric and protohistoric Mississippian chiefdoms in one region of the SAM culture area that I think supports Kowalewski's hypothesis.

Finally, I will touch on another subject that I think will shape future research on the late prehistoric archaeological period. Namely, how do we begin to identify coalescent aggregations in the archaeological record and how might we go about studying the strategies and processes of integration at the nested levels of household, settlement, and region? Once again, I draw on the mounting evidence of migration and cultural integration in the southwestern United States to frame this discussion, as theoretical and methodological approaches to this problem have been at the forefront of archaeological investigation in this geographical area for some time.

The Footprint of Megadrought and the Hohokam

Ignoring what the paleoclimate models are forced to do to square their chicken, it is with some appreciation for irony that I find the Hohokam render the squaring of the abandonment/drought scenario difficult to contain. I became familiar with the mystery of large-scale prehistoric collapse and abandonment events during my time working in the Hohokam Core Area in the 1980s, specifically in Canal System 2, which is now cemented over by the urban sprawl of Phoenix, Arizona (Figure 8.1). Over a multitude of centuries, the Hohokam developed a stable and complex society dependent on intensive irrigation agriculture, but beginning in the later half of the fourteenth century a poorly understood interlude of organizational decline is registered, culminating finally in the virtual abandonment of the region (Chenault 1996; Sires 1984). The hallmark of the collapse was the shift from a complex system of large towns composed of coursed adobe residential compounds, platform mounds, and great houses in the Civano phase to a dispersed and less populated system of free-standing pithouse villages, hamlets, and farmsteads during the Polvorón phase. Although the places where major Civano phase towns were built continued to be occupied, additions to existing monuments or initiations of new monuments are not documented. In addition, the residential compounds shrank in occupied space and many were abandoned entirely and left to deteriorate. In some instances, Polvorón phase pithouses were in fact built on top of older residential compound walls and on the tops of mounds. Current thinking is that, after a period of about one hundred years, the Polvorón phase ended with the virtual abandonment of the Core Area, sometime around AD 1450 or 1475 (see Fish et al. 1994; Hill et al. 2004). Whether the Core Area was ever completely abandoned is an open question, but it becomes increasingly difficult to identify villages or even farmsteads that might date much later than the fifteenth century.

If the +PDO/+AMO scenario accurately depicts climatic conditions in southern Arizona at the end of the MCA, however, the Core Area abandonment coincides not with a megadrought but with pluvial or drought-neutral conditions (see Figure 4.2c). McCabe and colleagues (2004) classify as pluvial (i.e., a wet regime) those isotherms in their maps that depict drought frequencies of less than 25 out of 100 years; if the chronology of megadrought variability during the MCA is correct, then wet conditions should have prevailed across almost all

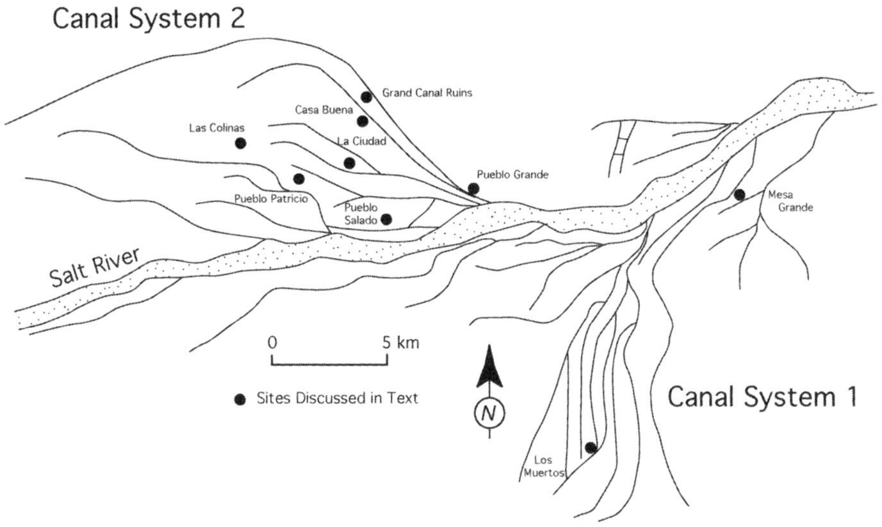

Figure 8.1. Generalized map of Canal System 2 in the Core Area of the Hohokam, Phoenix, Arizona.

of the Southwest in the fourteenth and fifteenth centuries. This would seem not to square with the megadrought-abandonment hypothesis, but before we can launch into the reasons why there would be such a contradiction, it is necessary to determine whether the +PDO/+AMO scenario accurately reflects the climatic conditions at this time.

There is a way to test this question by consulting the North American Drought Atlas, which provides access to summer (June–August) reconstructions of the self-calibrating Palmer Drought Severity Index (PDSI) on a 0.5° latitude/longitude grid centered over North America (E. Cook and Krusic 2004). The atlas was built by synthesizing 1,845 tree-ring chronologies and covers the period from AD 1 to 2005. PDSI measures the availability of soil moisture at a given time relative to what is regarded as normal or average conditions for a region and is calculated by incorporating temperature and precipitation data within a water balance model (E. Cook et al. 2007:98). The corresponding paleoclimatic variable represents an analog to this modern index, which primarily reflects tree-ring growth responses during the spring and summer. It should be noted, however, that a significant portion of tree-ring growth is linked to soil moisture buildup from precipitation that falls during the previous winter and spring. In practice, PDSI values range between -4 (severe drought) and +4 (extremely wet) in accordance with Palmer's (1965) classification of wet and dry conditions. Respectively, moderate and mild droughts prevail between PDSI values of -2.00 and -2.99 and -1.00 and -1.99. Incipient wet or dry and normal conditions range between -0.99 and +0.99, while wet conditions exceed values of + 1.00.

Figure 8.2 displays the time series graph for annual PDSI values between the years AD 1300 and 1500 as extracted from a window of graticule coordinates (32.5–34.0° N/110.5–113.5° W) corresponding roughly to the Hohokam Core Area (i.e., the lower Salt and middle Gila valleys). The dark line in the graph represents an eight-year smoothing spline, while the gray lines record actual tree-ring measurements. Splinning is a commonly used statistical technique aimed at reducing the effects of nonclimatic, localized variance registered in individual tree-ring samples (E. Cook and Peters 1981). Yearly values for the Core Area window vary considerably between dry and wet periods, but they alternate in a relatively regular cycle. The long-term average for this period indicates moderately wet conditions (Table 8.1). Although the fourteenth century was wetter than the fifteenth, the averaged 25-year increments displayed in the table indicate that the net negative averages were negligible in the fifteenth century, with the exception of the years between AD 1450 and 1474, where moderate drought conditions prevailed. Thus, the PDSI data are in line with the prediction that either wet or drought-neutral conditions predominated during the fourteenth and fifteenth centuries in the Core Area.

Rather than presenting a challenge to the +PDO/+AMO scenario, then, the paleo-PDSI data for southern Arizona actually help to confirm it. This particular juxtaposition of climatic conditions happens because circulation patterns are teleconnected to sea surface temperatures in the Pacific Ocean (PDO) by what is known as the El Niño-Southern Oscillation (ENSO). The ENSO is "a two-to-seven-year cycle of equatorial sea surface temperature (SST) anomalies in the eastern Pacific Ocean that comprise the leading mode of interannual precipitation variability in the western US" (Brown and Comrie 2004:1). During the winter months, cool conditions associated with ENSO (La Niña) are linked to below-normal precipitation in the Southwest, while warm conditions (El Niño) yield above-normal precipitation. Precisely how ENSO and PDO are related is not fully understood. It is known that the decadal pattern of SST (PDO) has a similar spatial structure to the ENSO, but the former is expressed in greater amplitude at higher latitudes (Zhang et al. 1997). Brown and Comrie (2004) have suggested that PDO represents the integration of interannual ENSO variability into the decadal SST signature, which is known to shift pronouncedly over phases of 20 to 30 years. The PDSI data presented in Table 8.1 suggest that these patterns may have even longer multidecadal signatures on the order of 50 to 75 years.

The model can be further evaluated by examining the alternative mega-

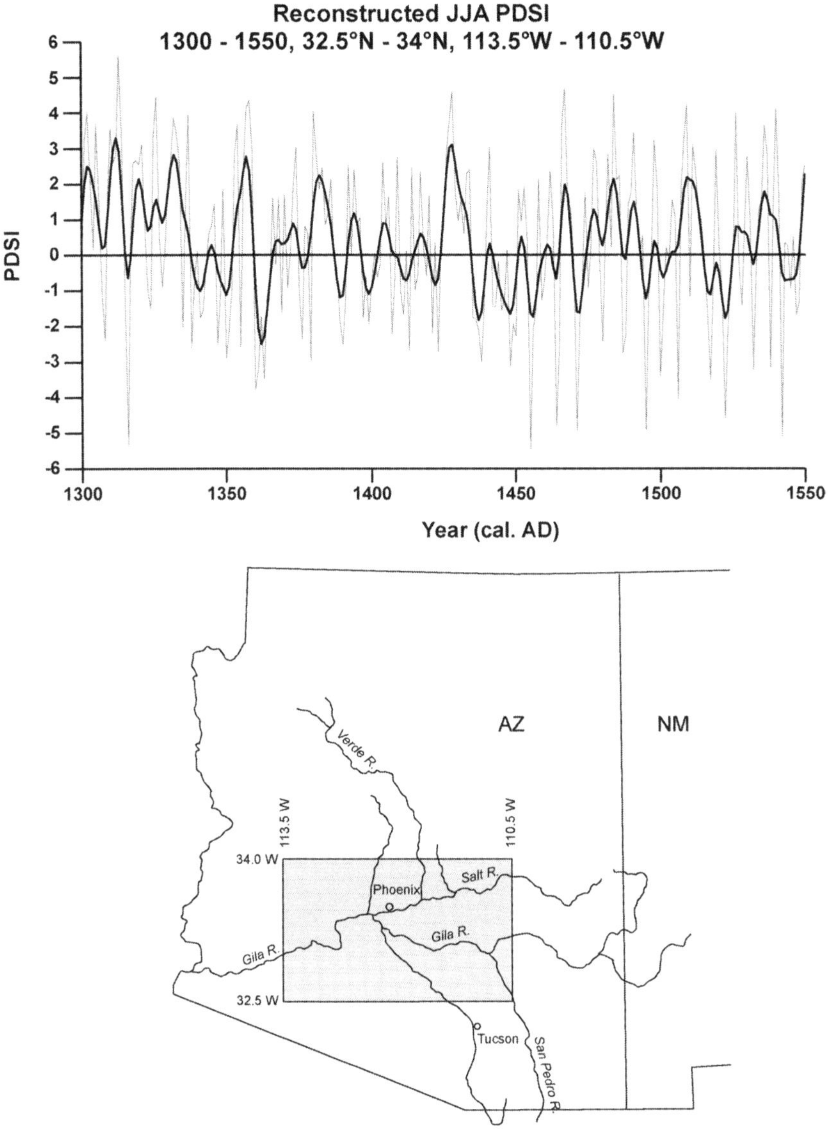

Figure 8.2. Palmer Drought Severity Index (PDSI) time series curve for the Hohokam Core Area between AD 1300 and 1550. JJA = June–July–August (data from the North American Drought Atlas, Cook and Krusic 2004).

Table 8.1. Twenty-Five-Year Increments for Mean PDSI Values in the Hohokam Core Area for the Period AD 1300 to 1499

25-Year Increment (cal AD)	Mean Measured PDSI	Mean PDSI (8-Year Spline)
1300–1324	1.33	1.44
1325–1349	0.80	0.72
1350–1374	0.19	0.19
1375–1399	0.28	0.38
1400–1424	-0.07	-0.06
1425–1449	-0.01	0.03
1450–1474	-0.21	-0.23
1475–1499	0.68	0.59
Grand Average	0.38	0.38

Note: Data from the North American Drought Atlas (Cook and Krusic 2004).

drought scenario in which cooler Pacific SST (-PDO) is coupled with warmer Atlantic SST (+AMO). In this scenario, much of the Southwest is predicted to have experienced much drier conditions coinciding with the timing of the abandonments in the Four Corners and Mogollon regions. Figure 4.2d indicates that relatively severe drought conditions are predicted to occur in both of these regions when Pacific SST is cooler, while the Core Area is predicted to support less severe, but still dry conditions. Instructing the North American Drought Atlas to extract PDSI measurements from the Core Area graticule window for the period between AD 1100 and 1300 confirms again these model expectations (Table 8.2). Over this period, there is a net negative PDSI, although relatively wet and dry phases on the order of 50 to 75 years duration can be identified. This suggests that the two scenarios of +AMO megadrought may have alternated in occurrence on a continent-wide basis during the MCA, on multidecadal time frames, and that there would have been interludes of more severe drought punctuated by pluvial conditions and perhaps increased flooding depending on geographic location and the patterns of teleconnected atmospheric temperature anomalies.

Even though the climatic implications of both megadrought scenarios are supported by the paleo-PDSI data for southern Arizona, we are still left with

Table 8.2. Twenty-Five-Year Increments for Mean PDSI Values in the Hohokam Core Area for AD 1100 to 1299

25-Year Increment (cal AD)	Mean Measured PDSI	Mean PDSI (8-Year Spline)
1100–1124	0.42	0.31
1125–1149	-0.73	-0.74
1150–1174	-0.29	-0.31
1175–1199	-0.30	-0.13
1200–1224	0.22	0.11
1225–1249	0.85	0.75
1250–1274	-0.35	-0.15
1275–1299	-1.01	-1.11
Grand Average	-0.15	-0.16

Note: Data from the North American Drought Atlas (Cook and Krusic 2004).

the contradiction that the Hohokam Core Area collapse and abandonment happened when either wet or drought-neutral conditions prevailed. Once the irrigation dependent economy of the Core Area Hohokam is appreciated, however, it is easy to see how pluvial conditions might pose a greater threat to survivability than drought. This is because it is during El Niño regimes, when Pacific SST is warm, that the frequency and magnitude of floods are highest in southern Arizona (Ely 1992; Ely et al. 1994; Harden et al. 2010; Huckleberry 2015a; Swetnam and Betancourt 1998). The most critical problem faced by irrigationists in arid environments is not so much reduced stream discharge under local drought conditions but the less predictable, destructive impacts to canal systems posed by large flood events. Damage to canal facilities and intakes along with shifts in river channel locations (avulsion) and streambed morphology (stable single channel versus braided pathways) could have potentially rendered the systems inoperable for extended periods.

Donald Graybill and colleagues (Graybill and Nials 1989; Graybill et al. 2006: 114–120; Nials et al. 1989) have proposed just such a hypothesis for the Core Area Hohokam. Deploying sophisticated paleo-discharge reconstructions based on tree-ring data, they have identified clustered, potentially catastrophic flood events in both river systems during terminal Civano times. On the lower Salt River, these events are inferred to have happened in consecutive sets of years between AD 1357 and 1359 and AD 1381 and 1384, while major floods on the middle Gila are thought to have occurred in five of the nine years between AD 1420 and 1428. Extreme magnitude floods of similar multiyear duration had not materialized for over 500 years in the Salt River and a comparable 9-year span of flooding had not been seen in over 400 years on the Gila. The several centuries prior had been characterized by stable, aggrading regimes conducive to channel stability and optimal conditions for canal irrigation. Graybill and colleagues argue, however, that the onset of more variable conditions wherein closely spaced extreme discharge episodes were sandwiched between prolonged periods of low streamflow exposed the river channels to greater risk of laterally erosive geomorphic processes, such as channel avulsion and widening. They ultimately conclude that this high amplitude variation exceeded the ability of the socioeconomic organization of the Core Area Hohokam to respond effectively to the devastating damages visited upon the physical facilities of the canal systems and their operation within the short span of time needed to reproduce Civano phase society. In keeping with Graybill and colleagues' general hypothesis, then, differential timing of catastrophic flood sequences in the two river systems would imply that the Civano phase collapse occurred earlier on the Salt than on the Gila. Accordingly, they propose that the Salt River Civano phase terminated as early as the AD 1350s and as late as the AD 1380s, but the Civano system on the middle Gila continued until the AD 1420s. The timing of these events has not yet been firmly established through absolute dating, but if they can be shown to coincide with the respective offset flood clusters in the two valleys, it would supply strong support for Graybill et alia's version of events.

Recent critiques of the model have come from a number of different directions. Kyle Woodson (2015) has presented a comprehensive summary and discussion of these, as well as a broader treatment of the relationship between flooding and the operation of canal systems across the larger span of the Hohokam occupation of southern Arizona. I have extracted from his overview three challenges that pose problems for accepting Graybill and colleagues' hypothesis. These are the reliability of streamflow retrodictions to predict the magnitude of flood events, the apparent absence of evidence for catastrophic flood damage to the river channels dating to the late fourteenth and early fifteenth centuries, and the problem of directly characterizing the extent of flood damage in canal systems and linking this evidence with specific flood events.

Historic data indicate that although a strong correlation exists between peak discharge (cubic feet per second) and tree-ring retrodicted annual discharge (acre-feet) years, some much smaller peak discharge years are associated with equally high, or even higher, retrodicted discharge from tree-ring reconstructions. Just because the relationship between large floods and retrodicted annual discharge is not linear, however, does not disprove that the catastrophic flood event sequences that Graybill and colleagues have identified happened in the way they specify. It does suggest, however, that the inferred relationship needs independent confirmation.

This leads to a second and more serious critique. Subsequent stratigraphic investigations have not found corroborative evidence that the geomorphic structure of either river channel was significantly modified at the times that these megaflood events were supposed to have occurred. The most extensive of these investigations was conducted by Waters and Ravesloot (2000, 2001) along the terraces of the middle Gila River. They concluded that only two major periods of channel modification are registered in the recent past: a major channel widening and cutting event that they date between 950–800 cal yr BP or AD 1000 to 1150 and a less extensive widening episode dating to the late nineteenth and early twentieth centuries. Between these events, they see only a regime of consistent low streamflow resulting in an extended period of channel stability and overbank sediment accumulation on the floodplain. This reconstruction runs counter to the hypothesis that Graybill and colleagues propose and leads Waters and Ravesloot (2001:293) to claim that the Civano phase collapse "in the middle Gila River valley cannot be linked to landscape changes." They further contend that the alluvial history of the lower Salt should mimic that of the middle Gila, because both stretches are responding to the same base-level conditions of the drainage catchment. Ergo, significant geomorphic modifications will not be found on the lower Salt relating to the Civano phase collapse either. As I explain below, neither of these inferences is necessarily supported by direct evidence.

Figure 8.3 displays the later two stages of the alluvial stratigraphy of the middle Gila as reconstructed by Waters and Ravesloot (2000:Figure 5). The lower profile represents what was actually observed in their trenching program, while the upper profile depicts the stable, gradually aggrading floodplain and narrowing channel that they infer to have been in place after circa AD 1150. A problem arises, however, when it is realized that the deposit representing the critical period of inferred flooding that Graybill and colleagues identify is not directly observable because it was removed by the downcutting and channel widening episodes accompanying the floods of the late nineteenth and early twentieth centuries. As Huckleberry (1993:76–83, 97; 1999:13) describes it, the historic modifications to the floodplain primarily occurred during a period of above-normal precipitation between 1905 and 1930, wherein an increased frequency of high magnitude as well as smaller flood episodes occurred. The 1905 floods succeeded an extended period of drought starting in 1899, which diminished vegetation cover and exposed the floodplain to greater risks of erosion, resulting in significant widening, channel avulsion, and braided stream morphology. Dur-

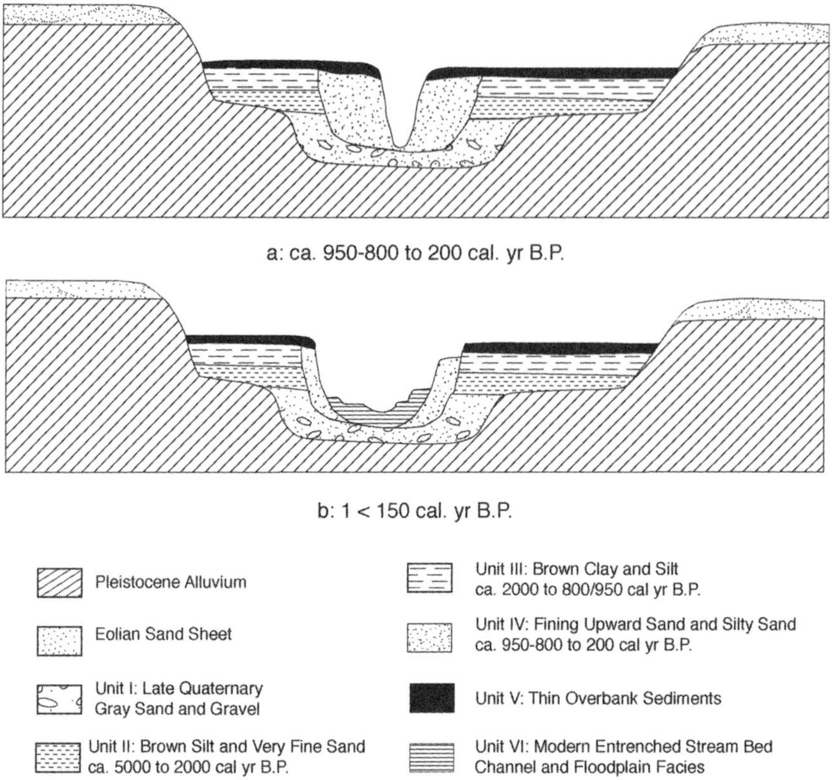

Figure 8.3. Stratigraphic sections of the geomorphic history of the middle Gila River: (a) ca. 950–800 to 200 cal yr BP; (b) 1 < 150 cal yr BP (adapted from Waters and Ravesloot 2000).

ing this period, the channel of the middle Gila widened by a factor of about 400 to 500% from its stable condition measured in 1869; subsequently, after 1930 low flow conditions resumed and a single, narrow channel conducive to canal irrigation reappeared. The floods of the AD 1420s, according to the retrodicted streamflow reconstruction (Graybill et al. 2006:98, 117), had a similar structure. Two flood years of magnitudes comparable to those of the 1905–1930 period are recorded along with a number of closely spaced smaller floods. In addition, this increased interlude of floods was preceded by a 10-year period of drought that could have exposed the floodplain to higher risks of hydraulic erosion. If the two periods were similar in structure and duration, then it is likely that their geomorphic impacts would have been similar as well. That the 1420s event is not present or easily discriminated from the effects of widening and cutting associated with the 1905–1930 period in the middle Gila stratigraphy cannot be used to disprove that it too resulted in major modifications to channel and stream geometry conforming closely in extent and breadth to the downcutting episode that Waters and Ravesloot associate with the historic period. According to the generalized profile (see Figure 8.2b), downward cutting of the historic floodplain contacts the Late Quaternary sand and gravel bed, so the remnant of the earlier event would only be preserved laterally, if at all, along the edges of the entrenched streambed. Such evidence has not yet been searched for adequately, however, as only one of the four trenches that Waters and Ravesloot excavated exposed even a portion of the modern floodplain.

With the exception of a single location not more than 4 km from its confluence with the middle Gila (Huckleberry et al. 2013; Onken et al. 2004), the lower Salt is *terra incognita* for geological stratigraphic studies. Consequently, direct evidence to support the proposition that the geomorphic history of the two rivers is the same is scanty and inconclusive. In fact, their contrasting catchment properties and hydroclimatologies (Huckleberry et al. 2013:49) suggest that they have distinctive flood histories and probably somewhat different geomorphic sequences. The majority of floods on the Salt are fed by winter frontal storms associated with large-scale low-pressure systems that butt against the topographic barrier formed by the Mogollon Rim escarpment, while floods on the middle Gila most frequently result from summer monsoons and tropical cyclones coming from the south. Ely and colleagues (1994:978–979) have shown that two-thirds of the large historic floods on the Salt River occurred during the winter; the same is true for only 14% of the floods on the Gila. From the record of historic floods on the middle Gila, Huckleberry (1993:95) observed that large, high discharge floods produced by summer cyclones had very little impact on channel geometry because of their generally short duration. By contrast, large winter floods tend to be of longer duration and, in combination with other factors related to lower sediment loads and water temperatures, they have the potential to exert much greater stress on channel geometry. Consequently, the two river systems should not be expected to have identical flood histories and, even when floods were coincident, they might not be expressed equally in changes to channel morphology, depending on specific atmospheric patterns of precipitation and runoff. This logic is further supported by the stark differences in annual discharge and floodplain widths. Mean annual discharge on the middle Gila is only about a third that of the lower Salt (Graybill et al. 2006:82–83) and the same magnitude of difference is reflected in the widths of the floodplains. The lower Salt floodplain extends across an area measuring about 4 km, while the floodplain width of the middle Gila is only about 1 km.

If the streamflow retrodictions of Graybill and colleagues (2006) are accurate, the flood histories of the two rivers are not fully synchronized, at least for high discharge episodes, and there is every expectation that these differences will be reflected in different geomorphic records. (See Loendorf and Lewis [2017] for an interesting hypothesis about how these differences might have created alternating flows of population between the two river systems). The major channel cutting and filling event on the middle Gila that Waters and Ravesloot date to the period between AD 1000 and 1150 may be registered in the geomorphic structure of both rivers, as this was an unprecedented period of large floods in both reconstructions. However, the subtler modifications associated with events like the 1905–1930 widening and entrenchment or the possibly similar 1420s flood sequence identified by Graybill and colleagues on the middle Gila may have resulted in smaller, unique imprints on the history of each river system. For instance, the major flood sequences proposed for the 1350s and 1380s on the lower Salt are not reflected in the streamflow reconstruction of the middle Gila and by the same token the 1420s event on the middle Gila is not registered on the lower Salt, at least in terms of retrodicted discharge magnitudes.

Based on this discussion, the claim advanced by Waters and Ravesloot (2003: 401) that climate and landscape modification are unimportant factors in explaining the Civano phase collapse cannot be substantiated with any degree of confidence. However, even if the proposed flood sequences that Graybill and colleagues say happened in the manner they specify, there is still the question of whether the impacts would have been sufficient to cause a societal collapse. Woodson's (2010) analysis of the earlier Sedentary to Classic period transition in the middle Gila Vlley, which correlates with the massive widening and cutting episode that Waters and Ravesloot date to sometime between AD 1000 to 1150 (see Figure 8.3), documents a dramatic settlement shift upstream in response to channel modifications. Even so, Woodson argues that complete recovery of the new irrigation system took only several crop cycles and entailed neither cultural collapse nor regional abandonment. By his own characterization, however, the settlement pattern changes he describes suggest that the developing hierarchical structure of Sedentary period society collapsed. Large, nucleated ballcourt villages were depopulated and residents dispersed into smaller settlements. In addition, population levels declined and it is hypothesized that some social groups migrated to the lower Salt. On a general level, then, these changes are not unlike those that have been proposed for the later Civano to Polvorón transition.

The question that remains is whether the geomorphic modifications of flood sequences hypothesized to have occurred in the late fourteenth and early fifteenth century would have had similar consequences to those attendant with the lower energy flooding events of the historic period. There is little argument that impacts to the early twentieth century Gila Pima canal systems were severe and resulted in the destruction of intakes, infilling of main trunks with flood deposit, repositioning of intakes to adjust to changes in channel morphology, and the establishment of new canal alignments (Huckelberry 1993, 1999). Darling and colleagues (2004:290) argue that these destructive and reconstructive events, which entailed shifts in channel locations and the loss of one mile or more of the southern riverbank, factored heavily in the relocation of settlements. It would seem, then, that even though the magnitude of the events was different, the impact of each had reached a critical threshold requiring significant societal reorganization. In the case of the Gila Pima, who were organized within the range of egalitarian or weakly ranked authority structures, reorganization might have resulted in only minor adjustments within a continuous cultural trajectory. This might not have been the case for Civano phase society, however. The existence of settlement ranking defined by differential distributions of monumental architecture suggests that the agricultural system would have been increasingly stressed by the need to support a sociopolitical hierarchy through surplus production. Disruption of long-term land tenure arrangements caused by the destruction and relocation of irrigation systems would likely have threatened the established system of social relationships and could have created the necessary conditions for the collapse of a ranked authority structure.

A full consideration of the factors that lead to the Civano phase collapse and the ultimate abandonment of the Core Area is beyond the scope this discussion (see Bayman 2001:291–292), but the hypothesis presented by Graybill and colleagues can be further evaluated through a consideration of absolute date associations from sites in Canal System 2. This task is complicated by the current confusion over the apparent temporal overlap of dates ascribed to the Civano and Polvorón phases, together commonly referred to as the late Classic period. Noting this feature of both the radiocarbon and archaeomagnetic databases, Henderson and Hackbarth (2000) argue that the Polvorón phase pithouses found at Civano phase towns represent the abodes of lower status families who had settled outside of the high-status compounds and platform mound complexes that constituted the residences of the original firstcomer families. This interpretation, or at least the idea of extreme chronological overlap between the phases, has grown to prominence in Core Area Hohokam studies, leading to potentially erroneous interpretations of population dynamics as a direct outcome of temporal and cultural conflation (cf. Henderson and Clark 2004; Hill et al. 2010, 2015a; Ingram 2008). Below I show how a Bayesian approach to analyzing absolute dates can clear up this confusion and can demonstrate the likelihood that the two phases are sequent with very little overlap. Moreover, I show how this interpretation synchronizes with Graybill and colleagues' reconstruction and discuss the implications of these findings for modeling the collapse and subsequent abandonment of the Core Area.

The radiocarbon database was collected during the 1980s and 1990s and is made up primarily of long-lived specimens of wood and agave charcoal that yield estimates that antedate the targeted events by unspecified numbers of years. Moreover, most of the error ranges are large, on the order of 50 to 80 years. The archaeomagnetic dating has similar problems of multiple date options and insufficient resolution in this time frame. There is little wonder, then, that absolute dates from closely adjacent phases would be difficult to distinguish. In recent years, several Bayesian analysis programs have been developed that provide a firmer basis for evaluating such problems with radiocarbon assays.

To further address phase overlap in the late Classic period, I apply the Oxcal version of this analysis program (Bronk Ramsey 2009a) to a select set of radiocarbon dates associated with architectural features from sites in Canal System 2 on the Salt River (Table 8.3). Classic period architectural styles are temporally sensitive (see Mitchell 1994) and when combined with ceramic evidence they provide a sound basis for selecting samples with reliable phase associations. The sample includes dates across the full range of the Hohokam Classic period because Bayesian modeling tends to yield better results when constrained by data from longer sequences. All but one of the radiocarbon assays that Henderson and Hackbarth (2000:Table 1) report in their study is included in the Canal System 2 sample. This particular specimen was recovered from an architectural feature (F. 8/9–40) at Pueblo Salado that they characterize as a compound adobe room. However, the excavators (Greenwald et al. 1995:55–57) identified it as a freestanding, adobe-walled surface structure containing three floors, each separated by 4 to 6 cm of mottled sand and silt indicating periods of abandonment and reoccupation. Moreover, the function of the structure appears to have changed from storage to residential through time, as the latest floor contained a hearth. The conventional age of the charcoal specimen (540 ± 80 BP) from the room would be positioned very late in the Canal System 2 sequence, but information concerning its precise stratigraphic context is not given. It seems likely that the dated charcoal was collected in conjunction with the archaeomagnetic sample taken from the hearth in the latest floor to aid in calibration, which would lead to the conclusion that the radiocarbon assay is as easily associated with a Polvorón phase reuse episode as it is with the Civano phase.

It is easy to see how one could argue in favor of extreme overlap between the Civano and Polvorón phases just by looking at the distribution of conventional radiocarbon assays in Table 8.3. In fact, there is no statistical difference ($t = 10.02$, $X^2_{.05} = 21$, $df = 12$) between the seven (#22–28) latest Civano ages

Table 8.3. Cleansed List of Radiocarbon Assays Associated with Architectural Features in Canal System 2, with the Addition of Polvorón Samples from the Queen Creek Drainage

Sample	Site	Phase	Conventional Age rcyB.P. (± 1-σ)	Feature No.	Architectural Type
1	Pueblo Grande	Sacaton	1040 ± 80	2782	S-1 Pithouse?
2	Pueblo Grande	Sacaton	1020 ± 60	575	S-1 Pithouse
3	Pueblo Grande	Sacaton	960 ± 60	2786	S-1 Pithouse
4	Pueblo Grande	Sacaton	950 ± 50	771	S-1 Pithouse
5	Casa Buena	Soho	1110 ± 70	150	True Pithouse
6	Grand Canal Ruins	Soho	1060 ± 60	42-16	Deep Pithouse
7	Pueblo Grande	Soho	1000 ± 45	1201	Other Pithouse
8	Pueblo Grande	Soho	950 ± 60	1550	Deep, Adobe-Lined Pithouse
9	Pueblo Grande	Soho	940 ± 70	782	Post-Reinforced Room
10	Pueblo Grande	Soho	905 ± 50	996	Post-Reinforced Pithouse
11	Pueblo Grande	Soho	845 ± 45	996	Post-Reinforced Pithouse
12	Pueblo Grande	Soho	790 ± 60	490	Post-Reinforced Pithouse
13	Pueblo Grande	Soho	760 ± 80	229	Deep Post-Supported Pithouse
14	Pueblo Grande	Soho	760 ± 80	3111	Post-Reinforced Room
15	Pueblo Grande	Soho	740 ± 60	2206	Post-Reinforced Room
16	Pueblo Grande	Soho	730 ± 70	144	Stone-Lined Pithouse
17	Pueblo Grande	Soho	700 ± 70	2838	indet. Pithouse
18	Pueblo Grande	Civano	840 ± 80	636	Compound Room
19	Pueblo Grande	Civano	810 ± 40	601	Compound Room

Continued on the next page

Table 8.3. *continued*

Sample	Site	Phase	Conventional Age rcyB.P. (± 1-σ)	Feature No.	Architectural Type
20	Pueblo Grande	Civano	735 ± 50	636	Compound Room
21	Pueblo Grande	Civano	710 ± 35	2975	Compound Room
22	Pueblo Grande	Civano	685 ± 60	2975	Compound Room
23	Pueblo Grande	Civano	680 ± 50	2947	Compound Room
24	Pueblo Grande	Civano	660 ± 50	641	Compound Room
25	Pueblo Salado	Civano	660 ± 70	8/9-9-1	Compound Room
26	Pueblo Grande	Civano	610 ± 60	2024	Compound Room
27	Pueblo Grande	Civano	590 ± 50	674	Compound Room
28	Pueblo Salado	Civano	580 ± 80	8/9-16-1	Compound Room
29	Dutch Canal Ruin	Polvorón	630 ± 70	8-3-1	Pit House
30	Pueblo Grande	Polvorón	610 ± 60		Jacal Structure
31	Pueblo Grande	Polvorón	580 ± 60	2428	Adobe-lined Pithouse
32	Pueblo Grande	Polvorón	580 ± 50	2508	Narrow-walled Adobe Pithouse
33	Pueblo Patricio	Polvorón	570 ± 50	805	Narrow-Walled Adobe Pithouse
34	Casa Buena	Polvorón	500 ± 60	118	Narrow-walled Adobe Pithouse
35	El Polvoron	Polvorón	390 ± 70	Structure 4	Narrow-walled Adobe Pithouse
36	Las Fosas	Polvorón	430 ± 60	Structure 11	Narrow-walled Adobe Pithouse
37	Las Fosas	Polvorón	370 ± 60	Structure 12	Narrow-walled Adobe Pithouse

Note: Samples 1 through 4, 8 through 24, 26, 27, 31, and 32 from Mitchell (1994); Samples 5 and 34 from Howard (1988); Samples 25 and 28 from Greenwald et al. (1995); Sample 29 from Greenwald et al. (1994); Sample 30 from Bostwick and Downum (1994); Sample 33 from Henderson (1995); Sample 35 from Sires (1984); and Samples 36 and 37 from Crown and Sires (1984).

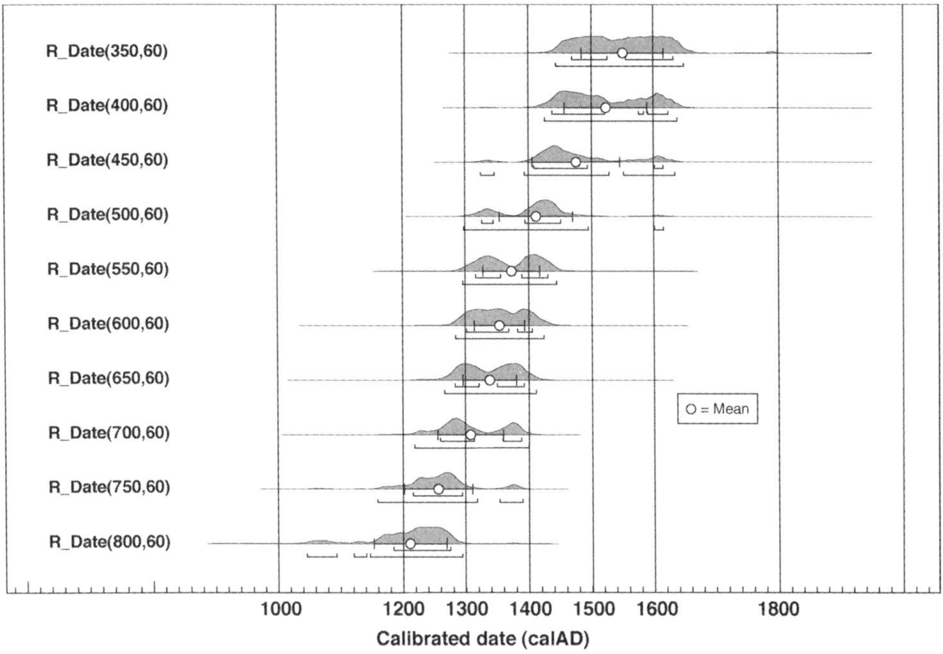

Figure 8.4. Heuristic demonstration of the bimodal structure of radiocarbon calibrations from AD 1200 to 1600 (adapted from the OxCal 4.3.2 [Bronk Ramsey 2017]; r.5 IntCal13 atmospheric curve [Reimer et al, 2013]).

and the six (#29–34) earliest Polvorón samples. As Thomas (1978) long ago warned, though, what is apparent is not always so in statistical analysis. The awful truth is that the calibration curve interjects variability into the interpretation of radiocarbon assays, often rendering our reliance on central tendency suspect. In the case of the late prehistoric time frame, as I have discussed previously for the SAM regional chronologies, calibration produces bimodal probability distributions with closely sequent date range options. Moreover, the chronological range of the modes shifts through time. Figure 8.4 heuristically displays this trend for late prehistoric sequences with arbitrary 50-year increments and common standard deviations of ± 60, which approximates the average uncertainty of the assays in Table 8.3. The bimodal couplet that covers the fourteenth and very early fifteenth centuries begins in earnest with conventional ages of about 700 BP and transitions to a mid-fifteenth- to early seventeenth-century couplet around 500 BP. The later one is represented in the sample by three Polvorón pithouses from the remote Queen Creek locality that is part of the outer reaches of the middle Gila drainage system, while the remaining Civano and Polvorón samples were obtained from Canal System 2 and belong to the earlier couplet. As can be seen, there is not a firm basis from which to evaluate the relative ages of the individual samples falling into either couplet because the means are tightly clustered and the modes have relatively equal probability densities throughout much of the trajectory.

It would seem that the extensive archaeomagnetic database collected over the years in the Hohokam Core Area would provide an alternative basis for evaluating the question of Civano/Polvorón phase overlap, but this method has its own unique limitations of discrimination in this time frame. Inspection of Table 1 in Henderson and Hackbarth (2000) indicates that the α95 uncertainty ellipses for individual Civano and Polvorón samples yield date ranges that cover the entire interval of relevance between the early fourteenth and the middle of the fifteenth century. This could be interpreted as evidence for extreme phase overlap, but there is a more fundamental reason that in fact explains this result. The length of the Secular Variation curve in this time range is extremely compressed (see Hagstrum and Blinman 2010) and therefore the resolution of most samples is not small enough to distinguish between the early and late ends of the trajectory. Standard uncertainty intervals would need to be reduced on the order of two to three times what individual samples commonly achieve before ages in this trajectory could be effectively discriminated. This might be accomplished through averaging groups of individual dates or by developing Bayesian methods for the analysis of archaeomagnetic samples. Neither of these approaches, however, has yet been sufficiently developed or applied in the southwestern United States, although Lanos and colleagues (2005) have made strides in developing a Bayesian modeling program that they have tested on sixth millennium BC Bulgarian archaeomagnetic data sets. For the moment, though, any further progress in evaluating phase overlap using absolute dates will come from the Bayesian analysis of radiocarbon assays.

Arguments in support of extreme overlap, then, should become more tempered when the limitations of absolute dating in this time frame are appreciated. It is certainly quite possible to construct a superficial chronological model that specifies extreme overlap between the Civano and Polvorón phases if the range of central tendencies is accepted at face value, but the question that has not been adequately addressed is whether the absolute date associations can alternatively support an interpretation of discrete phases without much overlap. If they can, then the timing of the phase transition might hold important implications for the Graybill et alia hypothesis and the pace of culture change. Addressing such questions is particularly well suited to Bayesian analysis, which can generate statistical tests of these inter-related propositions using a multiphase Markov chain Monte Carlo simulation model. The test statistic, the Agreement Index (A), pro-

Table 8.4. Estimated Bayesian End Dates for Phases within the Hohokam Classic Period Chronology, Results for Runs 1 and 2

Hohokam Phases	Phase End Dates First Run (All) (Mean ± SD) (cal AD)	Phase End Dates Second Run (w/o Samples 5 and 6) (Mean ± SD) (cal AD)
Sacaton	1085 ± 35	1117 ± 32
Soho	1260 ± 14	1259 ± 14
Civano	1324 ± 27	1324 ± 27
System 2 Polvorón	1416 ± 37	1417 ± 37
Queen Creek Polvorón	1525 ± 72	1522 ± 70
A_{model}	44.0%	80.1%

Note: Results generated from the OxCal 4.3.2 program (Bronk Ramsey 2009a).

vides a basis for evaluating goodness of fit between a hypothesized phase model and a set of calibrated radiocarbon dates. Indices that fall below a critical value of $A'c$ = 60%, which approximates a null hypothesis rejection region of $X^2_{.05}$, indicate a poor fit between the two.

Several different runs were undertaken to address outliers and to modify the underlying structure of the data set to contend with the offset problem that emerges from dating long-lived plant species. The first run included all of the assays within the sample and returned an Agreement Index (A_{model}) of only 44% (Table 8.4). This nonsignificant result, however, was not created by the Civano/Polvorón phase boundary, which individually yielded a significant separation. It was caused by three assays (#s 5, 6, and 17) associated with the early and late ends of the early Classic period (Soho) distribution. The former two had such low individual Agreement Indices (A_I) that they would qualify as true outliers, either due to the problems of old wood or because of the presence of an unconfirmed Sedentary-to-Classic transitional phase (the Santan phase) that should logically bridge this time frame (Doyel 2000; Gladwin et al. 1937). These dates were eliminated, then, in the second run, while the late end assay (#17) was retained because its A_I approached 60%. Removal of the two outliers achieved a good fit with the model (A_{model}=80.1%), although the projected end dates for the phases remained nearly unchanged (see Table 8.4).

Taken at face value, the estimated end date for the Civano phase of AD 1324 ±

27 is too early to be connected to the flood sequences of the late 1350s and early 1380s that Graybill and colleagues suggest were instrumental in the Civano phase collapse. However, the individual density plots for the Civano phase calibrations reveal that the group with the latest conventional ages (#22–28) have calibrated late mode probability densities that are much later than this estimated boundary. The pooled conventional age for this group of 641 ± 22 rcybp yields a calibrated late mode range of AD 1346 to 1394 at 95% confidence (95% CI). This suggests that there must be one or more factors acting on the model algorithm to cause it to exclusively emphasize the earlier mode, which ranges between AD 1285 and 1324 (95% CI). One factor that I think plays a major role in this modeled result of the first two runs is the wide range of ages within the Soho and Civano phases, which could bias the formation of temporal boundaries by emphasizing the within-phase averages. Inspecting the individual density distributions suggests that the calibrated dates in each of these phases can be separated into early and late groupings, within each of which the conventional ages are statistically the same (Table 8.5). A third run, then, was performed in which only the pooled means and standard deviations of the conventional ages within each group or phase were modeled so that each occupation period could be given equal weight. This did not reduce the error range of projected phase boundaries, but it did bring the Civano phase end date into closer alignment with the late mode distribution identified above. It also achieved an even higher agreement with the model assumptions of contiguous phases with only minimal overlap.

It is still a concern that the third run may be overestimating the time depth of late Classic phase boundaries because of the age offsets between the targeted events and the older ages of the dated long-lived wood and agave specimens. One solution would be to emphasize the later half of the standard deviations when estimating phase or subphase boundaries. Another method would be to add a universal age offset to the model (Bronk Ramsey 2009b). A fourth run was performed on the pooled conventional ages, adding a standard 25-year offset to each. This shifted the mean boundaries of all of the phases to slightly later positions, which may represent a modestly more accurate chronology. The accuracy and precision of phase boundaries could be considerably enhanced by implementing a policy in the future of dating only annuals or short-lived plant and faunal species, but the chronologies produced by runs 3 and 4 likely contain the actual boundaries within their reported error ranges. Bayesian analysis, then substantiates that the extant radiocarbon database from Canal System 2 can support the alternative hypotheses that the Civano and Polvorón phases are discrete with only slight overlap and that the Civano phase ended during the period of increased flood activity that Graybill and colleagues predict to have occurred between AD 1357 and 1384.

Table 8.5. Estimated Bayesian End Dates for Phases within the Hohokam Classic Period Chronology, Results for Runs 3 and 4

Phases/ Groups	Difference of Means[a]	Sample #s	Pooled Conventional Age rcyB.P. (± 1-σ)	End Dates, Third Run (Pooled, No Offset) cal AD (± 1-σ)	End Dates, Fourth Run (Pooled, 25 Offset) cal AD (± 1-σ)
Late Sacaton	t=1.47, $X^2_{.05}$=7.81, df=3	1 thru 4	983 ± 30	1095 ± 39	1102 ± 36
Ealy Group Soho	t=2.05, $X^2_{.05}$=7.81, df=3	7 thru 10	953 ± 30	1186 ± 46	1203 ± 51
Late Group Soho	t=4.45, $X^2_{.05}$=12.6, df=6	11 thru 17	775 ± 24	1255 ± 14	1276 ± 13
Early Group Civano	t=4.82, $X^2_{.05}$=7.81, df=3	18 thru 21	756 ± 22	1300 ± 27	1315 ± 27
Late Group Civano	t=3.25, $X^2_{.05}$=12.6, df=6	22 thru 28	641 ± 22	1359 ± 27	1376 ± 30
System 2 Polvorón	t=2.55, $X^2_{.05}$=11.1, df=5	29 thru 34	576 ± 23	1426 ± 42	1450 ± 39
Queen Creek Polvorón	t=0.51, $X^2_{.05}$=5.99, df=2	35 thru 37	397 ± 36	1474 ± 35	1536 ± 68
A_{model}				100.9%	100.9%

Note: Results generated from the OxCal 4.3.2 program (Bronk Ramsey 2009a).
[a]Calculated using Calib 6.1.0 calibration program, Stuiver and Reimer 1986–2011.

The mean boundaries that bracket the Canal System 2 Polvorón phase in runs 3 and 4 (see Table 8.5) suggest that this phase may have lasted 70 to 75 years in the Salt River. However, it is possible that the longevity of the phase was actually much shorter. Individual assays associated with the phase are tightly clustered in time but are widely distributed in Canal System 2, as they come from four different sites (see Table 8.3). If the earlier mode, which spans AD 1307 to 1363 (95% CI), is eliminated, then the later mode suggests that the Polvorón occupation in Canal System 2 may have lasted only about 30 years, between AD 1385 to 1416 (95% CI). If a 25-year offset is applied, the range of the later mode increases to AD 1385 and 1431, a potential 45-year span in which the original start date is retained. Both options are within 1-σ of the terminal phase boundary that is estimated with either Bayesian scenario (i.e., runs 3 and 4) and in this respect achieve consistency with the model. The conventional ages of the Queen Creek Polvorón sample are clearly much later, but this result too would be compatible with the claim of Graybill and colleagues that the Civano phase lasted up until the AD 1420s on the Gila River. The mean projected boundaries for the Queen Creek phase are somewhat more variable between the two scenarios, but if the probability distribution of the pooled mean is examined it suggests that the run 3 option is closer to the actual age when the early seventeenth-century mode is eliminated from consideration. Without an offset, the pooled mean yields a range of AD 1436–1525 (95% CI); with the offset, the range increases to AD 1455–1552 (95% CI). Using these results as guides, then, the midpoint of the Queen Creek Polvorón occupation is situated somewhere between AD 1480 and 1503.

Henderson and Hackbarth (2000:294) recognized that the resolution of individual absolute dates would be unsuccessful in segregating the temporal positions of "earlier from later or Civano from Polvorón structures." In spite of this realization, they still believed the overall pattern of date overlap was sufficient to demonstrate that Civano phase compounds "persisted throughout most of the time span defined by Polvorón." They then turned to stratigraphy and patterns of intrasite structure to develop a model that would be compatible with extreme phase overlap. First, they challenged the argument advanced by Sires (1984) and Chenault (1993) that instances of stratigraphic superposition always showed Polvorón pithouses above Civano compound walls, which would suggest that the two phases were discrete with only a limited period of transition. Examples from Mound 8 at Las Colinas, Compound F at Casa Grande, the Pueblo Salado Compound, and Habitation Areas 4 and 7 at Pueblo Grande were examined to verify this contention. They concluded that all of these instances, whether they involved earlier freestanding structures or compound walls, were associated with the earlier Soho phase and consequently there were no confirmed instances where Civano phase compound walls underlay Polvorón pithouses. Second, they proposed a settlement trajectory model that would accommodate virtually contemporaneous phases. In their view, the process of aggregation into large population centers climaxed between AD 1300 and 1350, accompanied by a shift from residential compounds characterized by freestanding rooms to ones with contiguous sets of rooms. After AD 1320 to 1350, growth in residential units involved almost exclusively accretions to existing compounds rather than the foundation of new ones. This process appears to have marked the beginning of population decline, which they argue accelerated in the fifteenth century. Declining population set the stage for a need to attract immigrants into Civano phase towns to sustain population levels and satisfy labor demands. These immigrants were the constructors of the Polvorón pithouses, which are commonly, although not exclusively, found in small clusters around the periphery of Civano phase residential compounds. The best examples of this pattern are seen at Las Fosas, Pueblo Salado, and Habitation Areas 4 and 7 at Pueblo Grande. The less substantial construction of the pithouses is further explained by cultural differences and the suggestion that many of these immigrants were only part-time, seasonal occupants. This dynamic of a migratory workforce attached to first-comer Civano families residing in room block compounds, then, may explain how these two very different forms of residential architecture existed contemporaneously.

There are problems with this model. The stratigraphic argument is at best equivocal. It is true that the Polvorón pithouses built at ground level in the Mound 8 precinct at Las Colinas superimpose only Soho phase structures, but the contention that the same relationship obtains on the top of the mound is unconfirmed and probably false. As Gregory and colleagues (1988:36–39) indicate, the top of the last building stage of the mound (Stage VII) was extensively disturbed by historic construction activities, including leveling for modern house construction. It was with Stage VII that massive coursed caliche retaining walls were extensively deployed to buttress the final mound expansion and several remnants of rooms with this type of wall were identified on the modern mound top. Gregory (1987:196) argues that these were part of "multiple room spaces." He did not specifically state that they formed elements of an adobe room block compound, but it is likely that the final stage arrangement would have been similar to the Civano phase room block built on top of the nearby Mound A at La Ciudad (see Wilcox 1987:93–106). By virtue of their deeper foundations, the two Polvorón pithouse floors escaped the destructive leveling activities that removed most of the earlier Civano surface structures. Clearly, it cannot be substantiated that the latest rooms that once stood on top of Mound 8 were associated with the Soho phase or that the Polvorón pithouses did not superimpose a Civano phase room block.

The proposed Soho phase association of Compound F at Casa Grande, where three Polvorón pithouses superimpose compound walls, is equally ambiguous. Henderson and Hackbarth (2000) infer that the presence of Casa Grande red-on-buff and the absence of polychrome indicates that the room block was constructed and abandoned prior to the Civano phase. Such ceramic evidence, if sufficiently representative of the entire and undefined occupation sequence of Compound F, may support the argument that the room block was abandoned prior to AD 1350, but it would not eliminate the possibility that it was occupied well into the early Civano phase. Finally, Henderson and Hackbarth (2000) observe that although there are two instances of Polvorón pithouses cutting through compound walls at Pueblo Grande (Abbott et al. 1994:196), it is unclear whether these features are associated with the Civano or Soho phases. Point taken, but, again, this concession does nothing to further advance their insistence that all Polvorón superimpositions involve only Soho phase occupation. Even if these few instances of superimposition involved only Polvorón above Soho architectural features, this fact does not disprove the contention that the Civano and Polvorón phases were sequent with little overlap.

One relationship, however, is given short shrift in their discussion, and that is the fact that there have not yet been any instances of compound walls intruding Polvorón pithouses. As Chenault (2000:281) cogently mused, why would this be the case if the two phases coexisted for any length of time? The obvious inference to draw is that the phases represent distinctly different chunks of time with only a short period of transition, just as the Bayesian chronology proposed herein stipulates. The other inference is not so obvious but is just as significant. And that is the implication that the compound room blocks ceased to grow during the brief transitional period, marking a rather abrupt, perhaps cataclysmic, social disruption.

This reinterpretation would not necessarily contradict the settlement trajectory model that Henderson and Hackbarth propose; it would only compress its time depth. Central to their trajectory are the small clusters of Polvorón pithouses that occur closely adjacent to the exterior compound walls of Civano phase room blocks. They reasonably argue that this pattern would only materialize if the compounds were still occupied. In addition, they observe that the architectural and ceramic contrasts between the two cultural expressions suggest that the Polvorón residents were newcomers not formally or only distantly integrated into the kinship systems of the Civano occupants of the room blocks. It is tacitly assumed, however, that the occupants of the room blocks represented fully functional Civano phase corporate lineages or lineage segments. I suggest alternatively that the architectural variability and late reoccupations of some rooms within the compounds in the late Classic/Polvorón occupation at Pueblo Grande (Mitchell 1994:65–71, 274–277) evidence a closely sequent and slightly earlier period of decline in the trajectory prior to and converging with the process of accretion that Henderson and Hackbarth identify. Mitchell's general conclusion was that just a small portion of the rooms in the larger room block complexes had extended spans of occupation, based on rebuilding sequences and the presence of ceramic assemblages in close agreement with those found in the Polvorón pithouses at the site. This in-place decline, I suggest, represents the period of overlap between the phases. Based on the Bayesian chronological model, this period appears to be circumscribed by the time of increased streamflow variation and high magnitude flooding that Graybill and colleagues date between AD 1357 and 1384 on the Salt River. The end of this transition would mark the point at which a fully formed Polvorón trait complex of narrow-walled adobe jacals and ceramic assemblages characterized by extremely high redware and polychrome percentages achieved widespread uniformity across the Salt River valley. I would further surmise that Civano phase room blocks would have undergone significant decomposition from discontinued use by this time, an inference that is supported by the fact that some Polvorón pithouses overlie compound walls but not Civano rooms. Exterior compound walls would have posed fewer hindrances to house construction than clearing away the rubble piles formed by crumbling adobe rooms. This might also explain the presence of Polvorón pithouses in the former plaza areas of room block compounds, which would probably not have been a proper option under the constraints that would have accompanied the accretion process described by Henderson and Hackbarth.

Resolving why there was such a dramatic shift in the architectural habit of the two phases is pivotal to understanding the collapse. Polvorón pithouses are of less sturdy construction than solid walled adobe room blocks, and because Henderson and Hackbarth assumed extreme phase overlap, they interpreted these differences as a reflection of stark contrasts in the residential permanence of the two social groups. Polvorón immigrants maintained a connection to their homeland through seasonally transhumant residential moves, choosing to visit Civano towns during periods of stress in their native communities or when enticed to relocate during the agricultural season to supply additional laborers in the face of regional population declines. Alternatively, I suggest that the shift to less permanent architecture became a quickly universal response to deteriorating local conditions. If indeed the Salt River Hohokam were subjected to destructive floods at the time of transition, then these episodes would have entailed a protracted and profound reorganization of the canal system, including relocation of the main trunk canals and abrupt shifts in agricultural land use patterns. Failure to achieve sustainable production for a number of seasons might have severely reduced populations through out-migration, famine, internal conflict, and other disruptive factors. Under such unpredictable conditions, an adaptive advantage would be realized by settlement systems emphasizing greater residential mobility and coincidentally less costly labor inputs for home construction. Polvorón immigrants were more likely dislocated local groups and not distantly connected foreigners, an inference that is strongly supported by continuity in Civano and Polvorón ceramic assemblages.

Correspondence between the timing of the Civano/Polvorón transition and the suite of drought-flood sequences that Graybill and colleagues date to the last half of the fourteenth century is not in itself proof that these episodes had catastrophic impacts on Canal System 2. Confirmation can only come from direct stratigraphic investigations of the canals. Much progress has been made in mapping the various canal systems of the Salt River Hohokam (see Howard 2006; Howard and Woodson 2018), but stratigraphic investigations are limited and the technology is not yet refined enough to provide precise chronologies of when individual canals were operated or the timing of flood events. An important exception for the purposes of the current discussion is Gary Huckleberry's (2015b) description of a cross section of one of the main trunks of Canal System 2 near the Park of the Four Waters below Pueblo Grande. He found a massive deposit in the upper two-thirds of this canal (Canal Bueno) linked to a single uncontrolled flood event that caused its abandonment. The flood deposit overtopped the canal's banks at this point, approximately 1.5 km upstream from the former head gate, and Huckleberry was able to trace the deposit in a correlated cross section located another 2.3 km upstream. An experimental, optically stimulated luminescence (OSL) date was recovered from a single sample in the lower portion of the flood deposit, which returned a 1-σ range of AD 1221–1341. Although the range corresponds to the Classic period, the central tendency of the sample is too early to represent the flood sequence of interest here, although it does fall within the 2-σ range. In a later adjunct to this study, however, Huckleberry and colleagues (2018) report that the excavation of a head gate feature belonging to a distribution canal branching from Canal Bueno, and containing the same flood deposit, has produced ceramic evidence (i.e., polychrome pottery) that would date the flood to the last half of the fourteenth century. Consequently, there is strong evidence that an important section of the Civano segment of Canal System 2 heading at Pueblo Grande was blown out and abandoned at a time consistent with the reconstruction proposed by Graybill and colleagues.

A later model of Core Decay offered by Hill and colleagues (2015a) to explain important factors contributing to the late Classic period collapse in the Salt River valley identifies a number of archaeological patterns that are of particular pertinence to the reconstruction I have developed here. Their model assumes extreme overlap between the Civano and Polvorón phases and effectively collapses the entire period of interest under the generalized rubric of the late Classic period. Yet, if the time depth of their model is collapsed in a manner similar to how I reinterpreted the Settlement Trajectory model of Henderson and Hackbarth, the archaeological patterns they discuss coincide just as well with a chronology of limited overlap and discrete phases. At the heart of the Core Decay model is the assumption that the Civano phase system of towns survived up until AD 1450 or slightly later. The architectural and material culture contrasts with the Polvorón phase are left unaddressed. The model's premise is that long-term operation of the canal systems resulted in a gradual buildup of unsuitably dense clay deposits in the areas closest to the traditional locations of the intakes for Canal Systems 1 and 2, which caused imbalanced distributions of productive agricultural land and hence a deterioration of cooperation between communities. The remaining arable lands were situated at the distal ends of the systems; in the case of Canal System 2 this resulted in a shift of power to the town of Las Colinas at the expense of the downstream communities, especially Pueblo Grande. Moreover, the rise of Las Colinas was facilitated by the aggregation there of immigrants, perhaps from the Kayenta region of northern Arizona, evidenced by much greater representations of late Roosevelt Redware types whose manufacturing dates begin after AD 1350 and high densities of perforated plates.

It is hypothesized that these immigrants represented an important factor in increasing local tensions, along with developing disparities in the proxemics of productive agricultural land. This circumstance, then, resulted in the early abandonment of downstream towns as is evidenced by lower representations of late Roosevelt Redware types at these locations. A review of native oral accounts and linguistic studies supports this interpretation and suggests that small groups of northern immigrants entered the Core Area at a late date, were given ritual authority, and were subsequently driven away by populations with a longer local history and stronger southern desert affiliations. Although acknowledging the impact of floods in the late fourteenth century, Hill and colleagues see these events as only contributing causes of increasing environmental degradation. In their view, the Hohokam collapse and abandonment of the Salt River valley was a product of mismanaged resource allocation under conditions of environmental degradation, declining populations, and differential coalescent integration between distal and proximal towns (Hill et al. 2015a:646).

Environmental degradation and declining fecundity are both factors that should be integrated into a more complete model of Hohokam collapse and abandonment. However, a radically different scenario can be developed from the distributional patterns Hill and colleagues identify if the Bayesian chronological model is applied. Lyons (2004; 2013) assigns respective start dates for the late types of Roosevelt Redware polychromes (i.e., Cliff, Phoenix, Nine Mile, and Los Muertos) to AD 1360, 1375, 1375, and 1390. If the end date for the Civano phase and room block architecture is likely no later than AD 1385 according to the Bayesian chronology, then these late types should not be expected to achieve significant proportions in Civano phase contexts. Their primary association should be with the Polvorón phase. Accordingly, the concentrations of late polychrome types at the distal ends of Canal Systems 1 and 2 that define the hypothesized loci of coalescent integration and power transfer from the towns situated closer to the intakes in the Core Decay model can alternatively be viewed as the markers of a post-collapse, Polvorón phase settlement shift.

I suggest that this settlement shift was tied to the conditions that brought about the Civano collapse and more specifically to the consequences of flood damage. In the cases of both Canal Systems 1 and 2, the main intakes were positioned at specific points along the river (see Figure 8.1), and if these locations were rendered inoperable by catastrophic flood events, then irrigation agriculture may have remained viable only at other intake locations. In the case of System 2, the downstream canals that ran past Pueblo Salado, Pueblo Patricio, and Las Colinas might have escaped the ravages that befell the intakes at Pueblo Grande, thus concentrating the post-collapse population at the distal end of the former system. A similar scenario could explain the distal concentration of late polychrome types at Los Muertos in System 1. Valley-wide agricultural acreage would have been severely reduced by continued flooding and it can be surmised that the local tensions that Hill and colleagues (2015a) allude to happened during the Civano/Polvorón transition, bracketed by the high discharge flood events of the late AD 1350s and early AD 1380s. Reduced irrigated acreage would have also forced populations at the proximal ends of the systems to supplement subsistence with other types of agricultural strategies, including runoff catchment systems along upland ranges at the edge of the valley and the buttes adjacent to the river (see Schaafsma and Briggs 2007). Additionally, numerous settling basins have been identified during archaeological investigations of canal segments at various locations within Canal System 2 that appear to have been designed to capture and retain water from runoff and overbank flooding (Huckleberry 2015b); these could have provided temporary "oasis-like" opportunities for small agricultural plots. Under such conditions, higher residential mobility would have imparted adaptive advantages and could well explain the shift from solid walled adobe room blocks to less labor-intensive jacal pithouse structures. A higher capacity to subsist on irrigation agriculture may have extended the longevity of the Civano phase around Las Colinas, but it would appear that the transition to a Polvorón phase adaptation was complete in the valley by AD 1385 to 1400, based on the clustering of calibrated dates associated with the Salt River Polvorón pithouse sample.

The reasons why the Salt River valley was eventually abandoned are not clear, but continued intervals of low and high discharge throughout the fifteenth century (Graybill et al. 2006:90–91) could have ultimately rendered the cost-benefit relationship between agricultural production and the labor needed to maintain canal systems untenable. By contrast, the Gila River valley may never have been abandoned after the Civano collapse, which is consistent with O'odham oral tradition as presented by Hill and colleagues (2015a), wherein many residents of the Salt River valley migrated southward while others may have returned to northern Arizona. The much later dates associated with the Polvorón phase occupations at Queen Creek seems to support this larger hypothesis. Graybill and colleagues (2006:110) show that discharge fluctuations were always much less extreme on the Gila due to differences in precipitation patterns and catchment size, and it appears that these differences supplied the conditions necessary to sustain irrigation agricultural practices there into the historic period.

An examination of the Hohokam response to increased flooding during the MCA demonstrates that collapses of hierarchical systems and regional abandonments can be protracted and complex processes. Did the collapse and abandonment of the mound sites along the middle and lower Savannah River happen overnight? That would seem unlikely. The same degree of material culture contrast may not exist to help segregate pre- and postcollapse habitations in much of the Mississippian Southeast, but refined dating methods like the Bayesian chronological model presented earlier could well supply the tools necessary to effectively investigate this question.

Disease Epidemics and Town Abandonments

There are many examples of abandoned towns in the Mississippian World, but their relationship to disease epidemics is clouded by the larger complications of megadrought in the fifteenth century. There is one example of such an event, however, that happened much earlier at Moundville, which provides an opportunity to evaluate the hypothesis that aggregation into large towns exposed some Mississippian societies to greater risks of epidemic disease during cyclical dry and wet periods. Sometime around AD 1200, the residents of the Black Warrior Valley initiated an ambitious preplanned construction project at Moundville that established a large-scale configuration of ranked residential and burial mound couplets arranged around a massive plaza (Knight 1998:52–53). Unprecedented influxes of residents from the hinterlands accompanied this event, resulting in the formation of crowded residential zones ringing the ritual precinct and the construction of a massive perimeter palisade wall. However, sometime around AD 1300 most of the town's residents dispersed into the hinterlands and the function of the site shifted to a nearly vacant ceremonial center and necropolis.

Several hypotheses have been advanced to explain this out-migration. Knight and Steponaitis (1998:18–19) suggest that it was engineered by the elite to socially distance themselves from commoners in an attempt to strengthen their hegemony. Primary evidence in support of this hypothesis is the appearance during this stage of lavishly adorned superordinate burials placed in the high-status mounds at the north end of the site, with grave furnishings indicative of an exclusive chiefly cult symbolism. Additionally, although the residential population of Moundville peaked between AD 1200 and 1300, the burial population peaked later, between about AD 1300 and 1450, which indicates that the power relations of the elite continued. They add that other factors might also have contributed, including soil depletion, exhaustion of wood resources in the site catchment area, and perhaps a lessening of regional hostilities as evidenced by the discontinued maintenance of the perimeter palisade wall. Blitz (2008:67–68) alternatively proposes that the dissolution of Moundville was a product of decentralizing forces brought about by factional disputes leading to segmentary fissioning and the dispersal of elite interest groups. He argues that a decline in long-distance exchange after AD 1300 and wider societal access to elite iconography as reflected in the engraved designs found on common utilitarian pottery might evidence a weakening of the elite authority structure.

Recourse to the North American Drought Atlas (E. Cook and Krusic 2004) indicates that there was a dry-wet cycle in the Black Warrior Valley spanning the late thirteenth and early fourteenth centuries that is structured similarly to the cycle coincident with the epidemic that the casica of Cofitachequi claimed was responsible for the abandonment of Talimeco in 1538. Figure 8.5 displays the time-series graph of PDSI values for this general time frame for the Black Warrior Valley. The solid line represents an eight-year smoothing spline. The cycle that is most pertinent to the depopulation of Moundville consists of an extended dry period spanning the years between AD 1270 and 1307, followed by a wet period of approximately twelve years ending at AD 1319. Average annual PDSI during the dry period was -0.975; during the ensuing wet period it was +0.495. If a hemorrhagic fever outbreak accompanied the wet period, it would most likely have occurred on the downward slope of the curve between AD 1314 and 1319. This would have been the period most favorable for an explosion of the resident rodent population and hence it would have been the nexus of greatest risk for the spread of rodent-borne diseases. Just as is the case for Talimeco, out-migration here involved a scattering of population into rural settlement zones of dispersed farmsteads and hamlets. Even the secondary mound sites supported only sparse populations. Intentional or not, such a settlement strategy would have lessened the impact of continued disease transmittal.

The fact that such cycles converge is no guarantee that a rodent-borne disease will accompany abrupt town abandonments in every instance, however. A study by Brown and Ernest (2002) concludes that a number of intervening factors create a complicated epidemiological pathway of infection. Rodent population responses to wet periods are complex and sometimes nonlinear and depend on temporal as well as quantitative patterns of precipitation. Too much rain, in fact, can have equally deleterious impacts on rodent population growth as droughts. Heavy and prolonged rains, for instance, dampened and spoiled stored seed caches in some of Brown and Ernest's control studies, resulting in population crashes. Floods can also decimate rodent populations. Moreover, not all population explosions in rodent reservoirs result in disease transmittal to associated human populations, and other microenvironmental factors can affect

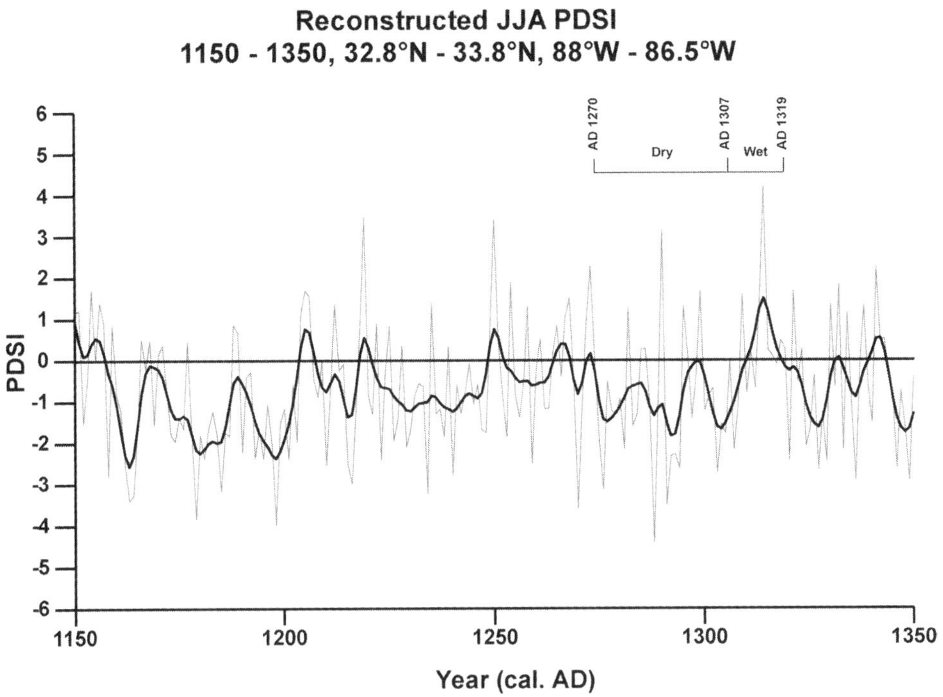

Figure 8.5. Palmer Drought Severity Index (PDSI) time series curve for the Black Warrior River valley between AD 1150 and 1350. JJA = June–July–August (data from the North American Drought Atlas, Cook and Krusic 2004).

the severity and longevity of individual outbreaks (Yates et al. 2002). Mitigating factors include landscape heterogeneity, microclimatic variation, local food access, and the degree of intraspecies competition, all of which can influence rodent densities. Yet, it is just the kind of convergence seen in the Moundville PDSI time-series at the beginning of the fourteenth century that would foster a rodent-borne disease eruption of epidemic proportions, especially in situations where agriculturally dependent human populations aggregate.

The correspondence between a dry-wet cycle at the time Moundville experienced a logistical abandonment does not constitute direct proof that a rodent-borne disease epidemic infected the population. In fact, Powell's (1998) analysis of the lesions found on individuals within the Moundville burial sample identified only treponematosis and tuberculosis as likely disease candidates and she doubts that either would have ever reached frequencies indicative of epidemics. However, many viral and bacteriological diseases that manifest themselves as epidemics do not affect bone in ways that allow their indications to be detected by simple macroscopic examination (cf. Antoine 2008:110; Wiktorowicz et al. 2017). This is particularly true of hemorrhagic fevers and plagues, which attack soft tissue and are of sufficiently short duration to not leave traces on bones before the victim dies. Recent advances in biomolecular DNA analysis of human skeletal remains are beginning to provide a basis for identifying a range of diseases that would otherwise go undetected in skeletal populations, and it is possible that such evidence is awaiting discovery in the Moundville burial sample. Viral epidemics of this sort were most probably a chronic but discontinuous feature of Misissippian systems during the MCA. The next section, in contrast, describes a common and continuous process leading to the reformulation of societal relationships and the denudation of chiefly authority structures.

Coalescence and the Collapse of Chiefly Authority

Kowalewski (2006) has suggested that there are mechanisms inherent in the integration of coalescent societies that would have weakened Southeastern chiefdoms and gradually transformed them into confederacies. In the Southeast this concept has been almost exclusively applied to describe the multiethnic aggregations of refugees forced into reserves or hidden backwater locales by the disruptions of European contact in the seventeenth and eighteenth centuries (Beck 2013; Ethridge and Hudson 2002; Galloway 1995). By this time, however, chiefdoms were a distant memory to Native Americans in the Southeast. To understand the collapse of chiefdoms, it is necessary to turn the dial back to the mid-sixteenth century, when "time's arrow" still preserved a trajectory of continuity with the powerful paramountcies of the thirteenth and fourteenth centuries. It is in this context that Kowalewski's model of political change becomes most relevant, because this was a time when the effects of the first wave of coalescent reorganization in the Southeast was well underway and the mechanisms of integration that he hypothesizes would have had sufficient time to disrupt long-standing social networks and sources of chiefly authority.

Based on the iconography of the Southeastern Ceremonial Complex, it is Kowalewski's (2006:119–120) thesis that Southeastern chiefdoms were organized around warrior cults rather than ranked lineal descent groups or conical clans as some have suggested (Knight 1990; Widmer 1994). Coalescent societies, however, require different institutions to integrate disparate ethnic groups and he suggests that this may have been accomplished through a resurrection of the latent clan system. Clans are macroregional in their scale of interaction and Kowalewski proposes that these connections may have extended across the entire span of eastern North America, because the same four naming conventions for clans (i.e., bear, deer, beaver, and wolf) are shared among most of the resident tribes. Such relationships would have cut across the authority structure of

Southeastern chiefdoms and eroded control of the restricted long-distance exchange networks that chiefs relied on to extend and enhance their powers. Thus, what appears to represent paramount chiefdoms in the sixteenth century when reading Spanish accounts may actually describe emerging confederacies with the outward symbolic trappings of the traditional fourteenth-century warrior cults.

An example of one of these hybrid social formations is the sixteenth-century Tascalusa paramountcy on the Alabama River alluded to in chapter 6. During the fifteenth century, this region received large influxes of people from three different cultural traditions (i.e., Pensacola, Moundville, and Lamar). I would suggest that these migrations were in response to forces set in motion by the effects of megadrought at the same time similar events were reshaping the demographic shifts I have described for the GDO and RDR. Each of these Alabama River cultures occupied different stretches of the valley, but crowding appears to have shrunk customary buffer zones, bringing them into intimate contact at territorial boundaries. Based on ceramic style representation, Amanda Regnier (2006) has demonstrated that individual towns contained unique mixes of people from these different cultures. Furthermore, many of the larger towns were situated in defensible positions and some were fortified, suggesting that internecine hostility was a common feature of social interaction. In spite of this, the cultural heterogeneity of the towns Regnier was able to identify suggests strongly that some overriding social institutions allowed individuals to move relatively freely across ethnic boundaries.

Based on her own findings, and reacting to the expectations of Kowalewski's model, Regnier argues that the polity over which Tascalusa held powerful influence at the time of the Soto entrada was more likely a confederacy as opposed to a paramount chiefdom (2006:250–251). According to her thesis, the towns under Tascalusa's purview had very different ethnic compositions and were widely dispersed, suggesting centralized control might have been difficult to maintain. Proof of this can be gleaned from the Soto accounts, which reveal that secondary chiefs maintained a good deal of political autonomy and yielded to Tascalusa's will only when in close proximity. Tascalusa was able to assemble an impressive army of some 2,500 warriors for the trap he planned for the Spanish at Mabila, which she argues is evidence for some kind of macroregional alliance akin to the later confederacies of the historic period. Thus, the extensive geographic reach of Tuscalusa appears to have functioned more as a means to assemble warriors for defensive and offensive maneuvers over long distances rather than a strictly administrative technique to collect tribute and administer social relations.

Subtle differences between chiefdoms and confederacies may be difficult to identify archaeologically, especially if these early social formations were hybrid expressions of an uneasy union of opposing institutions. There is ample evidence, however, that the power of sixteenth-century chiefdoms in the Southeast was on the wane. This is most succinctly demonstrated by a comparison of the sheer magnitude of difference in mound building between the prehistoric and protohistoric chiefdoms. Below I will present a comparison of mound-building efforts between the prehistoric Etowah and protohistoric Coosa chiefdoms of northwestern Georgia, which I think demonstrates the differences in power relations between the two periods.

The Little Egypt site, which is the presumptive capital of the sixteenth-century paramount chiefdom of Coosa in northwest Georgia (see Hally 1978; 1994b; M. Smith 2000), is estimated to have covered an area of about 12.3 acres (Hally and Rudolph 1986:73) and contains two known platform mounds and possibly a third one now plowed down (Hally 1994b). The largest mound (Mound A) stood about 3.3 m high and covered an area of 2,440 m^2, while the smaller one (Mound B) measured 960 m^2 in area and attained a height of about 1.8 m. By contrast, the Late Wilbanks phase (AD 1325–1375) component of the Etowah site (King 2003) extended over an area of 31.5 acres within the confines of a palisade ditch and contained three mounds. The largest mound (Mound A) covered an area of 6,560 m^2 and reached a height of 21 m. The two smaller mounds (Mounds B and C) covered about 1,000 m^2 each and respectively attained heights of 6 and 7 m.

There is clearly a difference in scale between the two centers, but the extent of it is brought out best by comparing the total volume of earth moved to construct the mounds. Using the formula for the volume of a frustum, roughly estimated earth volumes for the three Etowah mounds (A, B, and C) are respectively 90,282 m^3, 5,095 m^3, and 4,776 m^3, a little over 100,000 m^3 of earth combined. These figures are based on modern surface exposure and underestimate volumes because the original bases of the mounds are buried. The Little Egypt mounds, by contrast, return modern surface volumes of only 4,737 m^3 for Mound A and 1,225 m^3 for Mound B. Together, then, the Etowah mounds contain about 17 times more earth volume than the mounds at Little Egypt. Even without the contribution of Etowah Mound A, the labor expenditure represented by volume of earth moved is nearly twice as large for the remaining two mounds at Etowah compared with those at Little Egypt. One has only to climb to the summit of Mound A at Etowah to appreciate just what a commanding and towering view it offers of the Etowah valley and just how much it dwarfs the other two Savannah period mounds at the site, each of which has comparable volumes to Mound A at Little Egypt.

Hally (1994b:241) contends that the larger Coosa chiefdom, as he defines it, contained a minimum of 21 platform mounds distributed across 12 sites, most of which were of comparable height to Mound A at Little Egypt. The two tallest mounds, one at Citico and one at Toqua, attained heights of about 8 m each. The reported dimensions for Toqua yield a volume estimate of around 11,600 m^3. If an average of 4,000 m^3 is used to represent the lesser mounds and the volume estimate for Toqua is applied to the mound at Citico, then it would take the entire combined mound volume of the Coosa chiefdom (ca 99,000 m^3) to approximate the effort expended to build just Mound A at Etowah. Moreover, the sites Hally has used to construct the Coosa paramountcy span a distance of 420 km from northeastern Alabama to eastern Tennessee, while the Etowah system consisted of at least 10 mound sites within the expanse of only about 120 km along the Etowah and upper Coosa rivers (King 2003:36).

Clearly, the Etowah chiefdom was characterized by a much denser concentration of mound volume and, by extension, probably greater population density. Moreover, the area covered by the Etowah chiefdom would be more in line with the effective area that could have been controlled by a singular complex chiefdom, while the nearly four-times-greater areal extent of Coosa is more analogous to the hybrid confederacy that Regnier proposes for the Tascalusa paramountcy. In fact, although there is not direct evidence, it is likely that the relationships established between the disparate polities in the Coosa alliance were probably forged through similar processes of coalescence that Regnier describes for the Tascalusa confederacy. An objection could be raised that the Etowah mounds took much longer to build and, therefore, the comparison is invalid. However, according to King (2003:63–81), nearly all of the Savannah period mound stages at Etowah were built during the Early and Late Wilbanks phases, a period lasting only about 125 years. Mound A at Little Egypt contains both Early and Middle Lamar building stages (Hally and Rudolph 1986:70), which indicates a span of at least 125 years.

The apparent positive correlation between mound volume and the relative power of Mississippian polities has long been recognized (see Carneiro 1967; Haas 1982; Knight 1986; Milner and Schroeder 1999; Peebles and Kus 1977; Steponaitis 1978). Late twentieth-century studies of architectural energetics, however, have caused archaeologists to modify their appraisals of the labor force needed to build mounds, their estimates of population density, and, by extension, their views on the power of Mississippian chiefdoms (see Lindauer and Blitz 1997:187–189; Milner 1998:144–147). As Milner (1998:120–121) observes, nineteenth-century claims that the Cahokian system supported a population comparable to the cradles of civilization (e.g., Egypt, Mesopotamia, Peru, and Mesoamerica) were based on the assumption that the extensive array of mounds must have required a huge full-time labor force to construct. Population estimates of Cahokia, however, have gradually declined in magnitude from as many as 40,000 individuals to much lower levels on the order of 10,000 to 15,000 (Pauketat and Lopinot 1997) and 3,000 to 8,000 (Milner 1998:123). Milner (1998:148) projects that the entire system of mounds (1,180,000 m^3) at Cahokia could have been constructed with a labor force of only 470, 490, 310, and 68 laborers working 10 five-hour days annually during, respectively, the Lohmann (100 years), Stirling (100 years), Moorehead (50 or 75 years), and Sand Prairie (100 or 125 years) phases. Similarly, Muller (1986) estimates that the mounds at the Kincaid site (93,278 m^3) could have been erected by 1,000 people in 130 to 228 years, with an annual labor contribution of only 4 days per household of 5 individuals each.

Thus, mound building does not appear to have placed onerous demands on household labor during the annual cycle, although it is likely that the scheduling of building stages was not equally distributed and that some years required more extensive efforts than others. This finding, however, does not negate the inferred relationship between mound volume and the relative power of Mississippian polities. Mound building was not the only demand placed on the polities' subjects. They were also compelled to intensify agricultural productivity, construct plazas and community structures, erect defensive fortifications, yield to conscription for defense and war campaigns, and contribute to ritual ceremonies and feasts. The "thermodynamic cheapness" of mounds, in fact, suggests to Milner and colleagues (2010:110–111) that they represented a cost-effective, visually impressive, and permanent architectural symbol for conveying the power and legitimacy of the new social order (see also Pauketat 2000:123). In this light, mound volume remains a viable measure of the effectiveness of chiefdoms to compel labor for community and intercommunity projects.

The picture that emerges from the interior Southeast during the fifteenth century is one of reorganization, regional abandonment, collapse of at least some powerful chiefdoms like Etowah, and the rise of others that were less strongly integrated and probably more limited in their ability to conscript labor for building projects and defense and to collect tribute. Although Anderson (1999:228) has asserted that the sixteenth-century chiefdoms described by the Spanish were as powerful as those of the thirteenth and fourteenth centuries, the sheer magnitude of difference in the scale of mound building—a contrast that is consistently manifest across the Southeast (see Milner 1998:146)—suggests otherwise. M. Williams and Shapiro (1996:147–148) have argued that mound volume is a poor indicator of chiefly power and that this quality is better explained by the duration of construction and use. Blitz and Livingood (2004), though, have shown that longevity of building sequence explains only 11% (total sample) and 41% (major mound centers excluded) of the variation in mound volume from a sample of 35 excavated mounds in the Southeast and American Bottom. The remainder of the variability is unexplained, but the fact that the pace and magnitude of mound construction at regional centers was overwhelmingly more intense suggests that the lion's share of the variability is directly tied to chiefly power.

Certainly, the contrast drawn between the Etowah paramountcy and the Coosa chiefdom confederacy, which had similar durations, supports this interpretation. In general, then, I would state that the chiefdoms that the Soto chroniclers encountered in AD 1540 were not complex chiefdoms similar to those of the thirteenth and fourteenth centuries but examples of a new social formation wherein simple and devolved chiefdoms experimented with long-distance confederations to mobilize large aggregations of warriors for defensive and offensive actions. DePratter (1983) provides a detailed description of existing ethnohistoric accounts of the sophisticated military organization of sixteenth-century Southeastern chiefdoms. Initially this was a response to the uncertainties of environmental degradation directly linked to drought related migrations and the accompanying regional tensions and hostilities this process engendered. As such, these extensive but fragile political alliances were preadapted to respond militarily to the threat of Spanish incursions. Whether drought or coalescence played the major role in this transformation is not yet determinable, but it is likely that both factors were important and interconnected.

The Archaeological Correlates of Coalescence

In her review of the broader topic of human movement, Catherine Cameron (2013) has argued that there are many different reasons why people vacate settlements and regions and move to others and that there are numerous ways in which this act is accomplished. Not every displacement results from an environmental catastrophe and such instances are not necessarily planned or conducted out of free will. Often movements that result in the coalescence of unrelated peoples happen in the normal course of dynamic social relations and demographic processes; indeed, we are gradually coming to realize that coalescence is a normal feature of the human condition and no doubt a prime mover in most cases of material culture change. Under the culture-historic paradigm, dramatic changes to material culture were commonly interpreted as evidence of population replacement by a superior intruding culture at the expense of a technologically or ideologically inferior indigenous population. Yet, they could never perfectly match up the newly arriving set of traits they recognized at a site or in a region with a distant culture that they could satisfactorily demonstrate was the source of the intrusion. This led Phillips and colleagues (1951:454) to posit a concept that coincides closely with what modern researchers regard as the general process of coalescent culture invention: "Introduced and local cultural traits were quickly welded together to produce traits that appear unlike the items from which they were derived and thus it is difficult for the archaeologist to trace them. This is what we have designated as the X-factor, the contributions made by the culture to its own development."

As Cameron observes, archaeologists have made great advances in conceptualizing and identifying the correlates and characteristics of long-distance migration and understanding the dynamic between diasporic radiations and host communities (cf. Anthony 1990; Clark et al. 2013; Nelson and Strawhacker 2011). Her larger point, however, is that there are other types of movement that are less planned or subject to random processes that can assert just as dramatic an impact on social and material culture change. One such process is captive-taking, of which she (Cameron 2016) has undertaken a lengthy cross-cultural study. Once incorporated into the social fabric, captives play important roles in transforming their captors' society, especially in the realm of new technologies and practices of material culture innovation. In particular, female captives appear to contribute significantly to changes in pottery production, decorative styles, and to the organization of household space. Sometimes women captives are folded quickly into society through marriage, while in other situations both male and female captives remain social outcasts while still making important contributions to labor and craft production. In either case their participation results in significant modifications to their captors' society and material culture.

There is reason to believe that migration/emigration and captive-taking were both prevalent features of coalescent reformations during the fifteenth century across the SAM landscape. However, we do not yet have a conceptual framework or a sufficient archaeological database to identify and explore such detailed and sequent archaeological patterns. The SAM culture area is especially challenging in this regard because stylistic contrasts in architecture and ceramic assemblages between possibly interacting communities are exceedingly subtle. Coalescent settlements in the region, in fact, might go completely unnoticed given our present classification systems. Below I review two different examples of coalescent assimilation that can be used as models to inform the study of coalescence in the SAM culture area. The first examines approaches to identifying and studying the processes of migration and negotiation in southern Arizona during the late prehistoric period, where contrasts in material culture can be readily seen. The second example concerns the fifteenth-century convergence of three Mississippian cultures on the Alabama River, which I think might be appropriately identified as an example of captive-propelled assimilation and trait recombination. I conclude with some advice about how we might confront the problem of low stylistic variability in situations where we suspect coalescence might be found.

During the thirteenth century, small groups of Puebloan and Mogollon migrants from northern and east-central Arizona streamed into the river valleys of southern Arizona along the eastern flank of the Hohokam Core Area and negotiated resettlements among Hohokam hosts (Clark 2011; Clark et al. 2013; Huntley et al. 2016). The impetus for this movement, what might be called the push factor in migration parlance (Anthony 1990), was an extensive megadrought that correlates with the -PDO/+AMO megadrought scenario. Early response to drought-related stress in the north consisted of population aggregation along watercourses and a virtual abandonment of many portions of the previously occupied hinterland, followed by fission and straggling out-migration. Here I discuss the archaeological patterning found in two regions of north-south coalescence, the Tonto Basin on the upper Salt River and the San Pedro River, a tributary of the middle Gila River.

Migrants from unspecified and probably multiple homelands in northern Arizona began filtering into the Tonto Basin between AD 1250 and 1300, at a time when the local Hohokam residents were undergoing changes in architecture and ceramic production (Clark 2001; Stark et al. 1995). The pre-Classic architectural pattern of small clusters of pithouses arranged around informal courtyards was transformed into one of surface adobe masonry rooms surrounded by compound walls. However, the spatial arrangement of the earlier pithouse courtyard groups was maintained in the room configurations. In addition, platform mounds appeared in several of these compound complexes. This architectural shift has never been explained, but because it was universally repeated across the larger Core Area it is assumed to have emerged from internal dynamics and diffusionary assimilation. Other surface adobe masonry complexes of the same age, however, resemble Puebloan room blocks and were found in settlements without platform mounds. These are difficult to visually distinguish from the contiguous walled compound complexes, but Stark and colleagues (1995) have devised an index of contiguity that successfully differentiates them. Specifically, room blocks are composed of rooms with a high incidence of wall-sharing like those typical of Pueblos on the Colorado Plateau, while the masonry compounds exhibit low contiguity in keeping with a traditional Hohokam domestic arrangement. Species identification of architectural wood also exhibits patterns that strongly suggest cultural differences. The compound complexes were constructed with commonly available lowland species, while the room blocks contained primarily high-elevation species such as Douglas and white fir and Ponderosa pine, indicating procurement from sources closer to a Puebloan homeland. In addition, the two types of architectural configuration were geographically segregated. The compound complexes were situated adjacent to prime irrigable land, while the room blocks were located in marginal areas, consistent with a negotiated latecomer status.

Other evidence suggests that this marginal economic position motivated the latecomers (i.e., Puebloan/Mogollon immigrants) to diversify their economic pursuits to include specialized pottery manufacture and perhaps secondary crop production. Although detail is lacking, it would appear that the early Classic Hohokam manufactured ceramic assemblages composed mostly of plain and red ware just like the rest of the regional populations in the Core Area, but these utilitarian types were replaced by plain and red slipped corrugated wares soon after the arrival of the Puebloan migrants (Stark and Heidke 1998). Much of the red slipped corrugated pottery appears to have had a non-local manufacturing source, while the plain type was produced with local clays. The association of corrugation with Puebloan and Mogollon ceramic traditions strongly indicates that the migrants had assumed the role of pottery production specialists and that the firstcomers were the consumers of these products. True to the X-factor, a precise extraregional source for the corrugated types could not be found, supporting the idea that trait drift and recombination takes only a short time when separation by distance from a homeland is maintained. Flotation samples also suggest that at least some of the room block communities specialized in the cultivation of agave for exchange.

Economic specialization by the latecomers would have established a symbiotic exchange relationship between them and the host community, which may have compensated for the migrants' marginal position in relation to irrigation agriculture. This economic interaction would have fostered integration, which appears to have been progressing apace as several instances of small enclaves of room blocks began to pop up in the hosts' village complexes. Intermingling such as this suggests that the two communities were forming affinal ties, possibly reinforced by eventual consanguineal relations. Another indication that the groups were assimilating is provided by the pattern of room construction on the top of the platform mound at the principal firstcomer village of Meddler Point. Here, two rooms stood, one constructed with local wood species, while the other contained high elevation species, suggesting that the migrants had achieved parity in the local administrative and ritual networks just before the end of the Roosevelt phase, at which time the eastern part of the Tonto Basin was temporarily abandoned. Widespread evidence of burned structures dated to around AD 1325 in this settlement zone suggests some kind of hostile relations had erupted, but the role migrants played in the disruption, if any, is not known.

The reconstructed sequence in the San Pedro valley is more detailed and provides a clearer picture of the relationship between material culture change and coalescence. Clark and colleagues (2013:400) link the late thirteenth-century migration stream along the eastern periphery of the Core Area to the diasporic phenomenon known as "Salado" that emerges in the fourteenth century. Its archaeological signature is recognized by the distribution of a family of polychromes (Roosevelt Redware) manufactured by the coil-and-scrape technique common to Puebloan ceramic traditions. They argue that the stream was populated by multiple small groups moving southward in serial episodes from the Tusayan (Hopi) and Kayenta (northeast Arizona and southeast Utah) regions, resulting in numerous leapfrogging coalescent settlement negotiations along the way. Despite this fissioning, the groups maintained an extensive diasporic community of widespread interaction that not only preserved their own group identities but also integrated those of their hosts into a powerful Salado meta-identity. It did not transcend the more traditional community identities of the hosts, however, which is postulated to have created a source of tension within this inter-regional network.

The story of the Kayenta migration into the San Pedro valley begins in the late thirteenth century, when a pristine and fully self-contained migrant group settled in an uninhabited district at the edge of the central economic zone of the host Hohokam irrigationists (Clark 2011; Clark et al. 2013). As was true in the Tonto Basin case, the most fertile agricultural land was already taken. The Kayentas' status as migrants can be archaeologically detected by the presence of masonry-walled room blocks organized in the Puebloan tradition, with the addition of lower visibility architectural features such as entry boxes. At two of the migrant settlements, distinctive ceremonial structures of Puebloan affiliation called kivas were erected. Moreover, the migrants made pottery using the coil-and-scrape technique instead of the paddle-and-anvil technique common to the earlier southern desert manufacturing tradition. Perforated plates of Puebloan affiliation were made on local clays, as well as decorated pots with Puebloan-inspired designs (Maverick Mountain series).

Clark and colleagues suggest that the intrusion of these migrants motivated the host community to make a series of changes. Manufacture of the traditional decorated pottery in the region, San Carlos Red-on-Brown, was revived and platform mounds were constructed. In addition, settlement patterns shifted from dispersed farmsteads and hamlets to planned aggregated villages. It is further suggested that these changes were a conscious attempt by the Hohokam to distinguish and maintain their separate firstcomer identity and adopt defensive postures. In keeping with this interpretation, adjacent villages of the two communities were fortified and built in defensible positions.

By the mid-fourteenth century, the social boundaries between the groups began to fade. Maverick Mountain series and San Carlos Red-on-Brown were universally replaced by Roosevelt Redware, which appears to have been exclusively manufactured by the migrant community. According to Crown (1994: 164–173), the decorative tradition of Roosevelt Redware is derivative from the Kayenta region, but the iconography was new and probably held meaning for both communities. Similar to the Tonto Basin, the migrant community may have supplemented its inferior agricultural productivity by diversifying economic strategies. Although assimilating to some degree, the communities appear to have remained ostensibly segregated until the late fourteenth century. Clark and colleagues (2013:411) indicate that population was in decline by this time and only a few villages remained, concentrated at the northern end of the valley. These villages, however, exhibited evidence of mixed ethnic associations. Domestic architecture consisted of immigrant-type room blocks constructed with firstcomer rock-reinforced adobe walls, and both cremated and inhumated burial customs were practiced. Neither platform mounds nor kivas were associated with these settlements.

Frequencies of Roosevelt Redware, however, were much higher in these mixed settlements than previously and vessel size increased along with the incidence of exterior decoration. According to Crown (1994), these features may be related to increasing reliance on communal feasts (see also B. Mills 2007). Clark and colleagues (2013) suggest that the increased popularity and distribution of Roosevelt Redware in the face of declining populations in southern Arizona signifies an attempt to maintain sustainable communities by melding the iconographic symbolism of the remaining immigrant and host populations into a new unified cultural expression. They further suggest that the integral role played by the latecomers in the long-distance trade network that emerged from the Kayenta diaspora bestowed upon them a dominant socioeconomic position in the mixed settlements of the terminal occupation of the valley. Support for this proposition is supplied by the correspondence between the distribution of Roosevelt Redware and high frequencies of obsidian after AD 1300 obtained, in particular, from the upper Gila, where other migrant settlements affiliated with the Kayenta diaspora were to be found (Hill et al. 2015b; B. Mills et al. 2013). On the other hand, the immigrant stream of these groups to southern Arizona is seen to have forced aggregations of both immigrant and host communities, which led to declining fertility and/or increased mortality. Clark and colleagues suggest that integration attempts ultimately failed in the wake of demographic collapse.

I earlier suggested a different scenario for these patterns based on the probability that the Roosevelt Redware (i.e., polychromes) phenomenon is primarily associated with the post-collapse Polvorón phase. I would suggest that the Salado meta-identity, as represented by the distribution of Roosevelt Redwares, was a response to the need to even out production disparities through communal redistributive feasts in the context of high residential mobility brought about by a degraded environment, as opposed to an appeal to a functionalist argument that populations naturally decline through normal demographic processes brought about by aggregation. As an alternative hypothesis, I submit that the mixed settlements that Clark and colleagues (2013) place at the end of the San Pedro occupation in the late fourteenth century conform quite well to the period of transition between the Civano and Polvorón phases that I propose for the Core Area. These terminal San Pedro settlements exhibited higher frequencies of polychrome, yet there is no evidence of platform mounds or other architectural indications of hierarchical administrative institutions. That high residential mobility may have become the dominant organizing principle of settlement in southern Arizona during the Polvorón phase may in fact be demonstrated in the network analysis that B. Mills and colleagues (2013:4) have undertaken. They show that especially after AD 1350 there was an increasing disparity between spatial proximity and social connectivity and conclude that physical distance did not strongly structure social relations, even when pedestrian travel was the only mode of transportation. Such a pattern would fit well within a settlement system of high residential mobility where the consequences of fission and relocation driven by the need to efficiently exploit unpredictable resource patches would result in gradually increasing distances between close kin.

From the above examples of coalescent contact, three distinct stages of interaction in traditional types of small group migration can be discerned. The first stage consists of initial intrusion by migrants into a host community. Settlements are established at the periphery of the host's economic domain, generally resulting in the latecomers occupying inferior resource locations. Social and territorial boundaries are well defined and interaction is minimal. Intercommunity hostilities would be greatest at this time and defensive maintenance of territorial boundaries would be manifest in fortifications and strong social boundaries. The next stage is marked by migrant communities establishing peaceful economic exchange relationships with their hosts by developing strategies of symbiotic production specialization. In the cases reviewed here, the most prominent activity is specialized pottery manufacture, although diversification of crops and perhaps brokerage of long-distance trade relations would also represent important integrative strategies. It is during this period that small enclaves of migrant groups might settle within host communities, and perhaps *vice versa*, suggesting early assimilation through affinal marriage arrangements. A final stage would consist of gradually increasing integration in which a new cultural entity emerges (the X-factor), characterized by a unique set of melded social, ritual, and material culture patterns. In the case of the Salado meta-identity, the high-profile marker of this newly assimilated culture would be Roosevelt Redware. It is difficult to know whether tensions between migrant and host communities would have led to collapse and abandonment; at the very least, we must understand how the impacts of floods and droughts also figured into such processes.

The second example of coalescent assimilation to be discussed comes from the southeastern United States and appears to represent an altogether different process of amalgamation. Namely, Amanda Regnier (2006) demonstrates that during the last half of the fifteenth century the sparsely populated Alabama River valley was the focus of a dramatic migration event participated in by three different cultural groups (Pensacola, Moundville, and Lamar). These migrant communities formed the basis of the Tascalusa chiefdom confederacy that the Soto entrada contacted in the mid-sixteenth century. Exactly why this particular event would have occurred has not been answered, but as I suggested earlier in this chapter, it was likely a response to the stresses brought on by megadrought. Figure 4.2c indicates that although the fifteenth-century drought might not have been quite as intense in northern Alabama and Georgia as it was on the South Atlantic coast, it was nevertheless strong and might have provided the external push factor for southward migrations of Lamar and Moundville groups. The model predicts that the coastal Pensacola groups would have been situated outside of the high drought frequency isotherms, but it is probable that the Gulf Coast too was experiencing problems of declining estuarine productivity due to decreased streamflow and eutrophication from river systems originating in the Piedmont.

Owing to its location in the Coastal Plain, the Alabama River valley would seem an unlikely location for migrants to target as a refuge, given the low agricultural potential of the sandy, excessively drained soils of the Coastal Plain. The USDA, NRCS general soil map for the state of Alabama, however, indicates that much of the valley is situated in a unique soil belt called the Blackland Prairie, which is composed of clayey soils that would have been more favorable for intensive corn agriculture. Coalescent population packing in the Coastal Plain portion of the Alabama River valley appears to have been made possible by these unique soil characteristics. Certainly, the nearly unprecedented migratory convergence of three cultural traditions in such a small space argues that this situation came about as a response to a strong external push factor such as drought-related environmental degradation.

Each of these migrant communities occupied different stretches of the valley, but crowding appears to have shrunk customary buffer zones, bringing them into intimate contact at territorial boundaries. Regnier's analysis of ceramic samples from four sites in the drainage indicates that these towns contained different mixtures of pottery from each tradition, leading to the inference that the settlement areas housed mixes of people from each culture, although the traditional ceramic tradition associated with each territory remained dominant. There is no evidence that stylistic elements were blended on individual pots, as spatial association did not result in modifications to the original models of ceramic technology and stylistic decorative systems. In spite of this, the cultural heterogeneity of the towns suggests to Regnier that some overriding social institutions allowed individuals to move relatively freely across ethnic boundaries. Perhaps this mobility was made possible by marriage alliances arranged through the cross-cultural linkage of corresponding clans, as might be suggested by Kowalewski's model.

Low incidences of trait blending on individual pots further confirms for Regnier (2006: 246–247) that the operable residence pattern was matrilocal and that females, who it is assumed were the potters, remained in the households of their birth, as would be typical of matrilineal/matrilocal kinship organization. As she notes, it is generally held that matrilineal descent was the common kinship system of Mississippian societies based on analogies with Creek ethnology (see Widmer 1994). Although I take her point, there is a complicating factor. According to Divale's (1974:102–103) cross-cultural study, matrilineal-matrilocal societies tend to be locally endogamous. Divale hypothesizes that this pattern works to reduce regional warfare by breaking up localized fraternal interest groups while strengthening extra-village bonds. Such a system, however, would not encourage the introduction of new matrilines into a town, as Regnier's data seem to suggest happened. Consequently, it is difficult to see how such a pattern could have emerged organically as a function of traditionally formed matrilocal residence patterns in each of the communities under consideration here. The most likely explanation for the occurrence of minority ethnies in these towns would be that they arrived there because of some external factor not directly related to kinship. Two examples immediately come to mind, namely, refugee family enclaves and captives taken in the conduct of internecine warfare. I propose that the pattern of nonintegrated mixing that Regnier identifies is most likely a consequence of captive-taking, since many of the larger towns were situated in defensible positions and some were fortified, suggesting that internecine hostility was a common feature of community interaction.

It cannot be determined whether this example represents a different pathway to coalescence than the one described earlier for the Arizona examples. Captive-taking might constitute only one of several mechanisms that led to further assimilation. Before this question can be better addressed, though, considerably more archaeological investigation directed toward delineating the sequence of coalescent stages in the Alabama River valley would be needed. It does not appear that captive-taking was common in the Tonto and San Pedro valleys, but one could see how this mechanism might be a common element of societies coming into first contact with one another and future analysis may find that it played a role in coalescence scenarios in the Southwest as well. Regnier indicates that the three communities actually do appear to have assimilated later in the sixteenth century. The principal indication of this is the emergence of a new pottery tradition made up of blended elements belonging to each of the original traditions. Regnier (2006:247) attributes the emergence of a blended tradition to population decline caused by warfare, slave-raiding, political destabilization, and European introduced disease epidemics. Alternatively, I contend that this pattern of assimilation is an expected outcome of increasing contact in a coalescing multiethnic community and that Tascalusa's attempt to consolidate these groups prior to Spanish contact may have been instrumental in this process. If so, this would further suggest that captives played an important role in assimilation, probably due to the formation of matrimonial alliances that might also have included polygynous unions.

The example of Alabama River multiethnic communities, then, may point to the role coalescence played in the formation of the politically expansive protohistoric chiefdom confederacies that have been somewhat mistakenly interpreted as paramount or complex chiefdoms. More generally, however, had European contact been forestalled for several more centuries it is quite possible that the multisited instances of coalescence in southeast and southwest North America may have set the stage for the development of what Friedman and Rowlands (2008) would call "Asiatic" states. As they model it, these are relatively small, socially stratified formations extending over domains measuring minimally 20 to 30 km in radius with populations of around 10,000 at the low end. Such systems are fed by an engine of local subsistence intensification and the expansion of long-distance exchange networks controlled by central administrations for the purposes of obtaining exotic prestige goods that function in ritual, matrimonial alliances, and status symbolism. As regional competition increases, this exchange system can give rise to a unique method of societal control in which access to specific exotic goods that function as the requisite means of bridewealth payments, funerary rituals, and other vital activities of social reproduction are monopolized by central administrations. This form of exchange, which achieves a degree of independence from local production, has been referred to as the prestige-good system (Ekholm 1977). It is within this system that craft specialization becomes an institutionalized element of economic activity.

In the Arizona examples discussed above, economic diversification appears to have emerged as a response to differential access to productive land between migrant and host communities, suggesting that coalescence might be one of the most important factors in driving the emergence of craft specialization. Certainly, the diasporic system of long-distance exchange that appears to have been established through a network of Puebloan migrant connections in diverse environmental zones can be seen as a foundational mechanism for the development of a prestige-good system. Both the monopolized production of a ritualized and highly symbolic ceramic industry and the hegemonic control of long-distance exchange in obsidian appear to have cast Puebloan migrants in such a role. This is an important step in the evolution of state systems because it provides a basis for evening out local productivity swings that might be caused by external perturbations such as droughts and floods, thus increasing the security and longevity of complex systems that might otherwise collapse.

At the far end of the sequences I have discussed are region-wide depopulations and abandonments. In some cases, this may have been a consequence of early historic contact, but for the Kayenta diaspora, population decline occurred considerably before European contact. This is also true of the abandonment events associated with Cahokia, the Ohio River valley, the GDO, and the Colorado Plateau. The larger story of abandonments is set squarely within the late prehistoric period and is subsequent to the development of hierarchically structured societies that aggregated into towns of unprecedentedlly large and dense populations. Certainly, a number of factors may have contributed to collapse and later abandonment, including disease epidemics, factionalism, warfare, breakdowns in regional exchange networks, etc. To understand why these symptoms were manifested, however, it is imperative that we look to the paleoclimatic record and the environmental degradation that it visited upon highly concentrated, populous societies that were poorly equipped to maintain their hierarchical organization in the face of famine and disease. That the adjustments made to survive the collapses were not successful attests to the power of long-lasting, unpredictable, and extreme climatic regimes during the MCA to threaten the viability of human systems situated in regions where the effects were felt the strongest.

The archaeological patterns that were used in the two examples presented above are easily distinguished because the coalescent communities had very different material culture traditions and separate developmental histories. No such formula exists for the bulk of potential coalescent relationships in the SAM culture area. The degree of difference in the ceramic and architectural assemblages in this region is quite subtle. The ceramic assemblages have nearly indistinguishable technological and stylistic attributes, as I demonstrated in chapter 2. The same can be said of domestic architecture, which in the fifteenth century most generally conforms to what Lacquement (2007:65) describes as Type IV domestic structures.

To discern coalescent patterning in the SAM culture area, then, it will be necessary to focus on the less-visible attributes that exist at the analytical level below traditionally devised culture-historic typological systems. These are the features of material culture that possess very low visibility in terms of conveying messages about social or cultural identity. Yet, if they are subjected to detailed scrutiny, they can reveal valuable and diagnostic information about the differences from place to place in the customary ways mundane actions were performed. This would include, for example, habitual ways of felling a tree, organizing activity spaces in residents' houses, the specific production steps taken to make pots and projectile points, or food preparation idiosyncrasies. All these activities have readily identifiable attributes that are not intended by the actors to convey a stylistic message (Carr 1995), but they hold very real implications for archaeologists trying to detect different communities of practice within a region where coalescence is expected to have occurred. Clark (2011) has cogently observed that these attributes tell part of the story about what Bordieu (1977) calls the habitus, that is, the ingrained habits, skills, and dispositions of individuals as these are distributed within the social and economic fields of communities and cultures. Habitus is the group culture frame in which individuals are socialized.

To get us where we want to go, then, we will need to reconstruct the habitus of the individual communities of both the pre-abandonment cultures in the Greater Desert of Ocute and those of the surrounding Ring of Drought Resilience. Pauketat (2000) has advocated a similar approach for tracing the pace and meaning of Mississippianization, although his focus was on the more highly visible traits of shell tempering in pottery assemblages and wall-trench houses. But simply monitoring the progress of traits like these across geographic space would not answer the question of how they got there. On this basis alone one would not know whether shell-tempering and wall-trench construction emerged from the diffusionary acceptance of ideas or whether they were inspired by an infusion of migrants into a new area. A more satisfactory resolution to this question would require detailed study of the more mundane characteristics of material culture that inform on habitus and the different pathways cultures might take to accomplish the same task. For instance, are there differences in the way that interior space is organized in a set of wall-trench houses from a particular settlement or region, and might one of these patterns be linked to a pattern manifest in another region? If so, this might provide a basis for entertaining the hypothesis that the occupants of these houses represent recent immigrants. The other set of houses, which represent a blending of local and extralocal traits, then, could be conjectured to represent a later stage of coalescence in which firstcomers and latecomers were assimilating into a new cultural expression (i.e., the X-factor). Similar observations could be made on the technological and morphological characteristics of shell-tempered pottery. This is not to say that some attributes of higher visibility such as particular types of decorative elements would not be of value in most situations, only that when stylistic variability is low, or when the questions of cultural relatedness can be explained by different processes, we will benefit greatly from focusing on quotidian variation. Moreover, I think that such an enterprise will yield bountiful new revelations about the archaeological societies we study and the human condition we endeavor to understand. We learn by rearranging what we know.

Final Thoughts

One might ask why it was necessary to travel all the way to the Southwest to talk about drought-related abandonments on the opposite side of the continent. It was not. However, this is the journey I have taken over a span of 40 years in pursuit of answers to such questions and I would argue that an understanding of how the MCA impacted a single region cannot be fully achieved unless we see how this climate phenomenon played out across continents. Only then will we be able to develop the comparative and generalist perspective necessary to understand the range of responses to such events as they proceeded from region to region.

One important implication of this study is that the complex agricultural societies of the late prehistoric period living in the southerly latitudes of North America were exposed to the same suite of drought-related pestilences that modern climate and environment scientists are warning us will soon befall our own world. The difference between us and them is that they did not contribute to the severity of climate change as we are doing and they had no way of knowing that droughts and floods would wreak havoc on their world for centuries. Most importantly, the world healed itself at the end of the MCA and climatic stability returned. We are much more complicit in our own fate and yet we choose to ignore what modern science is telling us about what will happen in only a very short time. Indeed we see it happening all around us. The relationship is simple: the more fossil fuel we burn the more carbon enters the atmosphere and the hotter it gets. Some argue that this trajectory is solely a function of the natural cycle of climate oscillations as has been true in the past, but the implications of this attitude should not be lost on observant individuals. If there is a natural cycle of drought on the way, why would we want to make it worse by dumping unprecedented amounts of carbon into the atmosphere, which we know raises temperatures?

The atmosphere healed from the devastating changes that fueled global warming during the MCA and the world returned to a more hospitable regime within a short time. What we are doing today, however, is permanently altering the protective envelope of gasses that support life on our planet, inevitably moving us toward a bleaker and hotter world that could take millennia rather than a few centuries to restore a healthy balance. What happened to the complex societies that inhabited North America (and other regions of the world) during the MCA should serve as an urgent warning about the kinds of vulnerabilities we are exposing ourselves to today because of an unwillingness to change our lifestyles and direct our sophisticated technology toward solving the ultimate global calamity. We procrastinate in such epic proportions. The day will come, however, when the motivation to change will far outweigh the arguments to maintain the status quo and I predict that this decision will mark the beginnings of a new and better era in human history.

References

Abbott, David R., Douglas R. Mitchell, and J. A. Merewether
1994 Chapter 6. Chronology. In *The Pueblo Grande Project: Feature Descriptions, Chronology, and Site Structure*, edited by Douglas R. Mitchell, pp. 157–254. Soil Systems Publications in Archaeology No. 20, Vol. 2. Phoenix.

Acuña-Soto, Rudolfo, Leticia Calderon Romero, and James H. Maguire
2000 Large Epidemics of Hemorrhagic Fevers in Mexico 1545–1815. *American Journal of Tropical Medicine and Hygiene* 62(6):733–739.

Acuña-Soto, Rudolfo, David W. Stahle, Malcom K. Cleaveland, and Mathew Therrell
2002 Megadrought and Megadeath in 16th Century Mexico. *Emerging Infectious Diseases* 8(4):360–362.

Acuña-Soto, Rudolfo, David W. Stahle, Mathew D. Therrell, Sergio Gomez Chavez, and Malcom K. Cleaveland
2005 Drought, Epidemic Disease, and the Fall of Classic Period Cultures in Mesoamerica (AD 750–950). Hemorrhagic Fevers as a Cause of Massive Population Loss. *Medical Hypotheses* 65(2):405–409.

Acuña-Soto, Rudolfo, David W. Stahle, Mathew D. Therrell, Richard D. Griffin, and Malcom K. Cleaveland
2004 When Half the Population Died: The Epidemic of Hemorrhagic Fevers of 1576 in Mexico. *FEMS Microbiology Newsletters* 240(1):1–5.

Adams, Natalie P., Jennifer Langdale, Leslie Raymer, Mason Sheffield, Kate McKinley, and Arthur Cohen
2006 *Archaeological Investigations at Yourhaney Plantation (38GE18) Yauhannah Bluff, Waccamaw National Wildlife Refuge, Georgetown County, South Carolina*. Submitted to US Fish and Wildlife Service, Savannah, New South Associates Technical Report 1314. Columbia.

Agha, Andrew, and Charles F. Phillips Jr.
2006 *Cultural Resources Survey of Part One of the Country Road Environmental Analysis, Francis Marion National Forest, Berkeley and Charleston Counties, South Carolina*. Francis Marion and Sumter National Forest Cultural Resource Management Report No. 2005–01. Brockington and Associates, Inc., Atlanta.

Anderson, David G.
1982 The Mattassee Lake Ceramic Artifact Assemblage. In *Mattassee Lake: Archaeo-*

logical Investigations along the Lower Santee River in the Coastal Plain of South Carolina, by David G. Anderson, Charles E. Cantley, and A. Lee Novick, pp. 207–322. Atlanta Interagency Archeological Services Division, Special Publication 1982.

1990 Political Change in Chiefdom Societies: Cycling in the Late Prehistoric Southeastern United States. Unpublished PhD dissertation, Department of Anthropology, University of Michigan, Ann Arbor.

1994 Factional Competition and the Political Evolution of Mississippian Chiefdoms in the Southeastern United States. In *Factional Competition and Political Development in the New World*, edited by Elizabeth M. Brumfiel and John W. Fox, pp. 61–76. Cambridge University Press, Cambridge, United Kingdom.

1996 Chiefly Cycling and Large-Scale Abandonments as Viewed from the Savannah River Basin. In *Political Structure and Change in the Prehistoric Southeastern United States*, edited by John F. Scarry, pp. 150–191. University Press of Florida, Gainesville.

1999 Examining Chiefdoms in the Southeast: An Application of Multiscalar Analysis. In *Great Towns and Regional Polities in the Prehistoric American Southwest and Southeast*, edited by Jill E. Neitzel, pp. 95–107. University of New Mexico Press, Albuquerque.

Anderson, David G., David J. Hally, and James L. Rudolph

1986 The Mississippian Occupation of the Savannah River Valley. *Southeastern Archaeology* 5(1):32–51.

Anderson, David G., A. Lea Novick, and Charles E. Cantley

1982 *The Mattassee Lake Sites: Archeological Investigations along the Lower Santee River in the Coastal Plain of South Carolina.* Prepared for the US Department of the Interior National Park Service Interagency Archeological Services-Atlanta, Contract C54030(80), Final Report No. 2311.

Anderson, David G., David W. Stahle, and Malcom K. Cleaveland

1995 Paleoclimate and the Potential Food Reserves of Mississippian Societies: A Case Study from the Savannah River Valley. *American Antiquity* 60(2):258–286.

Anthony, D.

1990 Migration in Archeology: The Baby and the Bathwater. *American Anthropologist* 92(4):895–914.

Antoine, Daniel

2008 The Archaeology of "Plague." *Medical History, Supplement* 27:101–114.

Armelagos, George J., Alan H. Goodman, and Kenneth H. Jacobs

1991 The Origins of Agriculture: Population Growth during a Period of Declining Health. *Population and Environment* 13(1):9–22.

Arnold, Dean E.

1999 Advantages and Disadvantages of Vertical-Half Molding Technology: Implications for Production Organization. In *Pottery and People*, edited by James M. Skibo and Gary M. Feinman, pp. 59–80. University of Utah Press, Salt Lake City.

Bayman, James N.

2001 The Hohokam of Southwest North America. *Journal of World Prehistory* 15(3):257–311.

Beck, Robin A., Jr.
2003 Consolidation and Hierarchy: Chiefdom Variability in the Mississippian Southeast. *American Antiquity* 68(4):641–661.
2013 *Chiefdoms, Collapse, and Coalescence in the Early American South.* Cambridge University Press, New York.

Beck, Robin A., Jr., and David G. Moore
2002 The Burke Phase: A Mississippian Frontier in the North Carolina Foothills. *Southeastern Archaeology* 21(2):192–205.

Beck, Robin A., Jr., David G. Moore, and Christopher B. Rodning
2006 Identifying Fort San Juan: A Sixteenth-Century Spanish Occupation at the Berry Site, North Carolina. *Southeastern Archaeology* 25(1):65–77.

Beck, Robin A., Jr., Christopher B. Rodning, and David G. Moore
2016 *Fort San Juan and the Limits of Empire: Colonialism and Household Practice at the Berry Site.* University Press of Florida, Gainesville.

Bender, Margaret M.
1968 Mass Spectrometric Studies of Carbon 13 Variations in Corn and Other Grasses. *Radiocarbon* 10(2):468–472.

Benson, Larry V.
2011 Factors Controlling Pre-Columbian and Early Historic Maize Productivity in the American Southwest, Part 1: The Southern Colorado Plateau and Rio Grande Regions. *Journal of Archaeological Method and Theory* 18(1):1–60.

Benson, Larry V., Timothy R. Pauketat, and Edward R. Cook
2009 Cahokia's Boom and Bust in the Context of Climate Change. *American Antiquity* 74(3):467–483.

Benson, Larry, Kenneth Peterson, and John Stein
2007 Anasazi (Pre-Columbian Native-American) Migrations during the Middle 12th and Late 13th Centuries—Were They Drought Induced? *Climatic Change* 83(1/2):187–213.

Binford, Lewis R.
1962 A New Method of Calculating Dates from Kaolin Pipe Stem Samples. *Southeastern Archaeological Conference Newsletter* 9(1):19–21.
1964 Archaeological and Ethnohistorical Investigation of Cultural Diversity and Progressive Development among Aboriginal Cultures of Coastal Virginia and North Carolina. Unpublished PhD dissertation, Department of Anthropology, University of Michigan, Ann Arbor.
1965 Archaeological Systematics and the Study of Cultural Process. *American Antiquity* 31(2, Part 1):203–210.

Blanton, Dennis R.
2000 Drought as a Factor in the Jamestown Colony, 1607–1612. *Historical Archaeology* 34(4):74–81.

Blitz, John H.
2008 *Moundville.* University of Alabama Press, Tuscaloosa.

Blitz, John H., and Patrick Livingood
2004 Sociopolitical Implications of Mississippian Mound Volume. *American Antiquity* 69(2):291–301.

Bocquet-Appel, Jean-Pierre
2002 Paleoanthropological Traces of a Neolithic Demographic Transition. *Current Anthropology* 43(4):637–650.
2011 The Agricultural Demographic Transition during and after the Agriculture Inventions. *Current Anthropology* 52(S4):497–510.

Bocquet-Appel, Jean-Pierre, and Ofer Bar-Yousef (editors)
2008 *The Neolithic Demographic Transition and Its Consequences.* Springer Science & Business Media, New York.

Bocquet-Appel, Jean-Pierre, and Stephen Naji
2006 Testing the Hypothesis of a Worldwide Neolithic Demographic Transition: Corroboration from American Cemeteries. *Current Anthropology* 47(2): 341–365.

Boone, Stephanie A., and Charles P. Gerba
2007 Significance of Fomites in the Spread of Respiratory and Enteric Viral Disease. *Applied and Environmental Microbiology* 73(6):1687–1696.

Bostwick, Todd W., and Christian Downum
1994 *Archaeology of the Pueblo Grande Platform Mound and Surrounding Features.* Vol. 2: Features in the Central Precinct pf the Pueblo Grande Community. Pueblo Grande Museum Anthropological Papers No. 1. City of Phoenix.

Boudreaux, Edmond A.
2005 The Archaeology of Town Creek: Chronology, Community Patterns, and Leadership at a Mississippian Town. Unpublished PhD dissertation, Department of Anthropology, University of North Carolina, Chapel Hill.
2007 *The Archaeology of Town Creek.* University of Alabama Press, Tuscaloosa.

Bourdieu, Pierre
1977 *Outline of a Theory of Practice.* Cambridge University Press, Cambridge, United Kingdom.

Bourne, Edward G. (editor)
1904a *Narratives of the Career of Hernando de Soto*, Vol. 1. A. S. Barnes and Company, New York.
1904b *Narratives of the Career of Hernando de Soto*, Vol. 2. A. S. Barnes and Company, New York.

Brainerd, George W.
1951 The Place of Chronological Ordering in Archaeological Analysis. *American Antiquity* 16(4):301–313.

Braun, David P.
1985 Absolute Seriation: A Time-Series Approach. In *For Concordance in Archaeological Analysis: Bridging Data Structure, Quantitative Technique, and Theory*, edited by Christopher Carr, pp. 509–539. Westport, Kansas City, Missouri.
1987 Coevolution of Sedentism, Pottery Technology, and Horticulture in the Central Midwest, 200 BC–AD 600. In *Emergent Horticultural Economies of the Eastern Woodlands*, edited by William F. Keegan, pp. 153–181. Southern Illinois University at Carbondale, Center for Archaeological Investigations, Occasional Paper No.7.

Brockington & Associates

2011 *Archaeological Assessment of Selected Sites along the Intracoastal Waterway, Charleston County, South Carolina.* American Recovery and Reinvestment Act 2009, Section 110 Compliance, Report for the US Army Corps of Engineers, Charleston District. Brockington and Associates, Inc., Atlanta.

Bronk Ramsey, Christopher

2009a Bayesian Analysis of Radiocarbon Dates. *Radiocarbon* 51(1):337–360.

2009b Dealing with Outliers and Offsets in Radiocarbon Dating. *Radiocarbon* 51(3):1023–1045.

2017 Methods for Summarizing Radiocarbon Datasets. *Radiocarbon* 59(2):1809–1833.

Brown, David P., and Andrew C. Comrie

2004 A Winter Precipitation "Dipole" in the Western United States Associated with Multidecadal ENSO Variability. *Geophysical Research Letters* 31(9):L09203.

Brown, James H., and S. K. Morgan Ernest

2002 Rain and Rodents: Complex Dynamics of Desert Consumers. *Bioscience* 52(11): 979–987.

Bundy, L. G.

1998 *Corn Fertilization.* University of Wisconsin-Extension, Cooperative Extension, Publication A3340.

Burrell, Victor

1977 Mortalities of Oysters and Hard Clams Associated with Heavy Runoff in the Santee River System, South Carolina, in the Spring of 1975. *Proceedings of the National Shellfisheries Association* 47:35–43. Washington, DC.

Buzan, David, Wen Lee, Jan Culbertson, Nathan Kuhn, and Lance Robinson

2009 Positive Relationship between Freshwater Inflow and Oyster Abundance in Galveston Bay, Texas. *Estuaries and Coasts* 32(1):206–212.

Byrd, William

1966 A Journey to the Land of Eden, Anno 1733. In *Prose Works: Narratives of a Colonial Virginian*, edited by Louis B. Wright, pp. 379–415. Belknap Press, Harvard University, Cambridge, Massachusetts.

Cable, John S.

1991 *Prehistoric Adaptations to the Interior Uplands Environment of the Central South Carolina Coast: Test Excavations at Four Sites (38BK769, 38BK1176, 38CH1045 and 38CH1189) in the Francis Marion National Forest.* Francis Marion and Sumter National Forest Cultural Resource Management Report No. 1991–34. New South Associates, Inc., Stone Mountaion, Georgia.

1998 Ceramic Analysis: A Study of Taxonomy and Systematics in the North Coastal Zone. In *Data Recovery Excavations at the Maple Swamp (38HR309) and Big Jones (38HR315) Sites on the Conway Bypass, Horry County, South Carolina: Prehistoric Sequence and Settlement on the North Coastal Plain of South Carolina*, edited by John S. Cable, Kenneth F. Styer, and Charles E. Cantley, pp. 239–324. Report submitted to South Carolina Department of Transportation, Columbia. New South Associates Technical Report 385, Stone Mountain.

2002a *Intensive Cultural Resource Survey, Santee Timber Thinning Unit, Berkeley and Charleston Counties, Francis Marion National Forest.* Francis Marion and Sumter

National Forest Cultural Resource Management Report No. 2002–07. Palmetto Research Institute, Irmo, South Carolina.

2002b Ceramic Analysis. In *Shaw Air Force Base: Archaeological Data Recovery at Sites 38SU136/137 and 38SU141, Poinsett Electronic Combat Range, Sumter County, South Carolina*, by Charles E. Cantley and John S. Cable, pp. 163–234. Report submitted to US Army Corps of Engineers, Savannah District. New South Associates, Stone Mountain.

2008 *Archaeological Survey of the Honey Hill Sawtimber Harvest, Francis Marion National Forest.* Francis Marion and Sumter National Forest Cultural Resource Management Report No. 2008–10. Palmetto Research Institute, Irmo.

2010 *Cultural Resources Survey of Part One of the Wando-I'on Analysis Biomass Harvest, Francis Marion National Forest.* Francis Marion and Sumter National Forest Cultural Resource Management Report No. 2010–06. Palmetto Research Institute, Irmo.

2011a *Phase II Archaeological Testing and Evaluation of Six Precontact Sites in Hoke and Richmond Counties, Fort Bragg and Camp Mackall, NC.* Report prepared for the National Park Service, Southeastern Archeological Center, Tallahassee, Florida, and the Department of the Army, Fort Bragg, North Carolina. Palmetto Research Institute, Irmo.

2011b Implications of the Quewhiffle Site Ceramic Assemblages for Reconstructing Regional Patterns of Cultural Interaction in the North Carolina Sandhills. Paper presented at the 2011 Sandhills Conference: Across the Border—An Archaeological and Environmental Discussion of the Sandhills Physiographic Province: A View from North and South Carolina. Weymouth Woods State Park, Pinehurst, North Carolina. Sponsored by the North Carolina Department of Transportation and the National Forests in North Carolina and the North Carolina Office of State Archaeology.

2013 *Investigations at the Fogarty Creek Site (38BK552): Mississippian Chronology and Settlement in the Upper Wando River Valley.* Palmetto Research Institute, Irmo.

Cable, John S., Howard A. Gard, Patricia Evans-Shumate, and Mary B. Reed

1994 *Intensive Archaeological Survey of Salt Pond, South Tibwin, Bellfield, and Brick Church Tracts, Wambaw Ranger District, Francis Marion National Forest, Charleston South Carolina.* Francis Marion and Sumter National Forest Cultural Resource Management Report No. 1994–07. New South Associates, Inc., Stone Mountain.

Cable, John S., David R. Lawrence, and Leslie A. Raymer

1991 *Archeological Test Excavations on The Northeastern Perimeter of the Buck Hall Site (38CH644), Buck Hall Recreation Area, Francis Marion National Forest, South Carolina.* Francis Marion and Sumter National Forest Cultural Resource Management Report No. 1998–09. New South Associates, Inc., Stone Mountain.

Cable, John S., George D. Price, and Ray Talley

2013 *Cultural Resources Survey of the Proposed Land Exchange of Four Tracts with The Nature Conservancy, Francis Marion National Forest.* Report prepared for the Nature Conservancy, Charleston, South Carolina. Francis Marion and Sumter National Forest Cultural Resource Management Report No. 2013–11. Palmetto Research Institute, Irmo.

DePratter, Chester B.

1976 *The 1974–75 Archaeological Survey in the Wallace Reservoir, Greene, Hancock, Morgan, and Putnam Counties, Georgia.* University of Georgia, Laboratory of Archaeology Series, Report No. 66. Athens.

1983 *Late Prehistoric and Early Historic Chiefdoms in the Southeastern United States.* University of Georgia, Laboratory of Archaeology Series, Report No. 72. Athens.

1985 University of South Carolina Department of Anthropology/SCIAA 1985 Archaeological Field School: Mulberry Site (38KE12). *Institute of Archaeology and Anthropology Notebook* 17(3/4):31–36. South Carolina Institute of Archaeology and Anthropology, University of South Carolina, Columbia.

1989 Cofitachequi: Ethnohistorical and Archaeological Evidence. In *Studies of South Carolina Archaeology: Essays in Honor of Robert L. Stephenson,* edited by Albert C. Goodyear III and Glen T. Hanson, pp. 133–156. Anthropological Studies 9, Occasional Papers of the South Carolina Institute of Archaeology and Anthropology.

1991 *W.P.A. Archaeological Excavations in Chatham County, Georgia: 1937–1942.* University of Georgia Laboratory of Archaeology Series, Report 29. Athens.

1994 The Chiefdom of Cofitachequi. In *The Forgotten Centuries: Indians and Europeans in the American South, 1521–1704,* edited by Charles Hudson and Carmen Chaves Tesser, pp. 197–226. University of Georgia Press, Athens.

1998 Cofitachequi. Paper presented at the 55th Annual Meeting of the Southeastern Archaeological Conference, Greenville, South Carolina.

2007 Coastal Shell Midden Research. *Legacy* 2(1):8–9.

2009 Irene and Altamaha Ceramics from the Charlesfort/Santa Elena Site, Parris Island, South Carolina. In *From Santa Elena to St. Augustine: Indigenous Ceramic Variability (A.D. 1400–1700),* edited by Kathleen Deagan and David H. Thomas, pp. 19–48. Anthropological Papers of the American Museum of Natural History No. 90. New York.

DePratter, Chester B., and Christopher Amer

1988 Underwater Archaeology at the Mulberry Site (38KE12) and Adjacent Portions of the Wateree River. South Carolina Institute of Archaeology and Anthropology, Columbia. Research funded by USC Venture Fund.

DePratter, Chester B., Charles M. Hudson, and Marvin T. Smith

1983 The Route of Juan Pardo's Explorations in the Interior Southeast, 1566–1568. *The Florida Historical Quarterly* 62(2):125–158.

DePratter, Chester B., and Christopher Judge

1986 A Provisional Late Prehistoric and Early Historic Ceramic Sequence for the Wateree River Valley, South Carolina. Paper presented at the Lamar Institute Conference on South Appalachian Mississippian. Macon, Georgia.

Dickens, Roy S.

1976 *Cherokee Prehistory.* University of Tennessee Press, Knoxville.

Dickens, Roy S., H. Trawick Ward, and R. P. Stephen Davis Jr. (editors)

1987 *The Siouan Project: Seasons I and II.* Research Laboratories of Anthropology, University of North Carolina, Chapel Hill, Monograph Series No. 1.

Divale, William Tulio
1974 Migration, External Warfare, and Matrilocal Residence. *Behavior Science Research* 9(2):75–133.

Dixon, C.W.
1948 Smallpox in Tripoltania, 1946: An Epidemiological and Clinical Study of 500 Cases, Including Trials of Penicillin Treatment. *The Journal of Hygiene* 46(4): 351–377.

Dobyns, Henry F.
1983 *Their Number Become Thinned: Native American Population Dynamics in Eastern North America*. University of Tennessee Press, Knoxville.

Doyel, David E.
2000 The Santan Phase in the Phoenix Basin. In *The Hohokam Village Revisited*, edited by David E. Doyel and Suzzane K. Fish, pp. 221–244. Southwestern and Rocky Mountain Division of the American Association for the Advancement of Science, Glenwood Springs, Colorado.

Drennan, Robert D.
2010 *Statistics for Archaeologists: A Common-Sense Approach, Second Edition*. Springer, New York.

Duff, Andrew I.
1996 Ceramic Micro-Seriation: Types or Attributes? *American Antiquity* 61(1):89–101.

Earle, Timothy K.
1991 *Chiefdoms: Power, Economy, and Ideology*. Cambridge University Press, Cambridge.
2011 Chiefs, Chieftaincies, Chiefdoms, and Chiefly Confederacies: Power in the Evolution of Political Systems. *Social Evolution & History* 10(1):27–54.

Eastman, Jane M.
1999 The Sara and Dan River Peoples: Siouan Communities in North Carolina's Interior Piedmont from AD 1000 to AD 1700. Unpublished PhD dissertation, Department of Anthropology, University of North Carolina, Chapel Hill.

Eastman, Jane M., Victoria Saxe, Keith Seramur, Loretta Lautzenheiser, and Mary Ann Holm
1997 *Archaeological Salvage Recovery, Site 31SKI5, Stokes County, North Carolilla, NCDOT B-263–1 alld B-2635*. Prepared for the North Carolina Department of Transportation, Planning, and Environmental Branch, Raleigh. Coastal Carolina Research, Inc., Tarboro.

Egloff, Keith T., and Stephen R. Potter
1982 Indian Ceramics from Coastal Plain Virginia. *Archaeology of Eastern North America* 10:95–117.

Ekholm, Kajsa
1977 External Exchange and the Transformation of Central African Social Systems. In *The Evolution of Social Systems*, edited by Jonathan Friedman and M. Rowlands, pp. 115–136. Duckworth Publishers, London.

Elliott, Daniel
1984 *Archaeological Testing and Evaluation at 38BK552, Compartment 117, Witherbee District, Francis Marion National Forest*. Francis Marion and Sumter National

Forest Cultural Resource Management Report No. 1984–01. Palmetto Research Institute, Irmo.

Ely, Lisa L.
1992 *Large Floods in the Southwestern United States in Relation to Late-Holocene Climatic Variations*. PhD dissertation, Department of Geosciences, University of Arizona, Tucson.

Ely, Lisa L., Yehouda Enzel, and Daniel R. Cayan
1994 Anomalous North Pacific Atmospheric Circulation and Large Winter Floods in the Southwestern United States. *Journal of Climate* 7(6):977–987.

Enfield, David B., Alberto M. Mestas-Nuñez, and Paul J. Trimble
2001 The Atlantic Multidecadal Oscillation and Its Relation to Rainfall and River Flows in the Continental US. *Geophysical Research Letters* 28(10):2077–2080.

Engelbrecht, William
1980 Inferring Prehistoric Social and Political Organization in the Northeast. *Proceedings of the Conference on Northeastern Archaeology, Research Report* 19:109–122. University of Massachusettes–Amherst.

Epstein, Paul R.
2001 Climate Change and Emerging Infectious Diseases. *Microbes and Infection* 3(9):747–754.

Espenshade, Christopher T.
1989 *Archaeological Evaluation of 38CH146, 38CH426, 38CH1047, and 38CH1048, Francis Marion National Forest, Charleston County, South Carolina*. Francis Marion and Sumter National Forests, Cultural Resources Report 1989–54. Brockington and Associates, Inc., Atlanta.

Ethridge, Robbie
2009 Introduction. In *Mapping the Mississippian Shatter Zone: The Colonial Indian Slave Trade and Regional Instability in the American South*, edited by Robbie Ethridge and Sheri M. Shuck-Hall, pp. 1–50. University of Nebraska Press, Lincoln.

Ethridge, Robbie, and Charles M. Hudson (editors)
2002 *The Transformation of the Southeastern Indians, 1540–1760*. University of Mississippi Press, Jackson.

Evans, Clifford
1955 *A Ceramic Study of Virginia Archeology*. Smithsonian Institution, Bureau of American Ethnology, Bulletin 160. Washington, DC.

Feathers, James K.
1990 An Evolutionary Explanation for Prehistoric Ceramic Change in Southeast Missouri. Unpublished PhD dissertation, Department of Anthropology, University of Washington, Seattle.

Feldman, L.
1994 The Maize Root. In *The Maize Handbook*, edited by Michael Freeling and Virginia Walbot, pp. 29–36. Springer Lab Manuals Series XXVI. Springer-Verlag, Inc., New York.

Ferguson, Leland G.
1971 South Appalachian Mississippian. Unpublished PhD dissertation, Department of Anthropology, University of North Carolina, Chapel Hill.

1973a *Exploratory Archeology at the Scott's Lake Site (38CR1) Santee Indian Mound—Ft. Watson Summer 1972*. South Carolina Institute of Archaeology and Anthropology, Research Manuscript Series 36. University of South Carolina, Columbia.

1973b *Mulberry Plantation Exploratory Archeology*. South Carolina Institute of Archaeology and Anthropology, Research Manuscript Series 54. University of South Carolina, Columbia.

1974 Archaeological Investigations at the Mulberry Site. *Institute of Archaeology and Anthropology Notebook* 6(3/4):57–124. South Carolina Institute of Archaeology and Anthropology, University of South Carolina, Columbia.

Fish, Paul R.

1978 Consistency in Archaeological Measurement and Classification: A Pilot Study. *American Antiquity*: 43(1):86–89.

Fish, Paul R., Suzanne K. Fish, George J. Gumerman, and J. Jefferson Reid

1994 Toward an Explanation for Southwestern "Abandonments." In *Themes in Southwest Prehistory*, edited by George J. Gumerman, pp. 135–164. School of American Research Press, Santa Fe.

Fitts, Mary Elizabeth

2006 Mapping Catawba Coalescence. *North Carolina Archaeology* 55:1–59.

Ford, James A.

1962 A Quantitative Method for Deriving Cultural Chronology. *Pan American Union, Technical Bulletin* 1.

Friedman, Jonathan, and M. J. Rowlands

2008 Notes toward an Epigenetic Model of the Evolution of "Civilization." In *Historical Transformations: The Anthropology of Global Systems*, edited by Kajsa Ekholm Friedman and Jonathan Friedman, pp. 87–140. Altamira Press, Walnut Creek. First published in *The Evolution of Social Systems*, 1977, by Duckworth, London.

Gallay, Alan

2002 *The Indian Slave Trade: The Rise of the Indian Slave Trade in the American South, 1670–1717*. Yale University Press, New Haven.

Galloway, Patricia

1995 Book Review: The Savannah River Chiefdoms: Political Change in the Late Prehistoric Southeast. By David G. Anderson. *The Georgia Historical Quarterly* 79(3): 701–703.

1997 Conjoncture and Longue Durée: History, Anthropology, and the Hernando de Soto Expedition. In *The Hernando de Soto Expedition: History, Historiography, and "Discovery" in the Southeast*, edited by Patricia Galloway, pp. 283–294. University of Nebraska Press, Lincoln.

Gardner, Jeffrey W.

1992 *Sewee Fire Survey: Archaeological Survey of 135 Acres in the Sewee Historic Area, Wambaw Ranger District, Francis Marion National Forest, Charleston County, South Carolina*. Francis Marion and Sumter National Forests, Cultural Resources Report 1992–19. Brockington and Associates, Inc., Atlanta.

1993 *Archaeological Evaluation of Five Sites in the Sewee Historic Area, Wambaw Ranger District, Francis Marion National Forest, Charleston County, South Carolina*.

Francis Marion and Sumter National Forests, Cultural Resources Report 1993–15. Brockington and Associates, Inc., Atlanta.

Gardner, Paul S.

1980 An Analysis of Dan River Ceramics from Virginia and North Carolina. Unpublished PhD dissertation, Department of Anthropology, University of North Carolina, Chapel Hill.

Gladwin, Harold S., Emil W. Haury, E. B. Sayles, and Nora Gladwin

1937 *Excavations at Snaketown: Material Culture.* Gila Pueblo, Medallion Papers No. 25. Globe.

Goddard, Ives

2005 The Indigenous Languages of the Southeast. *Anthropological Linguistics* 47(1):1–60.

Gray, Jay W.

2012 *Phase II Testing of 12 Sites (31CD797, 31CD1840, 31CD1841, 31CD1846, 31CD1848, 31CD1849, 31CD1850, 31CD1854, 31SC142, 31SC159, 31SC176, and 31SC177) at Fort Bragg and Camp Mackall, Cumberland, and Scotland Counties, North Carolina.* Submitted to US Army Corps of Engineers, Savannah District, Wilmington Contracting Office. TRC Environmental Corporation, Chapel Hill.

Graybill, Donald A., and Fred L. Nials

1989 Aspects of Climate, Streamflow, and Geomorphology Affecting Irrigation Systems in the Salt River Valley. In *The 1982–1984 Excavations at Las Colinas: Environment and Subsistence*, by Donald A. Graybill, David A. Gregory, Fred L. Nials, Suzanne K. Fish, Robert E. Gasser, Charles H. Miksicek, and C. R. Szuter, pp. 5–24. Archaeological Series No. 162(5). Arizona State Museum, University of Arizona, Tucson.

Graybill, Donald A., David A. Gregory, Gary S. Funkhouser, and Fred L. Nials

2006 Long-Term Streamflow Reconstructions, River Channel Morphology, and Aboriginal Irrigation Systems along the Salt and Gila Rivers. In *Environmental Change and Human Adaptation in the Ancient Southwest*, edited by David E. Doyel and James S. Dean, pp. 69–123. University of Utah Press, Salt Lake City.

Greenwald, David H., Mark L. Chenault, and Dawn M. Greenwald (editors)

1995 *The Sky Harbor Project, Early Desert Farming and Irrigation Settlements, Archaeological Investigations in the Phoenix Sky Harbor Center, Volume 3: Pueblo Salado.* SWCA Anthropological Research Paper No. 4. Flagstaff.

Greenwald, David H., Mark Zyniecki, and Dawn M. Greenwald (editors)

1994 *Early Desert Farming and Irrigation Settlements, Archaeological Investigations in the Phoenix Sky Harbor Center, Volume 2: Dutch Canal Ruin.* SWCA Anthropological Research Paper No. 4. Flagstaff.

Gregory, David A.

1987 The Morphology of Platform Mounds and the Structure of Classic Period Hohokam Sites. In *The Hohokam Village: Site Structure and Organization*, edited by David E. Doyel, pp. 183–210. Southwestern and Rocky Mountain Division of the American Association for the Advancement of Science, Glenwood Springs, Colorado.

Gregory, David A., William L. Deaver, Suzanne K. Fish, Ronald Gardner, Robert W. Layhe, Fred L. Nials, Lynn S. Teague, and W. Bruce Masse

1988 *The 1982–1984 Excavations at Las Colinas: The Site and Its Features*. Cultural Resource Management Division Arizona State Museum, Archaeological Series 161, Vol. 2. University of Arizona, Tucson.

Griffin, James B.

1967 Eastern North American Archaeology. *Science*, New Series 156(3772):175–191.

Guerrero, E., Stephen Naji, and Jean-Pierre Bocquet-Appel

2008 The Signal of the Neolithic Demographic Transition in the Levant. In *The Neolithic Demographic Transition and Its Consequences*, edited by Jean-Pierre Bocquet-Appel and Ofer Bar-Yosef, pp. 57–80. Springer Science & Business Media, New York.

Gunn, Joel D.

2002 Steep Shore, Deadly Environment: The Case for a Cultural Anvil along the Unembayed Atlantic Coast. *North Carolina Archaeology* 51:1–33.

Haag, William G.

1958 *The Archaeology of Coastal North Carolina*. Louisiana State University Press, Baton Rouge.

Haas, Jonathan

1982 *The Evolution of the Prehistoric State*. Columbia University Press, New York.

Hagstrum, Jonathan T., and Eric Blinman

2010 Archeomagnetic Dating in Western North America: An Updated Reference Curve Based on Paleomagnetic and Archeomagnetic Data Sets. *Geochemistry, Geophysics, and Geosystems* 11(6):1–18.

Hakluyt, Richard (translator)

1609 *The Discovery and Conquest of Terra Florida by Hernando De Soto and Six Hundred Spaniards His Followers. Written by a Gentleman of Elvas, Employed in All the Action, and Translated out of Portuguese*. The Hakluyt Society, London.

Hally, David J.

1978 *Archaeological Investigation of the Little Egypt Site (9Mu102), Murray County, Georgia, 1969 Season*. University of Georgia, Laboratory of Archaeology Series, Report No. 18. Athens.

1986 The Cherokee Archaeology of Georgia. In *The Conference on Cherokee Archaeology*, compiled by David G. Moore, pp. 95–121. Warren Wilson College, Swannanoa, North Carolina.

1990 Phase Characteristics: Upper Coosa River. In *Lamar Archaeology*, edited by Mark Williams and Gary Shapiro, pp. 43–44. University of Alabama Press, Tuscaloosa.

1993 The Territorial Size of Mississippian Chiefdoms. In *Archaeology of Eastern North America: Papers in Honor of Stephen Williams*, edited by James B. Stoltman, pp. 143–168. Mississippi Department of Archives and History, Archaeological Report No. 25. Jackson.

1994a An Overview of Lamar Culture. In *Ocmulgee Archaeology, 1936–1986*, edited by David J. Hally, pp. 144–174. University of Georgia Press, Athens.

1994b The Chiefdom of Coosa. In *The Forgotten Centuries: Indians and Europeans in the American South, 1521–1704*, edited by Charles Hudson and Carmen Chaves Tesser, pp. 227–256. University of Georgia Press, Athens.

Hally, David J., and James B. Langford Jr.
1988 *Mississippi Period Archaeology of the Georgia Valley and Ridge Province*. University of Georgia, Laboratory of Archaeology Series, Report No. 25. Athens.

Hally, David J., and James L. Rudolph
1986 *Mississippi Period Archaeology of the Georgia Piedmont*. University of Georgia, Laboratory of Archaeology Series, Report No. 24, Athens.

Harden, Tess, Mark G. Macklin, and Victor R. Baker
2010 Holocene Flood Histories in South-Western USA. *Earth Surface Processes and Landforms* 35(6):707–716.

Harding, Juliana M., Roger Mann, and Melissa J. Southworth
2008 Shell Length-at-Age Relationships in James River, Virginia, Oysters (*Crassostrea virginica*) Collected Four Centuries Apart. *Journal of Shellfish Research* 27(5):1109–1115.

Hargis, William J., Jr., and Dexter S. Haven
1999 Chesapeake Oyster Reefs, Their Importance, Destruction, and Guidelines for Restoring Them. In *Oyster Reef Habitat Restoration: A Synopsis and Synthesis of Approaches*, edited by Mark W. Luckenbach, R. Mann, and J. A. Wesson, pp. 329–358. Virginia Institute of Marine Science Press, Gloucester Point.

Harrington, Jean C.
1954 Dating Stem Fragments of Seventeenth- and Eighteenth-Century Clay Tobacco Pipes. *Quarterly Bulletin of the Archeological Society of Virginia* 9(1):10–14.

Hart, John P.
2012 The Effects of Geographical Distances on Pottery Assemblage Similarities: A Case Study from Northern Iroquoia. *Journal of Archaeological Science* 39(1): 128–134.

Hatch, James W.
1995 Lamar Period Upland Farmsteads of the Oconee River valley, Georgia. In *Mississippian Communities and Households*, edited by Daniel Rogers and Bruce D. Smith, pp. 135–155. University of Alabama Press, Tuscaloosa.

Hayden, Brian
1995 Pathways to Power: Principles for Creating Socioeconomic Inequalities. In *Foundations of Social Inequality*, edited by T. Douglass Price and Gary M. Feinman, pp. 15–86. Plenum Press, New York.

Heath, Charles L., and E. Clay Swindell
2011 Coastal Plain Iroquoians before and after European Contact: An Interpretive Study of Cashie Phase Archaeological Research in Northeastern North Carolina. In *The Archaeology of North Carolina: Three Archaeological Symposia*, edited by Charles R. Ewen, Thomas R. Whyte, and R. P. Stephen Davis Jr., pp. 10–104. North Carolina Archaeological Council Publication No. 30. Raleigh.
2013 Notes on Contact–Early Postcontact Period Carolina Algonkian Pottery: Colington II (1584–1650) and Colington III (1650–1750) Subphase Ceramics. Workshop on North Carolina Native Ceramics in the Historic Era, 2013 Meet-

ing of the North Carolina Archaeological Council, Research Laboratories of Archaeology, University of North Carolina, Chapel Hill.

Helms, Douglas
2000 Soil and Southern History. *Agricultural History* 74(4):723–758.

Henderson, T. Kathleen (editor)
1995 *The Prehistoric Archaeology of Heritage Square*. Anthropological Papers, No. 3. Pueblo Grande Museum, Phoenix.

Henderson, T. Kathleen, and Tiffany C. Clark
2004 Changing Perspectives: Considerations of Past Agricultural Use of the Salt River Floodplain. In *Hohokam Farming on the Salt River Floodplain: Refining Models and Analytical Methods*, edited by T. Kathleen Henderson, pp. 167–186. Center for Desert Archaeology, Anthropological Papers No. 43, Tucson and City of Phoenix, Pueblo Grande Museum Anthropological Papers No. 10, Phoenix.

Henderson, T. Kathleen, and Mark R. Hackbarth
2000 What Is Going on at the Hohokam Village? A Fourteenth- and Fifteenth-Century Perspective. In *The Hohokam Village Revisited*, edited by David E. Doyel, Suzanne K. Fish, and Paul R. Fish, pp. 287–316. Southwestern and Rocky Mountain Division of the American Association for the Advancement of Science, Glenwood Springs, Colorado.

Herbert, Joseph M.
1999 Prehistoric Pottery Taxonomy and Sequence on the Southern Coast of North Carolina. *North Carolina Archaeology* 48:37–58.
2003 Woodland Ceramics and Social Boundaries of Coastal North Carolina. Unpublished PhD dissertation, Department of Anthropology, University of North Carolina, Chapel Hill.
2009 *Woodland Potters and Archaeological Ceramics of the North Carolina Coast*. University of Alabama Press, Tuscaloosa.
2011 Recent Woodland Archaeology of Coastal North Carolina. In *The Archaeology of North Carolina: Three Archaeological Symposia*, edited by Charles R. Ewen, Thomas R. Whyte, and R. P. Stephen Davis Jr., pp. 4-1 to 4-58. North Carolina Archaeological Council Publication No. 30. Raleigh.

Herring, D. Ann
1992 Toward a Reconsideration of Disease and Contact in the Americas. *Prairie Forum* 17(2): 153–165.

Herskovits, Mellville J.
1924 A Preliminary Consideration of the Culture Areas of Africa. *Amercian Anthropologist* 26(1):50–63.

Hill, J. Brett, Jeffery J. Clark, William H. Doelle, and Patrick D. Lyons
2004 Prehistoric Demography in the Southwest: Migration, Coalescence, and Hohokam Population Decline. *American Antiquity* 69(4): 689–716.
2010 Depopulation of the Northern Southwest: A Macro-Regional Perspective with Insights from the Hohokam. In *Time of Peril, Time of Change: Explaining Thirteenth-Century Pueblo Migrations*, edited by T. A. Kohler, M. D. Varien, and A. Wright, pp. 34–52. University of Arizona Press, Tucson.

Hill, J. Brett, Patrick D. Lyons, Jeffrey J. Clark, and William H. Doelle
2015a The "Collapse" of Cooperative Hohokam Irrigation in the Lower Salt River Valley. *Journal of the Southwest* 57(4):609–674.

Hill, J. Brett, Matthew A. Peeples, Deborah L. Huntley, and H. Jane Carmack
2015b Spatializing Social Network Analysis in the Late Precontact US Southwest. *Advances in Archaeological Practice* 3(1):63–77.

Hoard, Robert J., Michael J. O'Brien, Mohammad Ghazavy Khorasgany, and Vellore S. Gopalaratnam
1995 A Materials-Science Approach to Understanding Limestone-Tempered Pottery from the Midwestern United States. *Journal of Archaeological Science* 22(6): 823–832.

Hoffman, Paul E.
1984 The Chicora Legend and Franco-Spanish Rivalry in La Florida. *The Florida Historical Quarterly* 62(4):419–438.
1990 *A New Andalicia and a Way to the Orient: The American Southeast during the 16th Century*. Louisiana State University Press, Baton Rogue.
1994 Lucas Vasquez de Ayllón's Discovery and Colony. In *The Forgotten Centuries: Indians and Europeans in the American South, 1521–1704*, edited by Charles Hudson and Carmen Chaves Tesser, pp. 36–49. University of Georgia Press, Athens.
2014 "Until the Land Was Understood": Spaniards Confront La Florida, 1500–1600. In *La Florida: Five Hundred Years of Hispanic Presence*, edited by Viviana Díaz Balsera and Rachel A. May, pp. 69–82. University Press of Florida, Gainesville.

Holmes, William H.
1903 *Aboriginal Pottery of the Eastern United States*. Twentieth Annual Report of the Bureau of American Ethnology, pp. 1–237. Washington, DC.

Howard, Jerry B.
2006 Hohokam Irrigation Communities: A Study of Internal Structure, External Relationships and Sociopolitical Complexity. Unpublished PhD dissertation, School of Human Evolution and Social Change, Arizona State University, Tempe.

Howard, Jerry B., editor
1988 *Excavations at Casa Buena: Changing Hohokam Land Use along the Squaw Peak Parkway*. Soil Systems Publications in Archaeology No. 11. Phoenix.

Howard, Jerry B., and M. Kyle Woodson
2018 The Development of Prehistoric Irrigation Studies in Arizona Under the National Historic Preservation Act. *Journal of Arizona Archaeology* 5(2):114–126.

Huckleberry, Gary
1993 Late-Holocene Stream Dynamics on the Middle Gila River, Pinal County, Arizona. Unpublished PhD dissertation, Department of Geosciences, University of Arizona, Tucson.
1999 Stratigraphic Identification of Destructive Floods in Relict Canals: A Case Study from the Middle Gila River, Arizona. *Kiva* 65(1):7–33.
2015a Defining the Environmental Context of Indigenous Agriculture in the Southwest: What We Don't Know about Middle to Late Holocene Climate Change

and Floodplain Dynamics. In *Traditional Arid Lands Agriculture Understanding the Past for the Future*, edited by Scott E. Ingram and Robert C. Hunt, pp. 89–130. University of Arizona Press, Tucson.

2015b Geoarchaelogical Analysis of Prehistoric Canals near Park of Four Waters: Phx Sky Train Project. In *Hohokam Irrigation and Agriculture on the Western Margin of Pueblo Grande: Archaeology for the PHX Sky Train Project*, edited by T. Kathleen Henderson, pp. 37–82. Archaeology Southwest, Anthropological Papers No. 41. Tucson.

Huckleberry, Gary, T. Kathleen Henderson, and Paul R. Hanson

2018 Flood-Damaged Canals and Human Response, AD 1000–1400, Phoenix, Arizona, USA. *Journal of Field Archaeology* 43(8):604–618.

Huckleberry, Gary, Jill Onken, William M. Graves, and Robert Wegener

2013 Climatic, Geomorphic, and Archaeological Implications of a Late Quaternary Alluvial Chronology for the Lower Salt River, Arizona, USA. *Geomorphology* 185:39–53.

Hudson, Charles M.

1970 *The Catawba Nation*. University of Georgia Press, Athens.

1976 *The Southeastern Indians*. University of Tennessee Press, Knoxville.

1984 The Genesis of Georgia's Indians. In *Forty Years of Diversity: Essays on Colonial Georgia*, edited by H. H. Jackson and Phinizy Spalding, pp. 25–45. University of Georgia Press, Athens.

1990 *The Juan Pardo Expeditions: Exploration of the Carolinas and Tennessee, 1566–1568*. Smithsonian Institution Press, Washington, DC.

Hudson, Charles M., Marvin T. Smith, and Chester B. DePratter

1984 The Hernando De Soto Expedition: From Apalachee to Chiaha. *Southeastern Archaeology* 3(1):65–77.

Huntley, Deborah L.

2008 *Ancestral Zuni Glaze-Decorated Pottery: Viewing Pueblo IV Regional Organization through Ceramics Production and Exchange*. University of Arizona Press, Tucson.

Huntley, Deborah, Jeffry J. Clark, and Mary F. Ownby

2016 Movement of People and Pots in the Upper Gila Region of the American Southwest. In *Exploring Cause and Explanation: Historical Ecology, Demography, and Movement in the American Southwest*, edited by Cynthia L. Herhahm and Ann F. Ramenofsky, pp. 279–295. University Press of Colorado, Boulder.

Ingram, Scott E.

2008 Streamflow and Population Change in the Lower Salt River Valley of Central Arizona, ca. AD 775 to 1450. *American Antiquity* 73(1):136–165.

Irwin, Jeffrey D., Wayne C. J. Boyko, Joseph M. Herbert, and Chad Braley

1999 Woodland Burial Mounds in the North Carolina Sandhills and Southern Coastal Plain. *North Carolina Archaeology* 48:59–86.

Jackson, Jeremy B.

2001 What Was Natural in the Coastal Oceans? *Proceedings of the National Academy of Sciences* 98(10):5411–5418.

Jones, Douglas S., Irvy R. Quitmyer, and Chester B. DePratter

2012 Validation of Annual Shell Increments and Shifting Population Dynamics in

Modern and Zooarchaeological Hard Clams (*Mercenaria mercenaria*) from the Litchfield Beach Region, South Carolina. In *Seasonality and Human Mobility along the Georgia Bight*, edited by Elizabeth J. Reitz, Irvy R. Quitmyer, and David Hurst Thomas, pp. 149–164. Anthropological Papers of the American Museum of Natural History No. 97. New York.

Jones, Eric E., and Peter Ellis

2016 Multiscalar Settlement Ecology Study of Piedmont Village Tradition Communities, AD 1000–1600. *Southeastern Archaeology* 35(2):85–114.

Jones, Grant

1978 The Ethnohistory of the Guale Coast through 1684. *Anthropological Papers of the American Museum of Natural History* 55(2):178–210.

Judge, Christopher

1986 Ceramics from the Mulberry Mound Site: A Study in Form and Function. In *The Wateree Papers: Recent Prehistoric and Historical Research in the Wateree Valley*, compiled by Christopher Judge, pp. 16–20. Annual Papers of the University of South Carolina Anthropology Students Association. Columbia.

1987 Aboriginal Pottery Vessel Function in South Appalachian Mississippian Society: A Case Study from the Mulberry Site (38KE12). Unpublished Master's thesis, Department of Anthropology, University of South Carolina, Columbia.

Kalin, Robert M.

1993 Report of Radiocarbon Dating of Sample from 38CH134 Addressed to Jeff Gardner, February 18 and 23, 1993. Center for Applied Isotope Studies, University of Georgia, Athens.

Keel, Bennie C.

1970 Excavations at the Red Springs Mound RB°4, Robeson County 1971. *Southern Indian Studies* 22:16–22.

Keene, Deborah A.

2004 Reevaluating Late Prehistoric Coastal Subsistence and Settlement Strategies: New Data from Grove's Creek Site, Skidaway Island, Georgia. *American Antiquity* 69(4): 671–688.

Kelly, Arthur R.

1974 Excavation History at the Mulberry Site. *Institute of Archaeology and Anthropology Notebook* 6:67–87. South Carolina Institute of Archaeology and Anthropology, University of South Carolina, Columbia.

Kelton, Paul

2002 The Great Southeastern Smallpox Epidemic, 1696–1700: The Region's First Major Epidemic? In *The Transformation of the Southeastern Indians, 1540–1760*, edited by Robbie Ethridge and Charles Hudson, pp. 21–37. University Press of Mississippi, Jackson.

Kemp, W. M., W. R. Boynton, J. E. Adolf, D. F. Boesch, W. C. Boicourt, G. Brush, J. C. Cornwell, T. R. Fisher, P M. Gilbert, J. D. Hagy, L. W. Harding, E. D. Houde, D. G. Kimmel, W. D. Miller, R. I. E. Newell, M. R. Roman, E. M. Smith, and J. C. Stevenson

2005 Eutrophication of Chesapeake Bay: Historical Trends and Ecological Interactions. *Marine Ecology Progress Series* 303:1–29.

Kemp, Andrew C., Benjamin P. Horton, Jeffrey P. Donnelly, Michael E. Mann, Martin Vermeer, and Stefan Rahmstorf

2011 Climate Related Sea-Level Variations over the Past Two Millennia. *Proceedings of the National Academy of Science* 108(27):11017–11022.

Kerr, R. A.

2000 A North Atlantic Climate Pacemaker for the Centuries. *Science* 288(5473): 1984–1986.

Kie, John G., and Marshall White

1985 Population Dynamics of White-Tailed Deer (*Odocoileus virginianus*) on the Welder Wildlife Refuge, Texas. *The Southwestern Naturalist* 30(1):105–118.

Kilikoglou, Vassilis, George Vekinis, and Y. Maniatis

1995 Toughening of Ceramic Earthenwares by Quartz Inclusions: An Ancient Art Revisited. *Acra Metallurgica et Materialia* 43(S):2959–2965.

Kilikoglou, Vassilis, George Vekinis, Y. Maniatis, and Peter M. Day

1998 Mechanical Performance of Quartz-Tempered Ceramics: Part I, Strength and Toughness. *Archoeormetry* 40(2):261–279.

King, Adam

2003 *Etowah: The Political History of a Chiefdom Capital.* University of Alabama Press, Tuscaloosa.

2007 The Southeastern Ceremonial Complex: From Cult to Complex. In *Southeastern Ceremonial Complex: Chronology, Content, Context*, edited by Adam King, pp. 1–14. University of Alabama Press, Tuscaloosa.

Kintigh, Keith W.

2002 Tools for Quantitative Archaeology (TFQA). Department of Anthropology, Arizona State University.

Knight, Vernon James

1986 The Institutional Organization of Mississippian Religion. *American Antiquity* 51(4):675–687.

1990 Social Organization and the Evolution of Hierarchy in Southeastern Chiefdoms. *Journal of Anthropological Research* 46(1):1–23.

1998 Moundville as a Diagrammatic Ceremonial Center. In *Archaeology of the Moundville Chiefdom*, edited by Vernon James Knight and Vincas P. Steponaitis, pp. 44–62. University of Alabama Press, Tuscaloosa.

Knight, Vernon James, and Vincas Steponaitis

1998 A New History of Moundville. In *Archaeology of the Moundville Chiefdom*, edited by Vernon James Knight and Vincas P. Steponaitis, pp. 1–25. University of Alabama Press, Tuscaloosa.

Kohler, Timothy A.

2000 The Final 400 Years of Prehispanic Agricultural Society in the Mesa Verde Region. *Kiva* 66(1):191–204.

Kohler, Timothy A., Matt Pier Glaude, Jean-Pierre Bocquet-Appel, and Brian M. Kemp

2008 The Neolithic Demographic Transition in the US Southwest. *American Antiquity* 73(4):645–669.

Kopytoff, Igor

1987 The Internal African Frontier: The Making of African Political Culture. In *The*

African Frontier: The Reproduction of Traditional African Societies, edited by Igor Kopytoff, pp. 3–86. Indiana University Press, Bloomington.

Kovats, R. Sari, Menno J. Bouma, and Andy Haines
1999 *El Niño and Health*. Task Force on Climate and Health, Geneva.

Kowalewski, Stephen A.
2006 Coalescent Societies. In *Light on the Path: The Anthropology and History of the Southeastern Indians*, edited by Thomas J. Pluckhahn and Robbie Ethridge, pp. 94–122. University of Alabama Press, Tuscaloosa.

Kowalewski, Stephen A., and James W. Hatch
1991 The Sixteenth-Century Expansion of Settlement in the Upper Oconee Watershed, Georgia. *Southeastern Archaeology* 10(1):1–17.

Kristiansen, Kristian
1998 Chiefdoms, States, and Systems of Social Evolution. In *Social Transformations in Archaeology: Global and Local Perspectives*, edited by Kristian Kritiansen and Michael Rowlands, pp. 236–260. Routledge, London.

Krus, Anthony M., and Charles R. Cobb
2018 The Mississippian Fin de Siècle in the Middle Cumberland Region of Tennessee. *American Antiquity* 83(2):302–319.

Kuhn, Katrin, Diarmid Campbell-Lendrum, Andy Haines, and Martha Anker
2005 Using Climate to Predict Infectious Disease Epidemics. World Health Organization, Geneva.

Lacquement, Cameron Hawkins
2004 How to Build a Mississippian House: A Study of Domestic Architecture in West-Central Alabama. Unpublished Master's thesis, Department of Anthropology, University of Alabama, Tuscaloosa.
2007 Typology, Chronology, and Technological Changes of Mississippian Domestic Architecture in West-Central Alabama. *Architectural Variability in the Southeast*, edited by Cameron H. Lacquement, pp. 49–72. University of Alabama Press, Tuscaloosa.

Lake, P. S.
2003 Ecological Effects of Perturbation by Drought in Flowing Waters. *Freshwater Biology* 48(7):1161–1172.

Lallo, J. W., and J. C. Rose
1979 Patterns of Stress, Diseases, and Mortality in Two Prehistoric Populations from North America. *Journal of Human Evolution* 8(3):323–334.

LaMotta, Vincent M., and Michael B. Schiffer
2005 Formation Processes of House Floor Assemblages. In *Archaeology: The Key Concepts*, edited by Colin Renfrew and Paul Bahn, pp. 121–127. Routledge, London.

Lanos, Philippe, Maxime LeGoff, Mary Kovacheva, and Elizabeth Schnepp
2005 Hierarchical Modelling of Archaeomagnetic Data and Curve Estimation by Moving Average Technique. *Geophysical Journal International* 160(2):440–476.

Lansdell, Brent, Jon Marcoux, and Eric C. Poplin
2005 Investigations at Magnolia Mound and the Mississippian in the Coastal Zone in South Carolina. Paper presented at the 62nd Annual Southeastern Archaeological Conference, Columbia, South Carolina.

Larsen, Clark Spencer
2002 Bioarchaeology: The Lives and Lifestyles of Past People. *Journal of Archaeological Research* 10(2):119–166.
2006 The Agricultural Revolution as Environmental Catastrophe: Implications for Health and Lifestyle in the Holocene. *Quaternary International* 150(1):12–20.

Larson, Lewis
1978 Historic Guale Indians of the Georgia Coast and the Impact of the Spanish Mission Effort. In *Tacachale: Essays of the Indians of Florida and Southeastern Georgia during the Historic Period*, edited by Jerald Milanich and Samuel Proctor, pp. 120–140. University Press of Florida, Gainesville.
1980 *Aboriginal Subsistence Technology on the Southeastern Coastal Plain during the Late Prehistoric Period.* University Press of Florida, Gainesville.
1998 *The Georgia and South Carolina Coastal Expeditions of Clarence Bloomfield Moore.* University of Alabama Press, Tuscaloosa.

LeBlanc, Steven A.
1975 Micro-Seriation: A Method for Fine Chronologic Differentiation. *American Antiquity* 40(1):22–38.

Lefler, Hugh Talmage (editor)
1967 *A New Voyage to Carolina by John Lawson.* University of North Carolina Press, Chapel Hill.

Lindauer, Owen, and John H. Blitz
1997 Higher Ground: The Archaeology of North American Platform Mounds. *Journal of Archaeological Research* 5(2):169–207.

Livingston, R. J., F. G. Lewis, G. C. Woodsum, X. F. Niu, B. Galperin, W. Huang, J. D. Christensen, M. E. Monaco, T. A. Battista, C. J. Klein, R. L. Howell IV, and G. L. Ray
2000 Modelling Oyster Population Response to Variation in Freshwater Input. *Estuarine, Coastal and Shelf Science* 50(5):655–672.

Loendorf, Chris, and Barnaby V. Lewis
2017 Ancestral O'odham: Akimel O'odham Cultural Traditions and the Archaeological Record. *American Antiquity* 82(1):123–139.

Loftfield, Thomas C.
1976 A Brief and True Report. . . : An Archaeological Interpretation of the Southern North Carolina Coast. Unpublished PhD dissertation, Department of Anthropology, University of North Carolina, Chapel Hill.

Lyons, Patrick D.
2004 Cliff Polychrome. *Kiva* 69(4):361–400.
2013 By Their Fruits Ye Shall Know Them: The Pottery of Kinishba Revisited. In *Kinishba Lost and Found: Mid-Century Excavations and Contemporary Perspectives*, edited by John R. Welch, pp. 145–208. Arizona State Museum Archaeological Series No. 206. Arizona State Museum, University of Arizona, Tucson.

McCabe, Gregory J., Julio L. Betancourt, Stephen T. Gray, Michael A. Palecki, and Hugo G. Hidalgo
2008 Associations of Multi-Decadal Sea-Surface Temperature Variability with US Drought. *Quaternary International* 188(1):31–40.

McCabe, Gregory J., Michael A. Palecki, and Julio L. Betancourt

2004 Pacific and Atlantic Ocean Influences on Multidecadal Drought Frequency in the United States. *Proceedings of the National Academy of Sciences* 101(12):4136–4141.

MacCord, Howard A., Sr.

1966 The McLean Mound, Cumberland County, North Carolina. *Southern Indian Studies* 18:3–45.

McMichael, Anthony J.

2012 Insights from Past Millennia into Climatic Impacts on Human Health and Survival. *Proceedings of the National Academy of Sciences* 109(13):4730–4737.

Marcoux, Jon Bernard, and Eric C. Poplin

2012 Defining Wando: A Distinct Late Woodland Manifestation in the Charleston Harbor Region. *South Carolina Antiquities* 44:35–65.

Marquardt, William H.

1978 Advances in Archaeological Seriation. In *Advances in Archaeological Method and Theory, Vol. 1*, edited by Michael B. Schiffer, pp. 257–314. Academic Press, Orlando, Florida.

Meeks, Scott C., and David G. Anderson

2013 Drought, Subsistence Stress, and Population Dynamics: Assessing Mississippian Abandonment of the Vacant Quarter. In *Soils, Climate, and Society: Archaeological Investigations in Ancient America*, edited by John D. Wingard and Sue Eileen Hayes, pp. 61–84. University Press of Colorado, Boulder.

Merry, Carl, and Sharon Pekrul

1981 An Analysis of the 1981 Ditch Excavations at the Mulberry Site, 38KE12. Ms on file, Department of Anthropology, University of South Carolina, Columbia.

1983 A Summary of the 1981 Ditch Excavations at the Mulberry Site. *South Carolina Antiquities* 15(1/2):61–62.

Michie, James L.

1984 *An Initial Archeological Survey of the Wachesaw/Richmond Plantation Property, Georgetown County, South Carolina.* South Carolina Institute of Archaeology and Anthropology, Research Manuscript Series 191. University of South Carolina, Columbia.

Millis, Heather

2009 *Data Recovery Excavations at the Contentnea Creek Site (21WL37), Wilson County, North Carolina.* Report submitted to Project Development and Environmental Development Branch, North Carolina Department of Transportation, Raleigh. TRC Environmental Corporation, Chapel Hill.

2011 Broad Reach Revisited: Preliminary Results of the 2006 Data Recovery Excavations. In *The Archaeology of North Carolina: Three Archaeological Symposia*, edited by Charles R. Ewen, Thomas R. Whyte, and R. P. Stephen Davis Jr., pp. 6-1 to 6-13. North Carolina Archaeological Council Publication Number 30. Raleigh.

Mills, Barbara J.

2007 Performing the Feast: Visual Display and Suprahousehold Commensalism in the Puebloan Southwest. *American Antiquity* 72(2):210–239.

Mills, Barbara J., Jeffrey L. Clark, Matthew Peeples, William R. Haas Jr., John M. Roberts Jr., J. Brett Hill, Deborah L. Huntley, Lewis Borck, Ronald L. Breiger, Aaron Clauset, and M. Steven Shackley
2013 *Transformation of Social Networks in the Late Pre-Hispanic US Southwest.* Proceedings of the National Academy of Sciences, Washington, DC.

Mills, James N., and James E. Childs
1998 Ecologic Studies of Rodent Reservoirs: Their Relevance for Human Health. *Emerging Infectious Diseases* 4(4):529–537.

Milner, George R.
1980 Epidemic Disease in the Postcontact Southeast: A Reappraisal. *Midcontinental Journal of Archaeology* 5(1):39–56.
1986 Mississippian Period Population Density in a Segment of the Central Mississippi River Valley. *American Antiquity* 51(2):227–238.
1991 Health and Cultural Change in the Late Prehistoric American Bottom, Illinois. In *What Mean These Bones? Studies in Southeastern Bioarchaeology*, edited by M. L. Powell, P. S. Bridges, and A. M. W. Mires, pp. 52–69. University of Alabama Press, Tuscaloosa.
1998 *The Cahokia Chiefdom: The Archaeology of a Mississippian Society.* University Press of Florida, Gainesville.

Milner, George R., David G. Anderson, and Marvin T. Smith
2001 The Distribution of Eastern Woodlands Peoples at the Prehistoric and Historic Interface. In *Societies in Eclipse: Archaeology of the Eastern Woodlands Indians, AD 1400–1700*, edited by David S. Brose, C. Wesley Cowan, and Robert C. Mainfort Jr., pp. 9–18. Smithsonian Institution Press, Washington, DC.

Milner, George R., Scott W. Hammerstedt, and Kirk D. French
2010 Chert Hoes as Digging Tools. *Antiquity* 84(323):103–113.

Milner, George R., and Sissel Schroeder
1999 Mississippian Political Systems. In *Great Towns and Regional Polities*, edited by Jill Neitzel, pp. 95–107. University of New Mexico Press, Albuquerque.

Milton, Donald K.
2012 What Was the Primary Mode of Smallpox Transmission?: Implications for Biodefense. *Frontiers in Cellular and Infection Microbiology* 2(Article 150):1–7.

Mitchell, Douglas R. (editor)
1994 *The Pueblo Grande Project, Vol. 2: Feature Descriptions, Chronology, and Site Structure.* Soil Systems Publications in Archaeology No. 20, Phoenix.

Mooney, James
1894 *The Siouan Tribes of the East.* Bulletin of the Bureau of American Ethnology No. 22. US Government Printing Office, Washington, DC.

Moore, David G.
2002 *Catawba Valley Mississippian: Ceramics, Chronology, and Catawba Indians.* University of Alabama Press, Tuscaloosa.

Morse, Stephen S.
1995 Factors in the Emergence of Infectious Diseases. *Perspectives* 1(1):7–15.

Mozingo, Dea, Connie Huddleston, Alana Lynch, and Meagan Brady
2004 *Transitions: Archaeological Data Recovery at the Shell Crescent Site (38BU1791)*

Palmetto Bluff, Beaufort County, South Carolina. Report prepared for Palmetto Bluff LLC. Brockington and Associates, Inc., Atlanta.

Muller, Jon

1986 *Archaeology of the Lower Ohio River Valley*. Academic Press, Orlando.

1997 Native Eastern American Population Continuity and Stability. In *Integrating Archaeological Demography Multidisciplinary Approaches to Prehistoric Population*, edited by R. R. Paine, pp. 343–364. Center for Archaeological Investigations Occasional Paper No. 24. Southern Illinois University, Carbondale.

Müller, Noemi, George Vekinis, Peter M. Day, and Vassilis Kilikoglou

2015 The Influence of Microstructure and Texture on the Mechanical Properties of Rock Tempered Archaeological Ceramics. *Journal of the European Ceramic Society* 35(2):831–843.

Neff, Hector

1992 Ceramics and Evolution. *Archaeological Method and Theory* 4:141–193.

Nelson, Margaret C., and Colleen A. Strawhacker (editors)

2008 *Movement, Connectivity, and Landscape Change in the Ancient Southwest*. University Press of Colorado, Boulder.

Newell, Roger I. E.

1988 Ecological Changes in Chesapeake Bay: Are They the Result of Overharvesting the American Oyster, *Crassostrea virginica*? In *Understanding the Estuary: Advances in Chesapeake Bay Research*, edited by M. P. Lynch and E. C. Krome, pp. 536–546. Chesapeake Research Consortium Publication 129 (CBP/TRS 24/88), Gloucester Point.

Newell, H. Perry, Alex D. Krieger, and Volney H. Jones

1949 The George C. Davis Site: Cherokee County, Texas. *Memoirs of the Society for American Archaeology* 5.

Nials, Fred L., David A. Gregory, and Donald A. Graybill

1989 Salt River Streamflow and Hohokam Irrigation Systems. In *The 1982–1984 Excavations at Las Colinas: Environment and Subsistence*, by Donald A. Graybill, David A. Gregory, Fred L. Nials, Suzanne K. Fish, Robert E. Gasser, Charles H. Miksicek, and Christine R. Szuter, pp. 59–76. Archaeological Series 162, Volume 5. Cultural Resource Management Division, Arizona State Museum, University of Arizona, Tucson.

O'Brien, Michael J.

2000 The Legacy of Culture History in the Southeastern United States. *Reviews in Anthropology* 29(2):111–139.

O'Brien, M. J., and T. D. Holland

1990 Variation, Selection, and the Archaeological Record. *Archaeological Method and Theory* 2:31–79.

O'Brien, Michael J., Thomas D. Holland, Robert J. Hoard, and Gregory L. Fox

1994 Evolutionary Implications of Design and Performance Characteristics of Prehistoric Pottery. *Journal of Archaeological Method and Theory* 1(3):259–304.

O'Brien, Michael J., and R. Lee Lyman

1999 *Seriation, Stratigraphy, and Index Fossils: The Backbone of Archaeological Dating*. Kluwer Academic/Plenum, New York.

2002 The Epistemological Nature of Archaeological Units. *Anthropological Theory* 2(1):37–56.

O'Connell, James F.
1987 Alyawara Site Structure and Its Archaeological Implications. *American Antiquity*: 52(1):74–108.

O'Leary, Marion H.
1988 Carbon Isotopes in Photosynthesis: Fractionation Techniques May Reveal New Aspects of Carbon Dynamics in Plants. *BioScience* 38(5):328–336.

Oberg, Kalvero
1955 Types of Social Structure among the Lowland Tribes of South and Central America. *American Anthropologist* 57(3):472–487.

Onken, Jill, Michael R. Waters, and Jeffrey A. Homburg
2004 *Geoarchaeological Assessment for the Tres Rios Project, Maricopa County, Arizona.* Submitted to US Army Corps of Engineers, Los Angeles District, Contract No. DACA09–98-D-0011. Statistical Research, Inc., Technical Report 03–68, Redlands, California, and Tucson, Arizona.

Orton, Clive, Paul Tyres, and Alan Vince
1993 *Pottery in Archaeology.* Cambridge University Press, London.

Palmer, W. C.
1965 *Meteorologic Drought.* Weather Bureau Research Paper No. 45. US Department of Commerce, Washington, DC.

Pauketat, Timothy R.
2000 The Tragedy of the Commoners. In *Agency in Archaeology*, edited by Marcia-Anne Dobres and John E. Robb, pp. 113–129. Routledge, London and New York.
2007 *Chiefdoms and Other Archaeological Delusions.* AltaMira Press, Lanham, Maryland.

Pauketat, Timothy R., and Neal H. Lopinot
1997 Cahokian Population Dynamics. In *Cahokia: Domination and Ideology in the Mississippian World*, edited by Timothy R. Pauketat and Thomas E. Emerson, pp. 103–123. University of Nebraska Press, Lincoln.

Pearson, Charles E.
1984 Red Bird Creek: Late Prehistoric Material Culture and Subsistence in Coastal Georgia. *Early Georgia* 12(1/2):1–40.

Pechenkina, Ekaterina A., Robert A. Benfer Jr., and Wang Zhiljun
2002 Diet and Health Changes at the End of the Chinese Neolithic: The Yangshao/Longshan Transition in Shaanxi Province. *American Journal of Physical Anthropology* 117(1):15–36.

Peebles, Christopher S., and Susan M. Kus
1977 Some Archaeological Correlates of Ranked Societies. *American Antiquity* 42(3):421–448.

Peros, Matthew C., Samuel E. Munoz, Konrad Gajewski, and André Viau
2009 Prehistoric Demography of North America Inferred from Radiocarbon Data. *Journal of Archaeological Science* 37(3):1–9.

Peters, C. J., and Ali S. Kahn
2002 Hantavirus Pulmonary Syndrome: The New American Hemorrhagic Fever. *Emerging Infections* 34(9):1224–1231.

Petes, Laura E., Alicia J. Brown, and Carley R. Knight
2012 Impacts of Upstream Drought and Water Withdrawals on the Health and Survival of Downstream Estuarine Oyster Populations. *Ecology and Evolution* 2(7):1712–1724.

Phillips, Philip, James A. Ford, and James B. Griffin
1951 *Archaeological Survey in the Lower Mississippi Alluvial Valley, 1940–1947*. Papers of the Peabody Museum of Amercian Archaeology and Ethnology, Harvard University, vol. 25. Cambridge, Massachusetts.

Pietrusewsky, Michael, and Michele Toomay Douglas
2001 Intensification of Agriculture at Ban Chiang: Is There Evidence from the Skeletons? *Asian Perspectives* 40(2):157–178.

Plog, Stephen, and Jeffrey L. Hantman
1990 Chronology Construction and the Study of Prehistoric Change. *Journal of Field Archaeology* 17(4):439–456.

Pluckhahn, Thomas J., and David A. McKivergan
2002 A Critical Appraisal of Middle Mississippian Settlement and Social Organization on the Georgia Coast. *Southeastern Archaeology* 21(2):149–161.

Pluckhahn, Thomas J., and Victor D. Thompson
2018 *New Histories of Village Life at Crystal River*. University of Florida Press. Gainesville.

Poplin, E. C., C. T. Espenshade, and D. C. Jones
1993 *Archaeological Investigations at the Buck Hall Site (38CH644), Francis Marion National Forest, South Carolina*. Francis Marion and Sumter National Forests, CRM Report 92-08. Brockington and Associates, Inc., Atlanta.

Powell, Mary Lucas
1998 Of Time and the River: Perspectives on Health during the Moundville Chiefdom. In *Archaeology of the Moundville Chiefdom*, edited by Vernon James Knight and Vincas P. Steponaitis, pp. 102–119. University of Alabama Press, Tuscaloosa.

Power, A., B. McCrickard, M. Mitchell, E. Covington, M. Sweeney-Reeves, K. Payne, and R. Walker
2006 *Perkinsus marinus* in Coastal Georgia, USA, Following a Prolonged Drought. *Diseases of Aquatic Organisms* 73(2):151–156.

Pugh, Daniel C.
2010 The Swantek Site: Late Prehistoric Oneota Expansion and Ethnogenesis. Unpublished PhD dissertation, Department of Anthropology, University of Michigan, Ann Arbor.

Rao, A. R.
1972 Infected Inanimate Objects (Fomites) and Their Role in Transmission of Smallpox. No. WHO/SE/72.40. World Health Organization, Geneva, Switzerland.

Redmond, Elsa M.
1998 Introduction: The Dynamics of Chieftaincy and the Development of Chiefdoms. In *Chiefdoms and Chieftaincy*, edited by Elsa M. Redmond, pp. 1–17. University Press of Florida, Gainesville.

Regnier, Amanda L.
2006 An Examination of the Social Composition of Late Mississippian Towns in the

Alabama River Valley through Ceramic Styles. Unpublished PhD dissertation, Department of Anthropology, University of Alabama, Tuscaloosa.

2009 What Indian Pottery of Sixteenth-Century Central Alabama Looks Like and Why It Matters. *The Search for Mabila: The Decisive Battle between Hernando de Soto and Chief Tascalusa*, edited by Vernon James Knight Jr., pp. 83–93. University of Alabama Press, Tuscaloosa.

2014 *Reconstructing Tascalusa's Chiefdom: Pottery Styles and the Social Composition of Late Mississippian Communities along the Alabama River*. University of Alabama Press, Tuscaloosa.

Reid, J. Jefferson

1967 Pee Dee Pottery from the Mound at Town Creek. Unpublished PhD dissertation, Department of Anthropology, University of North Carolina, Chapel Hill.

Reily, F. Kent, III, and James F. Garber (editors)

2007 *Ancient Objects, Sacred Realms: Studies in Mississippian Iconography*. University of Texas Press, Austin.

Reimer, Paula J., Edouard Bard, Alex Bayliss and others

2013 IntCal13 and Marine13 Radiocarbon Age Calibration Curves 0–50,000 Years cal BP. *Radiocarbon* 55(4):1869–1887.

Reitz, Elizabeth J.

1982 Vertebrate Fauna from Four Coastal Mississippian Sites. *Journal of Ethnobiology* 2(1):39–61.

Reitz, Elizabeth J., Bruce M. Saul, J. W. Moak, Gwendolyn D. Carroll, and Charles W. Lambert

2012 Interpreting Seasonality from Modern and Archaeological Fishes on the Georgia Coast. In *Seasonality and Human Mobility along the Georgia Bight*, edited by Elizabeth J. Reitz, Irvy Quitmyer, and David Hurst Thomas, pp. 50–82. Anthropological Papers of the American Museum of Natural History No. 97. New York.

Reitz, Elizabeth J., and Daniel C. Weinand

1993 Vertebrate Fauna from Fogarty Creek (38BK552). Report on file, Francis Marion National Forest, Huger, South Carolina.

Ricciardelli, Taryn

2014 The Hinterlands of Town Creek: A Settlement Pattern Study of the Mississippian Occupation of the North Carolina Piedmont. Master's thesis, Department of Anthropology, East Carolina University, Greenville.

Rice, Jackie

2000 The Ceramics of an Artisan: An Analysis of the Ceramic Assemblage from Structure 1 at the Mulberry Site (38KE12). Manuscript on file, Department of Anthropology, University of South Carolina, Columbia.

Rice, Prudence M.

1987 *Pottery Analysis: A Source Book*. University of Chicago Press, Chicago.

Riggs, Brett H.

2010 Temporal Trends in Native Ceramic Traditions of the Lower Catawba River Valley. *Southeastern Archaeology* 29(1):31–43.

Riggs, Brett H., R. P. Stephen Davis Jr., and Mark R. Plane

2006 Catawba Pottery in the Post-Revolutionary Era: A View from the Source. *North Carolina Archaeology* 55:60–87.

Rights, Douglas L.

1931 The Trading Path to the Indians. *The North Carolina Historical Review* 8(4): 403–426.

Robertson, James Alexander (translator)

1933 *True Relation of the Hardships Suffered by Governor Hernando De Soto & Certain Portuguese Gentlemen during the Discovery of the Province of Florida. Now Newly Set Forth by a Gentleman of Elvas. 1557.* Florida State Historical Society, Deland.

Robinson, William S.

1951 A Method for Chronologically Ordering Archaeological Deposits. *American Antiquity* 16(4):293–301.

Rogers, Rhea J.

1993 A Re-Examination of the Concept of the Tribe: A Case Study from the Upper Yadkin Valley, North Carolina. Unpublished PhD dissertation, Department of Anthropology, University of North Carolina, Chapel Hill.

Rouse, Irving B.

1967 Seriation in Archaeology. In *American Historical Anthropology: Essays in Honor of Leslie Spier*, edited by C. L. Reily and Walter W. Taylor, pp. 153–195. Southern Illinois University Press, Carbondale.

1986 *Migrations in Prehistory: Inferring Population Movement from Cultural Remains.* Yale University Press, New Haven.

Rowe, John Howland

1961 Stratigraphy and Seriation. *American Antiquity* 26(3):324–330.

Rudes, Blair A.

2004a Place Names of Cofitachequi. *Anthropological Linguistics* 46(4):359–426.

2004b The First Description of an Iroquoian People: Spaniards among the Tuscaroras before 1522. Manuscript on file, Blair Rudes Indigenous Language Collection, Coastal Carolina Indian Center (www.coastalcarolinaindians.com).

Rudes, Blair A., Thomas J. Blumer, and J. Alan May

2004 Catawba and Neighboring Groups. In *Handbook of North American Indians, Volume 14: Southeast*, edited by Raymond D. Fogelson, pp. 301–318. Smithsonian Institution, Washington, DC.

Rudolph, James L.

1983 Lamar Period Exploitation of Aquatic Resources in the Middle Oconee River Valley. *Early Georgia* 11(1/2):86–103.

1994 *Population, Settlement, and Subsistence in the Oconee River Valley, Georgia.* PhD dissertation, Department of Anthropology, University of California at Santa Barbara. University of Georgia, Laboratory of Archaeology Series 68. Athens.

Runquist, Jeannette

1980 Appendix 1: Faunal Remains from Five South Carolina Indian Sites. In Investigation of the Woodland Period along the South Carolina Coast, by Michael B. Trinkley, pp. 464–486. Unpublished PhD dissertation, Department of Anthropology, University of North Carolina, Chapel Hill.

Sahlins, Marshall

1963 Poor Man, Rich Man, Big-Man: Political Types in Melanesia and Polynesia. *Comparative Studies in Society and History* 5(3):285–303.

Saunders, Rebecca

2001 Negotiated Tradition? Native American Pottery in the Mission Period in La Florida. In *The Archaeology of Traditions: Agency and History before and after Columbus*, edited by Timothy R. Pauketat, pp. 77–93. University Press of Florida, Gainesville.

2002 Seasonality, Sedentism, Subsistence, and Disease in the Protohistoric: Archaeological versus Ethnohistoric Data along the Lower Atlantic Coast. In *Between Contacts and Colonies Archaeological Perspectives on the Protohistoric Southeast*, edited by Cameron B. Wesson and Mark A. Rees, pp. 32–48. University of Alabama Press, Tuscaloosa.

2009 Stability and Ubiquity: Irene, Altamaha, and San Marcos Pottery in Time and Space. In *From Santa Elena to St. Augustine: Indigenous Ceramic Variability (AD 1400–1700)*, edited by Kathleen Deagan and David H. Thomas, pp. 83–112. Anthropological Papers of the American Museum of Natural History No. 90. New York.

Scarry, C. Margaret, and Vincus P. Steponaitis

1997 Between Farmstead and Center: The Natural and Social Landscape of Moundville. In *People, Plants, and Landscapes: Studies in Paleoethnobotany*, edited by Kristen J. Gremillion, pp. 107–122. University of Alabama Press, Tuscaloosa.

Scavia, Donald, John C. Field, Donald F. Boesch, Robert W. Buddemeier, Virginia Burkett, Daniel R. Cayan, Michael Fogarty, Mark A. Harwell, Robert W. Howarth, Curt Mason, Denise J. Reed, Thomas C. Royer, Asbury H. Sallenger, and James G. Titus

2002 Climate Change Impacts on US Coastal and Marine Ecosystems. *Estuaries* 25(2):149–164.

Schaafsma, Hoski, and John M. Briggs

2007 Hohokam Field Building: Silt Fields in the Northern Phoenix Basin. *Kiva* 72(3):369–395.

Schiffer, Michael B.

1986 Radiocarbon Dating and the "Old Wood" Problem: The Case of the Hohokam Chronology. *Journal of Archaeological Science* 13(1):13–30.

1987 *Formation Processes of the Archaeological Record*. University of New Mexico Press, Albuquerque.

Seager, Richard, Alexandrina Tzanova, and Jennifer Nakamura

2009 Drought in the Southeastern United States: Causes, Variability over the Last Millennium, and the Potential for Future Hydroclimatic Change. *Journal of Climate* 22(19):5021–5045.

Seavey, J. R., W. E. Pine III, P. Frederick, and M. Berrigan

2012 Decadal Changes in Oyster Reefs in the Big Bend of Florida's Gulf Coast. *Ecosphere* 2(10):1–14.

Service, Elman

1962 *Primitive Social Organization*. Random House, New York.

Sheridan, Susan Guise

2003 Childhood Health as an Indicator of Biological Stress. In *Centuries of Decline during the Hohokam Classic Period at Pueblo Grande*, edited by David R. Abbott, pp. 82–106. University of Arizona Press, Tucson.

Shott, Michael J.

1996 Mortal Pots: On Use Life and Vessel Size in the Formation of Ceramic Assemblages. *American Antiquity* 61(3):463–482.

Sires, Earl W., Jr.

1984 Excavations at El Polvoron. In *Hohokam Archaeology Along the Salt–Gila Aqueduct, Central Arizona Project, Vol. 4, Part II: Prehistoric Occupation of the Queen Creek Delta*, edited by Lynn S. Teague and Patricia L. Crown, pp. 221–326. Arizona State Museum Archaeological Series, No. 150. University of Arizona, Tucson.

Smith, Bruce D.

2001 Low-Level Food Production. *Journal of Archaeological Research* 9(1):1–43.

2009 Resource Resilience, Human Niche Construction, and the Long-Term Sustainability of Pre-Columbian Subsistence Economies in the Mississippi River Valley Corridor. *Journal of Ethnobiology* 29(2): 167–183.

Smith, Marvin T.

1983 The Development of Lamar Ceramics in the Wallace Reservoir: The Evidence from the Dyar Site (9GE5). *Early Georgia* 11(1–2):74–85.

1987 *Archaeology of Aboriginal Culture Change in the Interior Southeast: Depopulation During the Early Historic Period.* University Press of Florida, Gainesville.

1992 *Historic Period Indian Archaeology of Northern Georgia.* Georgia Archaeological Research Design Paper No. 7. University of Georgia Laboratory of Archaeology Series Report No. 30. Athens.

1994 *Archaeological Investigations at the Dyar Site, 9GE5.* University of Georgia Laboratory of Archaeology Series Report No. 32. Athens.

2000 *Coosa: The Rise and Fall of a Southeastern Mississippian Chiefdom.* University Press of Florida, Gainesville.

2002 Aboriginal Population Movements in the Postcontact Southeast. In *The Transformation of the Southeastern Indians, 1540–1760*, edited by Robbie Ethridge and Charles M. Hudson, pp. 3–20. University Press of Mississippi, Jackson.

Smith, Marvin T., and Mark Williams

1994 Mississippian Mound Refuse Disposal Patterns and Implications for Archaeological Research. *Southeastern Archaeology* 13(1):27–35.

South, Stanley A.

2002 *Archaeological Pathways to Historic Site Development.* Kluwer Academic/Plenum Publishers, New York.

2005 *Archaeology on the Roanoke.* Research Laboratories of Archaeology, University of North Carolina, Chapel Hill, Monograph No. 4.

Southerlin, Bobby, Dawn Reid, Connie Huddleston, and Joseph Sanders

1997 *The Grand Strand Frontier: Mississippian Period Occupation at the Tidewater Site (38HR254), Horry County, South Carolina.* Report submitted to Tidewater Golf Club and Plantation. Brockington and Associates, Inc., Atlanta.

Southerlin, Bobby, Rachael Tibbetts, Michael Keith O'Neal, Dawn Reid, Leslie Raymer, and MacKensie Cornelius

2005 *Woodland Adaptations in The Grand Strand: Native American Settlement along the Little River Estuary, Horry County, South Carolina: Excavations at Glen Dornoch Golf Course, Horry County, South Carolina.* Archaeological Consultants of the Carolinas, Inc. Clayton, North Carolina.

Spivey, L. D., Jr., W. J. Busscher, and R. B. Campbell

1986 The Effect of Texture on Strength of Southeastern Coastal Plain Soils. *Soil and Tillage Research* 6(4):351–363.

Squier, Ephraim, and E. G. Davis

1848 *Ancient Monuments of the Mississippi Valley.* Smithsonian Institution Contributions to Knowledge, No. 1. Washington, DC.

Stahle, David W., and Malcom K. Cleaveland

1992 Reconstruction and Analysis of Spring Rainfall over the Southeastern US for the Past 10000 Years. *Bulletin of the American Meteorological Society* 73(12): 1947–1961.

Stahle, David W., Malcolm K. Cleaveland, Dennis B. Blanton, Matthew D. Therrell, David A. Gay

1998 The Lost Colony and Jamestown Droughts. *Science* 280(5363):564–567.

Stahle, David W., Falko K. Fye, Edward R. Cook, and R. Daniel Griffin

2007 Tree-Ring Reconstructed Megadroughts over North America since AD 1300. *Climatic Change* 83(1/2):133–149.

Stark, Miriam T.

2003 Current Issues in Ceramic Ethnoarchaeology. *Journal of Archaeological Research* 11(3):193–242.

Stark, Miriam T., Jeffrey Clark, and Mark D. Elson

1995 Causes and Consequences of Migration in the 13th Century Tonto Basin. *Journal of Anthropological Archaeology* 14(2):212–246.

Stark, Miriam T., and James M. Heidke

1998 Ceramic Manufacture, Productive Specialization, and the Early Classic Period in Arizona's Tonto Basin. *Journal of Anthropological Research* 54(4):497–517.

Stark, Miriam T., and William A. Longacre

1993 Kalinga Ceramics and New Technologies: An Ethnoarchaeological Perspective. In *The Social and Cultural Contexts of New Ceramic Technologies*, edited by W. D. Kingery, pp. 1–32. Ceramics and Civilization, vol. 7. John Wiley and Sons, New York.

Steen, Carl, and James Errante

1996 *An Archaeological Survey of the Proposed Bridgewater Subdivision.* Diachronic Research Foundation, Columbia, South Carolina.

Steen, Carl, Christopher Judge, and Tariq Gaffar

1998 *Searching for the 18th Century Pee Dee Indian Town in Marion County, South Carolina.* Report submitted to Marion County Development Board, Marion County, South Carolina. South Carolina Department of Natural Resources Heritage Trust Program, Columbia.

Steponaitis, Vincas P.

1978 Locational Theory and Complex Chiefdoms: A Mississippian Example. In *Mississippian Settlement Patterns*, edited by Bruce D. Smith, pp. 417–453. Academic Press, New York.

Stine, S.

1994 Extreme and Persistent Drought in California and Patagonia during Medieval Time. *Nature* 369(6481):546–549.

Stocking, George W.

1974 *A Franz Boas Reader: The Shaping of American Anthropology 1883–1911.* University of Chicago Press, Chicago.

Stone, Glenn Davis

2001 Theory of the Square Chicken: Advances in Agricultural Intensification Theory. *Asia Pacific Viewpoint* 42(2/3):163–180.

Stuart, George E.

1974 Analysis of the Mulberry Site. *Institute of Archaeology and Anthropology Notebook* 6(3/4):100–110. South Carolina Institute of Archaeology and Anthropology, University of South Carolina, Columbia.

Swanton, John R.

1911 *Indian Tribes of the Lower Mississippi Valley and Adjacent Coast of the Gulf of Mexico.* Bureau of American Ethnology Bulletin No. 43. Smithsonian Institution, Washington, DC.

1922 *Early History of the Creek Indians.* Bureau of American Ethnology Bulletin No. 73. Smithsonian Institution, Washington, DC.

1946 *The Indians of the Southeastern United States.* Bureau of American Ethnology Bulletin No. 137. Government Printing Office, Washington, DC.

Swetnam, Thomas W., and Julio L. Betancourt

1998 Mesoscale Disturbance and Ecological Response to Decadal Climatic Variability in the American Southwest. *Journal of Climate* 11(12):3128–3147.

Thomas, David Hurst

1978 The Awful Truth about Statistics in Archaeology. *American Archaeology* 43(2):231–244.

1993 *Historic Indian Period Archaeology of the Georgia Coastal Zone.* University of Georgia Laboratory of Archaeology Series Report No. 31. Athens.

2008 Radiocarbon Dating on St. Catherines Island. In *Native American Landscapes of St. Catherines Island, Georgia. The Data*, edited by David Hurst Thomas, pp. 345–371. Anthropological Papers of the American Museum of Natural History 88, Part II. New York.

Tite, M. S., V. Kilikoglou, and G. Vekinis

2001 Strength, Toughness, and Thermal Shock Resistance of Ancient Ceramics, and Their Influence on Technological Choice. *Archaeometry* 43(3):301–324.

Trinkley, Michael B.

1980 Investigation of the Woodland Period Along the South Carolina Coast. Unpublished PhD dissertation, Department of Anthropology, University of North Carolina, Chapel Hill.

1981a *Archaeological Testing of the Awendaw Shell Midden (38CH300), Charleston*

County. Francis Marion and Sumter National Forests Cultural Resources Report 1981–22. Chicora Foundation, Columbia.

1981b *Archaeological Testing of the Walnut Grove Shell Midden (38CH260), Charleston County*. Francis Marion and Sumter National Forests Cultural Resources Report 1981–21. Chicora Foundation, Columbia.

1989 An Archaeological Overview of the South Carolina Woodland Period: It's the Same Old Riddle. In *Studies in South Carolina Archaeology: Essays in Honor of Robert L. Stephenson*, edited by Albert C. Goodyear III and Glen T. Hanson, pp. 73–87. Occasional Papers of the South Carolina Institute of Archaeology and Anthropology, University of South Carolina, Anthropological Studies No. 9. Columbia.

Trinkley, Michael B., S. Holmes Hogue

1979 The Wachesaw Landing Site: The Last Gasp of the Coastal Waccamaw Indians. *Southern Indian Studies* 31:3–19.

Trinkley, Michael B., S. Holmes Hogue, Martha Zierden, and Jack H. Wilson

1983 *Test Excavations at the Wachesaw Landing Site, Georgetown County, South Carolina*. North Carolina Archaeological Publication No. 20. Raleigh.

Van Gerven, Dennis P., and Susan Guise Sheridan (editors)

1994 *The Pueblo Grande Project, Volume 6: The Bioethnography of a Classic Period Hohokam Population*. Soil Systems Publications in Archaeology 20(6). Soil Systems, Inc., Phoenix.

Varner, J. G., and J. J. Varner (translators)

1951 *The Florida of the Inca: A History of the Adelantado, Hernando de Soto, Governor and Captain General of the Kingdom of Florida, and Other Heroic Spanish and Indian Cavaliers, Written by Garcilaso de la Vega, an Officer of His Majesty, and A Native of the Great City of Cuzco, Capital of the Realms and Provinces of Peru*. University of Texas Press, Austin.

Waddell, Gene

1980 *Indians of the South Carolina Low Country, 1562–1751*. Southern Studies Program, University of South Carolina, Columbia.

Wagner, Gail E.

1997 Late Woodland-Mississippian Plant Remains from the Fogarty Creek Site (38BK552), Berkeley County, South Carolina. Manuscript on file, Francis Marion National Forest, Huger, South Carolina.

Ward, H. Trawick

1987 Mortuary Patterns at the Fredricks, Wall, and Mitchum Sites. In *The Siouan Project: Seasons I and II*, edited by Roy S. Dickens, H. Trawick Ward, and R. P. Stephen Davis Jr., pp. 81–110. Research Laboratories of Anthropology, University of North Carolina, Chapel Hill, Monograph Series No. 1.

Ward, H. Trawick, and R. P. Stephen Davis Jr.

1988 Archaeology of the Historic Occaneechi Indians. *Southern Indian Studies* (36/37):1–127.

1993 *Indian Communities on the North Carolina Piedmont, AD 1000 to 1700*. Research Laboratories of Anthropology, University of North Carolina, Chapel Hill, Monograph No. 2.

1999 *Time before History: The Archaeology of North Carolina*. University of North Carolina, Chapel Hill.

Waselkov, Gregory A.

2012 Making a Case for Coastal Subsistence Seasonality. In *Seasonality and Human Mobility along the Georgia Bight*, edited by Elizabeth J. Reitz, Irvy Quitmyer, and David Hurst Thomas, pp. 199–206. Anthropological Papers of the American Museum of Natural History No. 97. New York.

Waters, Michael R., and John C. Ravesloot

2000 Late Quaternary Geology of the Middle Gila River, Gila River Indian Reservation, Arizona. *Quaternary Research* 54(1):49–57.

2001 Landscape Change and the Cultural Evolution of the Hohokam along the Middle Gila River and Other River Valleys in Southern Arizona. *American Antiquity* 66(2):285–299.

2003 Disaster or Catastrophe: Human Adaptation to High-and Low-Frequency Landscape Processes—A Reply to Ensor, Ensor, and De Vries. *American Antiquity* 68(2):400–405.

Welch, Paul D., and C. Margaret Scarry

1995 Status-Related Variation in Foodways in the Moundville Chiefdom. *American Antiquity* 60(3):397–419.

Wetmore, Ruth Y.

1978 Report on Excavations at the Buie Mound, Robeson County, North Carolina. *Institute of Archaeology and Anthropology Notebook* 10:30–71. South Carolina Institute of Archeology and Anthropology, University of South Carolina, Columbia.

Wheaton, Thomas R., Mary B. Reed, John S. Cable, Theresa Hamby, and Leslie A. Raymer

1993 *Archeological Site Testing of Willow Hall and Walnut Grove Plantations, Francis Marion National Forest*. Francis Marion and Sumter National Forests, Cultural Resources Report 1992–13. New South Associates, Inc., Stone Mountain.

Whittaker, John C., Douglas Caulkins, and Kathyrn A. Kamp

1998 Evaluating Consistency in Typology and Classification. *Journal of Archaeological Method and Theory* 5(2):129–164.

Widmer, Randolph J.

1994 The Structure of Southeastern Chiefdoms. In *The Forgotten Centuries: Indians and Europeans in the American South, 1521–1704*, edited by Charles Hudson and Carmen Chaves Tesser, pp. 125–155. University of Georgia Press, Athens.

Wiktorowicz, Conner J., Bettina Arnold, John E. Wiktorowicz, Mathew L. Murray, and Alexander Kurosky

2017 Hemorrhagic Fever Virus, Human Blood, and Tissues in Iron Age Mortuary Vessels. *Journal of Archaeological Science* 78:29–39.

Wilcox, David R.

1987 *Frank Midvale's Investigation of the Site of La Ciudad*. Ciudad Monograph Series, Vol. 4, OCR, Department of Anthropology, Arizona State University, Tempe. Arizona DOT Contract No. 82–11.

Willey, Gordon R.

1945 Horizon Styles and Pottery Traditions in Peruvian Archaeology. *American Antiquity* 11(1):49–56.

Willey, Gordon R., Charles C. Di Peso, William A. Ritchie, Irving Rouse, John H. Rowe, and Donald W. Lathrap

1955 An Archaeological Classification of Culture Contact Situations. *Memoirs of the Society for American Archaeology* 11:1–30.

Williams, Mark

1983 *The Joe Bell Site: Seventeenth Century Lifeways on the Oconee River*. PhD dissertation, Department of Anthropology, University of Georgia, Athens. University of Georgia, Laboratory of Archaeology Series 42. Athens.

1994 Growth and Decline of the Oconee Province. In *The Forgotten Centuries: Indians and Europeans in the American South, 1521–1704*, edited by Charles Hudson and Carmen Chaves Tesser, pp. 179–196. University of Georgia Press, Athens.

Williams, Mark, and Gary Shapiro

1990 *Lamar Archaeology: Mississippian Chiefdoms in the Deep South*. University of Alabama Press, Tuscaloosa.

1996 Mississippian Political Dynamics in the Oconee Valley, Georgia. In *Political Structure and Change in the Prehistoric Southeastern United States*, edited by John F. Scarry, pp. 128–149. University Press of Florida, Gainesville.

Williams, Marshall W., and Carolyn Branch

1978 Tugalo Site, 9ST1. *Early Georgia* 6(1/2):32–37.

Williams, Stephen

1990 The Vacant Quarter and Other Late Events in the Lower Valley. In *Towns and Temples Along the Mississippi*, edited by David H. Dye and C. A. Cox, pp. 170–80. University of Alabama Press, Tuscaloosa.

Wilson, Jack Hubert, Jr.

1983 A Study of the Late Prehistoric, Protohistoric, and Historic Indians of the Carolina and Virginia Piedmont: Structure, Process, and Ecology. Unpublished PhD dissertation, Department of Anthropology, University of North Carolina, Chapel Hill.

Woodall, J. Ned

1984 *The Donnaha Site: The 1973, 1975 Excavations*. North Carolina Archaeological Council Publication No. 22. Raleigh.

1990 *Archeological Investigations in the Yadkin River Valley, 1984–1987*. North Carolina Archeological Council Publication No. 25. Raleigh.

1999 Mississippian Expansion on the Eastern Frontier: One Strategy in the North Carolina Piedmont. *Archaeology of Eastern North America* 27:55–70.

2009 The T. Jones Site: Ecology and Agency in the Upper Yadkin Valley of North Carolina. *North Carolina Archaeology* 58:1–58.

Woodson, Kyle

2010 The Social Organization of Hohokam Irrigation in the Middle Gila River Valley, Arizona. Unpublished PhD dissertation, School of Human Evolution and Social Change, Arizona State University, Tempe.

2015 The Impact of Flooding on Hohokam Canal Irrigation Agriculture. In *Tra-*

ditional Arid Lands Agriculture: Understanding the Past for the Future, edited by Scott E. Ingram and Robert C. Hunt, pp. 180–216. University of Arizona Press, Tucson.

Worth, John E.

1994 Exploration and Trade in the Deep Frontier of Spanish Florida: Possible Sources for 16th-Century Spanish Artifacts in Western North Carolina. Paper presented in the symposium "A Consideration of Archaeological and Documentary Evidence for 16th- and 17th-Century Spanish and Native Contact in Western North Carolina" at the 51st Annual Meeting of the Southeastern Archaeological Conference, Lexington, November 9–12.

1995 *The Struggle for the Georgia Coast: An 18th-Century Spanish Retrospective on Guale and Mocama*. Anthropological Papers of the American Museum of Natural History 75. New York.

1999 Coastal Chiefdoms and the Question of Agriculture: An Ethnohistorical Overview. Paper presented at the 56th Annual Southeastern Archaeological Conference, Pensacola, November 12.

2003 An Ethnohistorical Synthesis of Southeastern Chiefdoms: How Does Coosa Compare? Paper presented in the symposium "Coosa: Twenty Years Later" at the 60th Annual Meeting of the Southeastern Archaeological Conference, Charlotte, North Carolina, November 15.

Wright J. Leitch, Jr.

1981 *The Only Land They Knew: The Tragic Story of the American Indians in the Old South*. Free Press, New York.

Yates, Terry L., James N. Mills, Cheryl A. Parmenter, Thomas G. Ksiazek, Robert R. Parmenter, John R. Vande Castle, Charles H. Calisher, Stuart T. Nichol, Kenneth D. Abbott, Joni C. Young, Michael L. Morrison, Barry J. Beaty, Jonathan L. Dunnum, Robert J. Baker, Jorge Salazar-Bravo, and Clarence J. Peters

2002 The Ecology and Evolutionary History of an Emergent Disease: Hantavirus Pulmonary Syndrome. *Bioscience* 52(11):989–998.

Yoffee, Norman

1993 Too Many Chiefs? (or, Safe Texts for the 90s). In *Archaeological Theory: Who Sets the Agenda?*, edited by Norman Yoffee and Andrew Sherratt, pp. 60–78. Cambridge University Press, New York.

Zhang, Y., J. M. Wallace, and D. S. Battisti

1997 ENSO-Like Interdecadal Variability: 1900–93. *Journal of Climate* 10(5):1004–1020.

Index

Page numbers in italics refer to figures and tables.